IET CONTROL, ROBOTICS, AND SENSORS SERIES 120

Imaging and Sensing for Unmanned Aircraft Systems

IET International Book Series on Sensing–Call for Authors

The use of sensors has increased dramatically in all industries. They are fundamental in a wide range of applications from communication to monitoring, remote operation, process control, precision and safety, and robotics and automation. These developments have brought new challenges such as demands for robustness and reliability in networks, security in the communications interface, and close management of energy consumption. This Book Series covers the research and applications of sensor technologies in the fields of ICTs, security, tracking, detection, monitoring, control and automation, robotics, machine learning, smart technologies, production and manufacturing, photonics, environment, energy, and transport.

Proposals for coherently integrated international multi-authored edited or co-authored handbooks and research monographs will be considered for this Book Series. Each proposal will be reviewed by the IET Book Series Editorial Board members with additional external reviews from independent reviewers. Please email your book proposal to: vmoliere@theiet.org or author_support@theiet.org.

Imaging and Sensing for Unmanned Aircraft Systems

Volume 1: Control and Performance

Edited by
Vania V. Estrela, Jude Hemanth, Osamu Saotome,
George Nikolakopoulos and Roberto Sabatini

The Institution of Engineering and Technology

Published by The Institution of Engineering and Technology, London, United Kingdom

The Institution of Engineering and Technology is registered as a Charity in England & Wales (no. 211014) and Scotland (no. SC038698).

First published 2020

The Institution of Engineering and Technology
Michael Faraday House
Six Hills Way, Stevenage
Herts, SG1 2AY, United Kingdom

www.theiet.org

British Library Cataloguing in Publication Data
A catalogue record for this product is available from the British Library

ISBN 978-1-78561-642-6 (Hardback Volume 1)
ISBN 978-1-78561-643-3 (PDF Volume 1)
ISBN 978-1-78561-644-0 (Hardback Volume 2)
ISBN 978-1-78561-645-7 (PDF Volume 2)
ISBN 978-1-78561-679-2 (Hardback Volumes 1 and 2)

Typeset in India by MPS Limited

Contents

About the editors

Vania Estrela is a faculty/researcher at Telecommunications Department, Universidade Federal Fluminense (UFF) and a visiting scholar at UNICAMP. Her research interests include biomedical engineering, electronic instrumentation, modelling/simulation, sustainable design, multimedia, artificial intelligence, remote sensing, STEM education, environment, and digital inclusion. She has served as a reviewer for IEEE, Elsevier, ACM, IET, Springer-Verlag, and MDPI. She has extensive experience as a project manager, post-graduate advisor (M.Sc. and D.Sc.), as well as an editor of books and special issues. ORCID 0000-0002-4465-7691

Jude Hemanth is an associate professor in the ECE Department of Karunya University (KU), India. He is a member of the IEEE task force on deep learning and serves as associate editor and editorial board member for several international refereed journals.

Osamu Saotome is a professor at the Instituto Tecnológico de Aeronáutica (ITA), Brazil. He has been involved in several international research and cooperation projects with the Brazilian Air Force, INPE, IEAv (France, Sweden, USA, and Japan).

George Nikolakopoulos is a professor in robotics and automation at the Department of Computer Science, Electrical and Space Engineering at Luleå University of Technology (LTU), Sweden. He is also a member of the ARTEMIS Scientific Council of the European Commission. He has significant experience in Managing European and National R&D&I projects funded by the EU, ESA, Swedish and the Greek National Ministry of Research.

Roberto Sabatini is a professor of aerospace engineering and aviation in the School of Engineering of RMIT University. He is the founding Chair of the RMIT Cyber-Physical Systems Group and the Director of the Autonomous and Intelligent Aerospace Systems Laboratory (Sir Lawrence Wackett Centre). He has over 25 years of experience in the aerospace, defence, and transport sectors, including a combination of advanced academic and military education, extensive research and flight test practice, and progressively more responsible technical and operational leadership positions in universities and R&D organisations in Europe, in the United States and in Australia. In addition to doctoral degrees in aerospace/avionics systems (Cranfield University) and in satellite navigation/geospatial systems

(University of Nottingham), he holds the qualifications of flight test engineer (fast jets), private pilot (fixed-wing aircraft) and remote pilot (multi-rotor unmanned aircraft). Throughout his career, he has led successfully numerous industrial and government-funded research programmes on aerospace, defence, and transport systems, and he has authored or co-authored more than 250 peer-reviewed international publications and over 100 research/flight test reports.

He is a Chartered Professional Engineer (CPEng), Engineering Executive (EngExec) and Fellow Member of the Institution of Engineers Australia (FIEAust). Additionally, he is a Fellow of the Royal Aeronautical Society (FRAeS), Fellow of the Royal Institute of Navigation (FRIN), Senior Member of the Institute of Electrical and Electronics Engineers (IEEE), Senior Member of the American Institute of Aeronautics and Astronautics (AIAA), and Life Member of Armed Forces Communications and Electronics Association (AFCEA). He received various scientific and professional awards including the ADIA Scientist of the Year Award (2019), the NATO Research and Technology Organisation Scientific Achievement Award (2008), the SAE Arch T. Colwell Merit Award (2015), the SARES Science Award (2016), and the Northrop Grumman Professorial Scholarship (2017). He is the Avionics Editor for Progress in Aerospace Sciences, Technical Editor for the IEEE Transactions on Aerospace and Electronic Systems, Senior Editor for the Journal of Intelligent and Robotic Systems, Associate Editor for Aerospace Science and Technology, and Associate Editor for the Journal of Navigation. Additionally, he serves as Vice-Chair of the IEEE Avionics Systems Panel (ASP), member of the NASA Unmanned Aircraft Systems (UAS) Traffic Management (UTM) Collaborative Testing Initiative, and Australian National Representative at the International Civil Aviation Organization (ICAO) Committee on Aviation Environmental Protection (CAEP), Impact and Science Group (ISG).

He performs research on Intelligent Automation and Autonomous Systems for Aerospace, Transport and Defence applications. His research interests include Avionics and Space Systems; Communications, Navigation and Surveillance/Air Traffic Management (CNS/ATM); Guidance, Navigation and Control (GNC); Global Navigation Satellite Systems (GNSS); Unmanned Aircraft Systems (UAS) and UAS Traffic Management (UTM); Defence C4ISR and Electronic Warfare Systems; Human-Machine Systems; and Trusted Autonomy. His research has led to significant discoveries, including innovative navigation and guidance techniques; optimal control and trajectory optimisation; GNSS integrity augmentation; laser/optoelectronics sensors; experimental flight test techniques and instrumentation; UAS sense-and-avoid; and cognitive human-machine systems (adaptive human-machine interfaces and interactions for trusted autonomy and enhanced human performance).

Preface

An unmanned aerial vehicle (UAV) – also known as a drone, unmanned aircraft system (UAS) or remotely piloted aircraft system (RPAS) – is an aircraft without a human pilot on board. Its flight can be controlled autonomously by computers in the vehicle, or by remote control. They can help perform a large number of tasks such as surveillance, disaster relief, health care in remote regions, to name a few. Drones are uniquely capable of penetrating areas, which may be too dangerous for piloted craft. Making a UAV autonomous requires addressing problems from different disciplines, such as mechanical design, aeronautics, control, computer science, sensor technologies, and artificial intelligence, to name only some.

The UAV cyber–physical system (CPS) comprises all the subsystems and interfaces, for processing and communication functions performed by the embedded electronic system (avionics) and the ground control station. To achieve the desired real-time autonomy, the avionics is highly tied with aerodynamics sensing and actuation. An entirely autonomous UAV can (i) obtain information about the environment, (ii) work for an extended period of time without human interference, (iii) move either all or part of itself all over its operating location devoid of human help, and (iv) stay away from dangerous situations for people, and their possessions.

Avionics play a decisive role in UAVs, and they comprise the electronics (hardware and software) used on aircraft, spacecraft, and satellites. Their costs, as well as widespread availability, have made the use of UAV increasingly viable for several types of potential users and applications.

Among the sensors, vision sensors including human visual spectral range and also multispectral sensors, and the so-called hyperspectral sensors are of interest, because of the extensive application possibilities, such as in navigation, obstacle detection, obstacle avoidance, and others, besides avionics challenges of computational complexity and objective-optimised algorithms.

A vision system (VS) entails the way output data will be utilised, appropriate architecture for total avionics integration, control interfaces, and operational utilisation. Since the VS core is its sensors, multi-sensor fusion, navigation, hazard detection, and ground correlation in real time are some of the most important flight operational aspects.

UAV avionics aim to boost flight visibility and situational awareness. This book intends to be a reference for vision and sensor integration. It will display fundamental aspects, ongoing research efforts, accomplishments, and challenges faced when it comes to the deployment of imaging capabilities and sensor integration in UAVs.

Vania V. Estrela
Jude Hemanth
Osamu Saotome
George Nikolakopoulos
Roberto Sabatini

Chapter 1

Introduction to advances in UAV avionics for imaging and sensing

Vania V. Estrela[1], *Jude Hemanth*[2], *Osamu Saotome*[3], *George Nikolakopoulos*[4] *and Roberto Sabatini*[5]

An unmanned aerial vehicle (UAV) – aka drone, unmanned aircraft system or remotely piloted aircraft system – is an aircraft without a human pilot on board. Its flight can be controlled autonomously by computers in the vehicle or by remote control. They can help perform a large number of tasks such as reconnaissance, disaster prevention/remediation, and health care in remote regions, to name a few. UAVs can uniquely penetrate areas, which may be too dangerous or too difficult to reach for piloted craft. Making a UAV autonomous requires addressing problems from different disciplines, such as mechanical design, aeronautics, control, computer science, sensor technologies, and artificial intelligence [1–5], to name a few.

The UAV cyber-physical system comprises all the subsystems and interfaces for processing and communication functions performed by the embedded electronic system (avionics) and the ground control station [6]. To accomplish the desired real-time autonomy, the avionics is highly tied with aerodynamics sensing and actuation. An entirely autonomous UAV can (i) obtain evidence about the environment, (ii) work for an extended period of time without human interference, (iii) move either all or part of itself all over its operating location devoid of human help and (iv) stay away from risky situations for people and their assets.

This chapter intends to introduce the material addressed in further chapters of this book. The next sections go through some concepts that are recurrent in the book.

1.1 Basic concepts

Unmanned aerial vehicle (UAV) mobility allows operation in more dynamic domains and requires the ability to sense and react to a changing environment.

[1]Universidade Federal Fluminense, RJ, Brazil
[2]Karunya University, Coimbatore, India
[3]DCTA-ITA-IEEA, Sao Jose dos Campos, SP, Brazil
[4]Luleå University of Technology, Sweden
[5]RMIT University, Bundoora, Australia

Issues such as localisation and mapping stem from using sensors to localise the objects and avoid obstacles on the way.

Size helps categorise airborne platforms. The smaller the airframe, the narrower the range of appropriate sensors and other hardware that can be used. The main restriction is the smaller payload, which is also restricted by limited onboard power, where the payload is the transport capacity of an aircraft usually considered in terms of weight.

Some important UAV categories are listed below (refer to [7,8]):

(i) High-altitude long-endurance (HALE): They carry out extremely long-range (trans-global) reconnaissance and surveillance being capable of flying more than 24 h over 15,000 m.
(ii) Medium-altitude long-endurance (MALE): Similar to the HALEs saved for generally operating at shorter ranges, but still more than 500 km, from 5,000 to 15,000 m of altitude, with 24 h endurance, and from fixed bases.
(iii) Tactical UAV or medium-range UAV: It has a range between 100 km and 300 km and it is smaller/operated than a HALE or MALE.
(iv) Vertical takeoff and landing: This type of UAV can take off, hover, and land vertically.
(v) Close-range UAV: It usually works at ranges up to 100 km, and it has possibly the most significant number of uses, including diverse tasks like reconnaissance, target designation, surveillance, crop-spraying, power-line inspection, and traffic monitoring, to name a few.
(vi) Mini-UAV: It can be hand-launched and operate at ranges of up to 30 km.
(vii) Micro-UAV (MAV): It is widespread in urban locations and within buildings. Besides flying slowly, it stops and sits on a wall or post (hover-and-perch mode). MAVs are, in general, manually launched, and for that reason, winged versions have shallow wing loadings, which make them vulnerable to atmospheric turbulence and precipitation [9].
(viii) Nano-air vehicle: It is an ultra-light UAV intended adequate to swarms for purposes such as radar confusion or conceivably, provided camera, propulsion, and control sub-systems can be made small enough, for ultra-short-range surveillance.

Another important taxonomy of UAV is the aerodynamical classification in fixed-wing and rotative-wing UAVs [10–13].

Fixed-wing (FW) UAVs need to develop a certain minimum translational velocity to maintain in the air. They comprise a much simpler structure than rotary-wing UAVs, which allows for less complicated maintenance and repair. Hence, they offer more operational time at a lower cost, greater aerodynamics efficiency, longer flight durations at higher, which enables larger survey areas per given flight. These aircraft can transport greater payloads for longer distances on less power, which gives room to carry more prominent and more expensive sensors as well as twin sensor configurations.

Rotative or rotary-wing (RW) UAVs normally can have different numbers of rotors: 1 (helicopter), 3 (tricopter), 4 (quadcopter), 6 (hexacopter), and 8 (octocopter) as well as more atypical arrangements with 12 and 16 rotors. When perfectly controlled, they can remain in the air. Once more, each arrangement has its unique characteristic advantages and disadvantages. Control of rotary UAVs derives from the variation in thrust and torque from its rotors. Their most significant advantage is the aptitude to take off and land vertically, which permits the operation within a smaller neighbourhood without a large fixed landing/take-off area. The RW capacity to hover and perform quick manoeuvring makes them well suited to applications such as inspections where precision manoeuvring and the ability to monitor a single target for extended periods is vital. RW UAVs' disadvantages are the greater mechanical and electronic complexities that result in more complicated maintenance and repair processes, shorter operational time, and higher operational costs than FW UAVs.

The faster dynamics of flying platforms impose strict requirements on the timely execution of tasks and decision making. Time dependency is paramount, especially for indoor UAVs because they handle small distances to obstacles. Moreover, the use of airborne vehicles calls for strict safety requirements, high dependability, and robustness.

There are numerous reasons for focusing on imaging sensors. Drones are usually equipped with cameras, so no additional hardware is needed. Furthermore, they are lightweight, low-cost and have low power requirements. For an outdoor UAV, the level of autonomy, as well as the system situational awareness increases with colour and thermal cameras.

Stimuli have four characteristics: location, modality, intensity, and duration. A stimulus modality (sensor modality) is a type of physical phenomenon that can be detected like temperature, taste, sound, and pressure [14–17]. A sensor is an entity intended to perceive events or changes in its surroundings and then present an equivalent output. The kind of sensor activated by a stimulus has a vital role in processing its modality.

Multisensory or multi-modal integration (MMI) studies how information from different sensory modalities, such as vision, sound, touch, taste, motion, and smell, may be combined with the control system. A coherent representation of objects combining modalities results in meaningful and adaptive knowledge. MMI also deals with how different types of sensors interact and modify each other processing.

A UAV demands three main features:

- the aircraft itself;
- the command, control, communication, and computer system (C4 system) also known as Ground Control Station; and
- the operator.

Figure 1.1 shows the basic entities comprising avionics architectures according to [18,19].

Payloads can be high- and low-resolution cameras, video cameras, night-and-day reconnaissance gear, high-power radar, gyro-stabilised, electro-optical signals, meteorological, chemical, biological, relay (communications and navigation

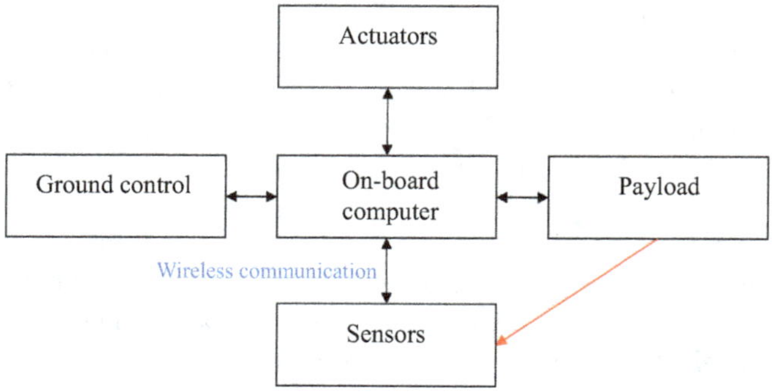

Figure 1.1 Basic UAV avionics architecture

signals), cargo (leaflets and supplies), and generally any equipment required for the UAV mission.

1.2 Navigation and intelligence

The C4 system comprises intelligence, procedures, organisational structures, ground personnel, equipment, facilities, and communications designed to support UAV command and control during all operational phases [20].

Advanced UAV sensor payloads gather an immense amount of data, where bandwidth is habitually restricted. At the same time, sharing, transmission, and display of mission-critical information need protection. These network limitations demand efficient information processing directly on the UAV.

Since the communication service demands are isolated from the functional model, the communications network can be specified in any preferred level of detail independently. As a consequence, the logical model becomes invariant concerning the physical model, which results in flexibility in designing a large-scale command, control, and communications system. The command, control, communications, and intelligence, surveillance and reconnaissance architectural structure specifies three architecture views: operational, systems, and technical. The proposed synthetic simulation technique and the invariant feature of the logical model allow collaborative work in developing that operational architecture and system architecture views.

An inertial navigation system (INS) aids UAV navigation using a computer, sensors (e.g., accelerometers, gyroscopes, magnetometers, and so forth) to estimate the position, the orientation, the direction, and speed of a moving item without external references. Other expressions that refer to the INS or closely associated devices include inertial instrument, the inertial guidance system, and inertial measurement unit (IMU), to name a few.

A satellite navigation system (SNS) provides autonomous geo-spatial positioning, which permits small receivers to control their location with high precision using satellite signals. Such a system can handle position, navigation, tracking the location of some object tracked by satellites and can operate independently of any telephony or Internet connection or concomitantly with these technologies for better positioning information. A global navigation satellite system (GNSS) is an SNS that affords global coverage.

Synthetic aperture radar (SAR) can be extremely helpful to aid UAV navigation when the INS measurements are not precise enough to get rid of drifts from a planned route [21]. This setback can influence MALE UAVs, which allows significant and spacious payloads (as necessary by SAR) gathering large drifts. The fundamental idea is to figure out the position and attitude of an aerial platform by inspecting both the amplitude and the phase of SAR images obtained onboard. For the amplitude-based method, matching the actual ground coordinates of landmarks with those automatically determined from the SAR image is a way of implementing the needed navigation corrections. When SAR amplitude analysis is impracticable, the phase data can be analysed via SAR interferometry and using a reference digital terrain model (DTM). A feasibility study can obtain adequate system requirements by exploring both radiometric and geometric parameters of the acquisition framework. MALE UAVs relying on particular commercial navigation sensors and SAR systems, with typical landmark position accuracy and classes, and existing DTMs allow available UAV technology using a SAR-based backup structure.

Indoor and outdoor vehicles pose different challenges. While some techniques work in both cases, some issues have to be solved independently. The biggest challenge for an indoor UAS is the lack of a ubiquitous positioning system such as the global positioning system (GPS). Even though its availability is never assured, GPS is an essential part of outdoor UAVs. Several commonly used sensors do not make sense or are not useful indoors, as is the case of the altitude sensor. Hence, autonomous UAV indoor operation is an open issue, and the detection of people can become challenging [22,23]. Because of imaging and other sensor types to facilitate the autonomy of a UAV operating in both outdoor and indoor environments, one may consider to tailor it to the application. In the case of the former, the input from cameras increases the number of tasks a UAV can perform autonomously. For the latter, a camera sensor can offer basic functionality, which is the ability to keep flight without human participation [24,25].

An embedded electronic system (sometimes called an onboard computer) controls the flight stability, the navigation, and the communication of a UAV. In the case of small UAVs' hardware, it is often called the flight controller, flight controller board, or autopilot, and it controls the communication with the base station, maintains the UAV floating and navigating into the air while performing the following actions:

1. Task distribution amongst a group of agents, with time and equipment constraints;
2. UAV path while meeting certain objectives and restrictions, such as obstacles [26];

3. Control manoeuvre to follow a given path or to go from one location to another; and
4. Activities sequence and their spatial distribution among agents to maximise the success chance for any given mission scenario.

Since UAV control is not in the cockpit, it has to take place by three possible means:

- ground-control or remote piloted;
- semi-autonomous; or
- autonomous.

Onboard and ground-based automated systems can control the UAV and its payload. The onboard autonomous management unit carries out the flight plan and performs other essential tasks related to flying, including payload-directed flight. Other desired characteristics are contingency management, in the event of unattainable or conflicting goals, and coordination with tactical and strategic intelligence. These clever manoeuvring (outer-loop) systems can incorporate planning and decision-making models to endow a UAV with goal-directed self-sufficient behaviour and facilitate time-critical re-planning and execution adjustments to compensate for unforeseen internal and external conditions or various mission-specific knowledge-related findings [27].

Challenges stem from real-time probing, computations, communication, requirements, environmental and operational uncertainty tracking, threats, and the growing need for enhanced UAV technology with more autonomy and reliability. There are significant challenges related to inter-UAS communications, links to command and control, and contingency management that can increase significantly when multi-vehicle coordinated control is considered. Meeting the lower echelon challenges has been moderately achieved, leaving open the whole field for subsequent developments. Hence, innovative, coordinated planning and control technologies such as distributed artificial intelligence, multi-agent system theory, computational intelligence, and soft computing, generalised system theory, game theory, optimisation, and refined hardware and software architectures will be required [28].

To examine large areas or simultaneously evaluate various points, multiple UAVs must move according to some formation. In a swarm, sensor data should be exchanged among UAVs via wireless communications to help correct and control flights. Most studies use numerical simulations to structure flight. An onboard system comprising sensors, processing units, and radiofrequency (RF) telemetry can handle flight formation algorithms for multi-platform UAV swarms [29].

1.3 Communications

Autonomy means that aerial robots can perform a wider range of tasks by interacting and operating safely in different kinds of environments without relying on external help via wireless transmission, with safety and self-sufficiency [30,31].

UAVs rely on communications for command and control and information spreading. Important issues to bear in mind are frequency and bandwidth availability, link security, link ranges, and network infrastructure to guarantee the availability for operational/mission support of unmanned systems. Planning and budgeting for UAV deployment must give room for realistic evaluations of projected Satellite Communications (SATCOMs) bandwidth and intensive use of onboard pre-processing to transmit only critical data [30,31].

Designing aeronautical wireless data links is much harder than other types of wireless links. The key challenges are long distances, aircraft high-speed, and limited availability of the radio frequency spectrum [30,31]. UAVs and human crewed aircraft demand the development of new data links because they will share the same airspace and have to be attentive to each other's presence [30,31].

Long distances are the primary challenge for aeronautical data links. The IEEE 802.11 wireless links (also called Wi-Fi) is the most common. IEEE 802.16 (WiMAX) wireless networks cover metropolitan areas. For longer distances, the signal strength decreases rapidly by the second to the fourth power of the distance. Long distances attenuate the power significantly along the path and result in a very low spectral efficiency (given in bits per second per Hertz or bps/Hz). Long distances also increase the round-trip delays that need big guard times. The increased guard times further decrease spectral efficiency even more. Aeronautical communications require links redundancy with overhead.

The second predicament is the speed of mobility. Wi-Fi supports minimal mobility. Data links based on WiMAX technology, designed for takeoff and landing applications, offer better performance.

Third, aeronautical communications systems have conventionally used high-frequency (HF), very high-frequency (VHF), as well as SATCOM bands. However, SATCOM systems are not always present during all phases of UAV flight. On the other hand, the HF and VHF bands are becoming very congested. Because of the increasing air traffic growth, new spectra for air-to-ground data links need to be found. Aeronautical communication already uses the L-band, which has become lately available for aeronautical mobile route service, has been cautiously designated as the next preferred band. Lower frequency bands are preferred. However, they are getting busy, and so the general trend is to move up in frequency. These higher frequency bands are more comprehensive to support the higher data rates needed [30,31].

1.4 Sensors

A sensor is a transducer that provides various types of output, most characteristically electrical or optical signals. For example, a thermocouple generates a voltage in response to its temperature [32].

Sensors support communication and coordination among multiple agents in the presence of incomplete and imperfect information.

With advances in technology, the uses of sensors have expanded beyond the most traditional fields of temperature, pressure, or flow measurement. Moreover, analogue sensors, e.g., potentiometers and force-sensing resistors, are still widely used. Applications include manufacturing and machinery, aeroplanes and aerospace, cars, medicine, and robotics.

Sensor sensitivity indicates how much its output changes when the input quantity being measured changes. Furthermore, some sensors disturb the phenomenon or characteristic they measure. For example, a room temperature thermometer can be affected by some failure in an indoor UAV. Hence, sensors must have a small effect on measurements. Often, making the sensor smaller improves this and may introduce other advantages [33–36].

Sensors provide basic functionality and aid in maintaining flight without human input, radar, cameras, Infra-Red (IR) scanners or Electronic Intelligence technology are most common. Other types may include a (laser) target designator to guide stand-off. Requirements for sensing payloads on UAV extend intelligence, data collection, reconnaissance/surveillance, tracking, and payload delivery since they assist detection and identification objects.

Sensors can represent one of the most costly items in an uncrewed aircraft and are necessary for navigation and mission achievement. Processing units allow UAVs to fly complete missions autonomously with little or no human intervention [37].

The fusion of multi-modal images can help investigate human body detection and geolocation for potential survivors in a search and rescue mission. The scanning result is a map of victim locations that can be used by first responders or a robotic system to, for example, deliver food or medical supplies [38,39]. The UAV tracking functionality can provide streams of low-level events such as geographical coordinates of targets on the ground, which allows for qualitative reasoning about the surroundings and assessment of vehicles overtaking, incoming, and leaving disaster sites besides improving the situational awareness mostly.

Position and movement sensors inform about the aircraft state. Exteroceptive sensors obtain data from the environment like distances to objects and deal with outside information. Expproprioceptive sensors measure values internal to the UAV (such as speed and heading), establishing a correlation between internal and outer states.

Non-cooperative sensors can detect targets autonomously, so they are used for separation assurance and collision avoidance.

The term degrees of freedom refers to both the amount and the quality of sensors onboard.

The critical communication technologies' issues are flexibility, adaptability, safety, robustness, and cognitive controllability of the bandwidth, frequency, and information/data flows [37]. A UAV data link usually consists of an RF transmitter and a receiver, an antenna, and modems to link these parts with the sensor systems. For UAS, data links serve three essential functions:

1. Uplinks from the ground station (GS) and/or a satellite to send control data to the UAV;
2. Downlinks to transmit records from the UAV onboard sensor and telemetry systems to the GS; and

3. Azimuth and range measurement schemes to maintain good communications among the GS, the satellite, and the UAV.

Efforts to standardise data links led to the common data link (CDL), which is typically full duplex wideband, jam-resistant and secure. These links connect the GS with the UAV through direct point-to-point links or use SATCOMs [40].

1.5 Computational aspects: image/video processing, computer graphics, modelling, and visualisation

The increasing use of UAVs in many civilian and military applications has generated considerable interest from the control community, primarily because the design of UAVs reveals some of the most stimulating control research issues. One of them is the use of computer vision (CV) as a sensor in the feedback control loop. Autonomous aircraft landing is particularly compatible with vision-based control, particularly for an indefinite or a moving landing site [41].

Many factors restrict the use of imaging to UAV systems such as the diffraction limit and resolution constraints from the optical elements involved. When it comes to image processing and understanding, several factors restrict design even further [1–3]. One can cite as examples of restrictions:

(a) The size, the shape, and the pitch of pixels that can lead to errors;
(b) Some tasks demand real-time response;
(c) CV algorithms may involve extreme computing conditions;
(d) The interaction among different levels of processing demand sophisticated control schemes; and
(e) Image acquisition hardware faces restrictions resulting from charge-coupled device cameras and environmental conditions.

A typical scenario in CV and control is for a given imaging degeneration model, to cast a problem as an optimisation task. Sparse reconstruction is a typical ill-posed inverse problem where measurement error and sparsity are used as conflicting objective functions to be handled at the same time. Classical optimisation and analysis techniques may perform disappointingly because of the amount of data, dimensionalities, and parallelism involved. These intensive computational problems motivate the application of computational intelligence methods and may benefit from advances from other areas like biomedical engineering.

Motion estimation (ME) algorithms help to determine position and orientation, linear and angular velocity and may work under unstable levels of image noise and unreliable camera altitudes, as well as different camera motion regarding the landing pad (assumed as being a planar surface) [42,43]. Commonly, the vision problem is equivalent to a particular case of the ego-ME problem, where all feature points lie on the planar surface. Both the discrete and the differential versions of the ego-ME help obtain the position and velocity of a UAV about the landing pad. The resulting algorithms allow vision sensors in the feedback loop as a state observer for landing control (LC). These methods are suitable for real-time implementation due to linearity, numerical robustness, and low computational burden. The resulting LC subsystem comes from a full UAV dynamic model. The UAV dynamics are equivalent to inner and outer systems combined when using nonlinear control theory. Control schemes may rely on the differential flatness of the outer system. If the overall closed-loop system exponential stability is guaranteed, then the controller can be tightly coupled with the vision-based state estimation and the auxiliary sensors such as accelerometers [41].

On average, a UAV vision system is a cheap way to augment the sensor suite of the aerial vehicle, including a GPS. Vision systems provide position data about the inertial frame and INSs [44]. As a passive and information-rich sensor, CV is gaining more and more importance in a UAV sensor set prompting growing interest in control design around vision sensors. Stereo vision systems expand a UAV multi-sensor suite, including laser range-finders for landing [45], estimate the location/orientation of the landing pad [46], and the camera regarding the landing pad.

Furthermore, the use of CV-based UAV control is more challenging than the classical approaches [47] since UAVs are underactuated nonlinear dynamical systems. To guarantee performance with stability for the overall closed-loop system, a comprehensive depiction of the UAV dynamics is necessary.

In various UAV imaging applications, the multisensory super-resolution (SR) method has turned out to be an unending setback and attracted escalating attention. Multi-sensor SR algorithms exploit the multispectral low-resolution (LR) pictures to create a high-resolution (HR) image to perk up the performance of the UAV image processing system. To restore image details without noise amplification or unnatural post-processing artefacts, regularised SR algorithms by combining the directionally adaptive constraints and a multi-scale non-local means filter are proposed. Thus, the physical limitation of multispectral sensors can be overcome by estimating the colour HR image from a set of multispectral LR images using intensity-hue-saturation image fusion [48].

UAVs can also be controlled based on 3D landmark observations and the landmark positions' predictions even under significant external perturbations [49].

Since incoming light to a UAV platform can be scattered by atmospheric haze and dust, the acquired subject image loses the original colour and brightness. The enhancement of hazy-affected images is an imperative undertaking to improve the visibility of various UAV images. Spatially adaptive de-hazing algorithms can rely on merging colour histograms with consideration of the wavelength-dependent atmospheric turbidity. Based on the wavelength-adaptive hazy image acquisition model, a de-hazing procedure consists of three steps: (i) image segmentation; (ii) creation of a context-adaptive communication map; and (iii) intensity transformation to enhance a hazy image. The degradation model must consider the wavelength of light sources, and the transmission map has to provide a theoretical basis to discriminate visually relevant regions from others using the turbidity and merged classification results [50].

Augmented reality (AR) comprises the technologies and methods to combine real images with virtually rendered scenes to improve the application of the UAV.

The high viewpoint and the combination of (multimodal) real and virtual scenes provide engineers with augmented views of the observed/analysed site. AR increases the possibility of uncovering problems. Important features for a successful UAV-AR platform are (1) stabilised image acquisition, (2) image/video transmission, (3) reliable communication, and (4) viewing interface. AR can lead to the (re)construction of scenarios. The UAV-AR solution can be used in construction, planning, inspection, and remediation [51,52].

Spatial information plays a decisive role in remote sensing (RS) and mapping applications like environment surveying and disaster monitoring [53,54]. A UAV-borne mobile mapping system (MMS) can achieve fast spatial data acquisition under limited conditions with better mobility and flexibility than other means. The accuracy of a direct geo-referencing (DG) system can be considerably improved by flying at a lower altitude using such hardware [55].

1.6 Security, health, and standards

Recently, significant research effort has been directed towards increasing operational aircraft safety through numerous approaches to fault-tolerant flight control schemes. Extensive research programs led by NASA have focused on designing and validating specific approaches for the design of some fault-tolerant flight control schemes [7,8], primarily focused on malfunctions of the actuators of the control surfaces or failures of the propulsion system. However, issues related to sensor failures have been considered of lower priority. Historically, triple or quadruple physical redundancy on the sensor suite, along with the implementation of built-in testing (BIT) and/or voting schemes, have been a reliable approach to cope with failures for the majority of the sensors of the flight control system [3.4]. On the other side, physical redundancy has clear limitations for light and inexpensive aircraft such as small UAVs. Nevertheless, recent events have shown that physical redundancy is not infallible in air data systems providing airspeed measurements since all the redundant sensors are subjected to the same environmental conditions in flight.

An alternative approach to physical redundancy is analytical redundancy. In general, an analytical redundancy-based sensor failure detection, identification, and accommodation scheme can be distinct and sequential tasks as follows: sensor failure detection and identification (SFDI) task and the sensor failure accommodation (SFA) task. The SFDI task is designed to monitor the sensor systems without physical redundancy to detect signs of incipient failures followed by the identification and isolation of such failures; instead, the SFA task is responsible for providing an alternative estimate in lieu of the measurements from the failed sensor. Several approaches implement analytical redundancy for detection, identification, and accommodation of failures on the rate gyros onboard an aircraft.

A flight termination system (FTS) is highly advantageous to work with all UAV hardware/software redundancies and independent functionalities besides to warrant the overall safety and predictability of the whole UAV cyber-physical system (CPS). If a UAV is found to be lacking in system redundancies, an independent FTS can safeguard the human subjects and critical resources.

When a UAV is designed to be completely autonomous, with no capability of pilot intervention, it will face the following onboard challenges: intelligence, teaming/swarming, health management using the software system "A Computational Logic for Applicative Common Lisp" (ACL 2), collision avoidance, affordability, and sensing.

The operating system (OS) along with the corresponding application programming interface (API) is essential because of the time-critical nature of flight control. The OS requires high reliability and real-time execution. The IEEE 1003.1 POSIX is the preferred OS interface standard because it is extensively supported and allows easy porting of applications between the various UNIX flavours and QNX. QNX is currently used widely for desktop and embedded computing since it provides an excellent set of features and performance.

The intent of sense and avoid equipment is to use sensors and other tools to find and maintain situational awareness of other traffic and to yield the right-of-way, by the rules, when there is a traffic conflict.

Automated recovery is crucial since pilot control is not possible. A UAV must have several fail-safes for the case of link losses to ensure that airborne operations can be kept in the event of a lost link.

1.7 Applications

Most UAVs are primarily used for intelligence, surveillance, and reconnaissance (ISR), which can use chemical, biological, radiological, and nuclear detection, or merely those tasks considered too dangerous or politically challenging for human crewed aircraft to undertake. UAVs are preferred over human-crewed aircraft not only because of downsizing risk and increasing confidence in mission success avoiding at the same time the human cost of losing lives if the mission is unsuccessful but also because UAVs have better and sustained vigilance over humans during monotonous operations. Moreover, many other technological, economic, and political factors have encouraged the development and operation of UAVs.

Operations that involve more than 30- or 40-h missions are best carried out using UAVs. Automation brings in low workload, low-intensity tasks are suited to unmanned aircraft.

UAVs are a perfect choice for operations in hostile environments, e.g., some fire brigades for reconnoitring fires in inaccessible locations or where smoke and flame would make human presence dangerous use small drones.

Operations like reconnaissance over unfriendly territory may result in loss of human lives. Thus, in such a scenario, multiple, cheap unmanned aircraft can be used to handle detection, command, and control systems.

A complete sensory platform for greenhouse monitoring (SPGM) consists of a sensory system onboard a small UAV to measure temperature, humidity, luminosity and CO_2 concentration and plotting maps of these variables. These SPGM features allow for climate control, crop monitoring, or vulnerability detection according to the climate and plant growth models as well as the requirements for their integration onboard [56,57].

Additional possible applications of UAVs include:

- reconnaissance, surveillance, and target acquisition;
- surveillance for SAR;
- support for maritime manoeuvres;
- meteorology missions;
- route and landing reconnaissance support;
- indirect fire adjustment and prevention;
- radio and data relay;
- nuclear cloud surveillance;
- relaying radio signals;
- contamination monitoring;
- long-range, high-altitude surveillance;
- radar system jamming and destruction;
- security;
- damage assessment;
- policing duties;
- traffic spotting;
- fisheries protection;
- pipeline survey;
- sports events film coverage;
- agricultural operations;
- powerline survey;
- aerial photography;
- border patrol;
- surveillance of coastal borders, road traffic, etc.;
- disaster and crisis management search and rescue;
- environmental monitoring;
- agriculture and forestry;
- public health;
- firefighting;
- communications relay and RS;
- aerial mapping and meteorology;
- research by university laboratories;
- law enforcement; and
- many other applications.

1.8 Book organization

The book contains the following chapters.

Chapter 1 gathers some important information about a UAV seen as a CPS.

Chapter 2 focuses on the problems related to acquiring, storing, processing, and compressing images and videos. It describes how CV software impacts tasks such as processing, communications, storage, and compression besides other applications specific to a UAV CPS while explaining the general architecture of the cloud-based UAV-CPS [58,59], its challenges, its necessities, its design goals, memory usage, specific requirements, limitations of onboard storage, general solutions, UAV DL, and protocol standardisation with examples.

Chapter 3 investigates the role of OF in visual situation awareness, detection, and obstacle avoidance systems, which interact with the environment through sensors and actuators. The use of cameras allows the integration of CV algorithms with the INS. The movement of characteristics of the image fused with the dynamic of the UAVs allows us to improve the process of remoting sense, avoid obstacles, or estimate the position and velocity of the UAV. In the literature, there are various algorithms to locate characteristics points between two consecutive images. This chapter shows how to integrate the movement of the pixel textures in the image with INS and compares the different algorithms that are used to determine points between consecutive images and to implement an algorithm that is used to determine points between consecutive images less expensive computationally speaking and with less memory consumption. A case study about using the FPGA as part of the visual servoing is discussed showing how to integrate results into the CV hardware system of a UAV and addressing the need to handle issues such as multi-resolution.

Chapter 4 explores intelligent vision-based sensor systems, future prerequisites for UAV performance assessment to unsegregated airspace for a given navigation sensor, quantitative assessment of visual sensors that are actualized in independent and incorporated frameworks, and much more so for INSs.

Chapter 5 addresses the need for modelling and simulation of UAV systems, reviews historical developments in the field, current techniques used, and future developments. The modelling of a system's dynamics and performing simulation are commonly used techniques and almost critical components in the modern development of manned and unmanned systems alike.

Chapter 6 provides an introduction to the integration of visual sensors in multi-sensor systems, along with a review of significant research efforts in the domain. The research area has matured owing to the need to operate UAS in GNSS-denied environments.

Chapter 7 describes the main visual-based pose estimation algorithms and discusses where they best apply and when each fails. Results of state-of-the-art are also presented to stimulate the development of new strategies that will overcome the remaining challenges of this research field.

As the use of UAVs increased, studies regarding their autonomous flight became an academic field of great interest for researchers. Until recently, most studies based their developments using an IMU and a GNSS as the main sensors to calculate and estimate the UAVs pose. These sensors, on the other hand, have several limitations, which can affect the navigation, therefore, the fully autonomous aspect of the system. Images captured during flight, computer vision algorithms, and photogrammetry concepts have become a core source of data to estimate the UAVs pose in real-time, therefore, composing new alternative or redundant navigation systems. Several algorithms have been proposed in the scientific community, each one working better in specific situations, and using different kinds of imaging sensors (active and passive sensors).

MAVs are a class of aircraft investigated in Chapter 8. MAVs have a large range of applications from commercial, research, government to military purposes. Recently, the applications of bio-inspired MAVs have an amount of ecological, geological, meteorological, hydrological, and human-induced disasters. Indeed, animals utilise different strategies of movement in a varied environment, which allows them to adapt themselves to get food, ran away from a threat, etc., by using the least energy dissipation. Hence, designing and modelling UAVs inspired from animals not only can decrease the energy consumption of the robot for long-time applications but also it helps us to provide some kinds of robots which can be utilised in different dangerous outdoors environment which even regular robots like UAVs cannot access them in disaster research. UAVs also have some useful advantages like safety, flexibility, relatively low cost of ownership, and ease of operation. Bio-inspired robots can be designed to both fly and move on land, either by jumping or by walking. One of the most critical parts of such vehicles (MAVs) is the vision (visual perception) system.

Chapter 9 discusses the robot operating system (ROS) with examples of its use in UAV deployed CV tasks with some references to the ROS 2.0. It also talks about the relationship between ROS and available software suites.

Chapter 10 extends some topics from Chapter 8 to the realm of security analysis. This chapter preliminarily presents typical, published, and on-going research works about security flaws on the UAV domain and possible scenarios that can arise from them. Following security strategies to the resilient operation are presented to support new UAV designs based on typical security approaches (e.g., authentication and cryptography), current research works and insights derived from aircraft design guidelines which relate intentional unauthorised interaction (cyber-threat) and its effects to safety. Finally, deployment of a robot operating system (ROS) on consumer UAV (Parrot AR.Drone 2) is performed, and a cybersecurity assessment is presented containing its findings, mitigations, and proposals to strengthen its operation (resiliency).

Chapter 11 explores how the type of environmental impacts the CV techniques, algorithms, and specific hardware to be used. Indoor environments, also known as controlled environments, generally rely on solutions based on beacons, proximity sensors, and image processing for data acquisition. In this case, as the environment is controlled, the illuminance of the scene is adjusted and sensors are previously positioned, which facilitates the development and execution of these systems. Outdoor environments, generally known for uncontrolled environmental variables, frequently require solutions based on image processing techniques to provide the data acquisition. In this environment, the non-constant variation of the illuminance of the scene and the great variation of the background of images are important complicating factors for the operation of the image processing algorithms. In addition, constructions and buildings block the signal of sensors and global positioning systems making it even harder to treat the exceptions caused by these factors. Each exception being treated in a CV system has a computational cost that can be high. If this is considered in applications using embedded hardware, some projects simply become infeasible. Researchers put great effort attempting to optimise the software for high performance and better use of the hardware resources so that less processing power is demanded as well as positively impacting the energy savings. This chapter presents a review of the main CV techniques currently used in the development of mission control software for the use in indoor and outdoor environments, providing autonomy navigation and interaction for these aerial robots.

In Chapter 12, contemporary research studies point towards increasing awareness of the importance of vision in the monitoring of UAV structural health. This chapter addresses algorithm-based solutions to real-time infrastructure damage recognition and damage accreditation for UAV systems. Damage detection and identification of effects are quantified for remediation. A vision-based UAV system may obtain visual evidence from a damaged surface of interest to detect faults, eliminate the irrelevant areas from image data, spot damage, measure the resulting consequences, record information, identify the type of fault, and point to the most noticeable kind of problem present. This chapter also discusses new approaches to sense and acquire vibration data and to pre-process these data on aeroelastic certification test flights. These new approaches aim to reduce the time to identify the aeroelastic phenomenon and to reduce the size of the hardware that must be boarded in the aircraft, thus minimizing the risks and costs of the vibration tests. The advance of computer vision systems enables the use of cameras as a motion tracker sensor with millimetre precision and accuracy. Non-contact sensors are suited to flutter analysis because they do not interfere with the dynamics of the aircraft. With the aid of computer vision algorithms, they allow the acquisition of 2D and/or 3D data, instead of one-dimensional data acquired by traditional vibration sensors, such as piezoelectric accelerometers. Nevertheless, it is worth noting that to capture the aeroelastic phenomena, these cameras must operate with at least five times higher frame rates than traditional ones. Therefore, this intelligent video sensor system that is capable of processing the obtained images and providing the user with the data of movements in a vector ready for use, at a reasonable cost, is an important topic under development. In addition, this work proposes modifications in the acquisition and pre-processing procedures of the signals traditionally used in flutter certification analytic, for example, modal analysis, suited to an embedded system, and near-real-time process.

Chapter 13 tackles ISR missions, persistent surveillance is commonly defined as the exercise of automatic intelligence discovery by monitoring a wide area coverage for hours of operation at a high altitude leveraging aerial platforms (manned or unmanned). The platform can be large enough to carry a matrix of high-resolution sensors and a rack of high-performance computing to process in real-time all sensors' feeds. With the current ISR growing in capability, engineering and optics-based aerial surveillance solutions have become a design challenge. More onboard processing is desired for an increasing fidelity/resolution sensors' feeds while matching a constraining SWaP (size, weight, and power) budget requirements in a bandwidth-constrained operating theatre. The advent in small

UAV (sUAV) technology, able to carry sophisticated optics payloads and to take aerial images from strategic viewpoints, has become unavoidable in nowadays battlespace contributing to moving forward the ISR capabilities. The constrained onboard processing power in addition to the strict limit in the flying time of sUAV are amongst the serious challenges that have to be overcome to enable cost-effective persistent surveillance based on sUAV platforms. All previous examples show that tailoring the sensors to match the platforms' environment is a challenging endeavour, and therefore, architects have shifted their design methodology to be based on hardware and software open architectures as a centrepiece of their approach in building cost-effective surveillance solution design. This chapter is a brief introduction to hardware and software building blocks for developing persistent surveillance systems. In our context, the focus is, in particular, on electro-optic (visual spectrum) and IR integrated solutions leveraging computer vision techniques for surveillance missions.

Chapter 14 closes the book with some important conclusions and more information on new research directions.

References

[1] Razmjooy, N., Mousavi, B. S., Soleymani, F., and Khotbesara, M. H. 2013. A computer-aided diagnosis system for malignant melanomas, *Neural Computing and Applications*, Vol. 23(7–8), 2059–2071.

[2] Moallem, P., Razmjooy, N., and Ashourian, M. 2013. Computer vision-based potato defect detection using neural networks and support vector machine. *International Journal of Robotics and Automation*, Vol. 28(2), 137–145.

[3] Mousavi BS, Sargolzaei P, Razmjooy N, Hosseinabadi V, and Soleymani F. 2011. Digital image segmentation using rule-base classifier. *American Journal of Scientific Research*, Vol. 35, 17–23.

[4] Hemanth, D.J., and Estrela, V.V. Deep learning for image processing applications. Adv. Par. Comp. IOS Press. ISBN978-1-61499-821-1 (print) 978-1-61499-822-8 (online) 2017.

[5] Jesus de M.A., Estrela V.V., Saotome O., and Stutz D. Super-resolution via particle swarm optimization variants. In: Biologically Rationalized Computing Techniques for Image Processing Applications, pp. 317–337, 2018.

[6] Estrela, V.V., Saotome, O., Loschi, H.J., *et al.* 2018. Emergency response cyber-physical framework for landslide avoidance with sustainable electronics, Technologies, 6, 42. doi: 10.3390/technologies6020042.

[7] Petricca, L., Ohlckers, P., and Grinde, C. 2011. Micro- and nano-air vehicles: state of the art. *International Journal of Aerospace Engineering*. Vol. 2011, Article ID 214549, 17 pages. http://dx.doi.org/10.1155/2011/214549

[8] Ezequiel, C. A. F., Cua, M., Libatique, N. C., *et al.*, UAV aerial imaging applications for post-disaster assessment, environmental management and

infrastructure development. In Proceedings of the 2014 International Conference on Unmanned Aircraft Systems (ICUAS 2014), 274–283.

[9] Zingg, S., Scaramuzza, D., Weiss, S., and Siegwart, R. 2010. MAV navigation through indoor corridors using optical flow. Proceedings of the 2010 IEEE International Conference on Robotics and Automation (ICRA 2010).

[10] http://www.questuav.com/news/fixed-wing-versus-rotary-wing-for-uav-mapping-applications (Accessed on November 20, 2016).

[11] Rangel, R. K., Kienitz, K. H., and Brandão, M. P. 2009. Development of a multi-purpose portable electrical UAV system, fixed & rotative wing. Proceedings of the 2011 IEEE Aerospace Conference, doi: 10.1109/AERO.2011.5747512.

[12] dos Santos, D. A., Saotome, O., and Cela, A. 2013. Trajectory control of multirotor helicopters with thrust vector constraints. Proceedings of the 21st IEEE Mediterranean Conference on Control & Automation (MED). doi: 10.1109/MED.2013.6608749.

[13] Figueiredo, H., Bittar, A., and Saotome, O. 2014. Platform for quadrirotors: analysis and applications. Proceedings of the 2014 IEEE International Conference on Unmanned Aircraft Systems (ICUAS). doi: 10.1109/ICUAS.2014.6842332.

[14] Krantz, J. 2013. What is Sensation and Perception? *Experiencing Sensation and Perception.*

[15] Stein, B. E., Stanford, T. R., and Rowland, B. A. 2009. The neural basis of multisensory integration in the midbrain: its organization and maturation. *Hear Res.* 258 (1-2): 4–15. doi:10.1016/j.heares.2009.03.012. PMC 2787841free to read. PMID 19345256.

[16] Lewkowicz, D. J., and Ghazanfar, A. A. 2009.The emergence of multisensory systems through perceptual narrowing. *Trends in Cognitive Science*, 13 (11): 470–8. doi:10.1016/j.tics.2009.08.004.

[17] Zmigrod, S., and Hommel, B. 2010. Temporal dynamics of unimodal and multimodal feature binding. *Attention, Perception, & Psychophysics,* 72 (1): 142–52. doi:10.3758/APP.72.1.142.

[18] Ellen, R. A., Roberts, P. J., and Greer, D. G. 2005. An investigation into the next generation avionics architecture for the QUT UAV project. In Proceedings of the Smart Systems 2005 Postgraduate Research Conference. Goh, Roland and Ward, Nick (Eds.), Brisbane. http://eprints.qut.edu.au.

[19] Bilbao, J.,Olozaga, A., Bravo, E., García, O., Varela, C., and Rodríguez, M. 2008. How design an unmanned aerial vehicle with great efficiency in the use of existing resources. *International Journal of Computers*, Vol. 2 (4), 442–451.

[20] Salkever, A. 2003. The Network is the Battlefield (Business Week). https://citizenlab.org/2003/01/the-network-is-the-battlefield-business-week/ (Accessed October 21, 2016).

[21] Nitti, D.O., Bovenga, F., Chiaradia, M.T., Greco, M., and Pinelli, G. 2015. Feasibility of using synthetic aperture radar to aid UAV navigation. *Sensors*, 15, 18334–18359.

[22] Zhu, Q., Avidan, S., Yeh, M.-C., and Cheng, K.-T. 2006. Fast human detection using a cascade of histograms of oriented gradients. In Proceedings of the 2006 Computer Vision and Pattern Recognition (CVPR06), 1491–1498.

[23] Zhou, J., and Hoang, J. 2005. Real time robust human detection and tracking system. In Proceedings of 2005 IEEE Computer Society Conference on Computer Vision and Pattern Recognition (CVPR05), 149.

[24] Vanegas, F., and Gonzalez, F. 2016. Enabling UAV navigation with sensor and environmental uncertainty in cluttered and GPS-denied environments. *Sensors*, 16, 666.

[25] Rudol, P., and Doherty, P. 2008. Human body detection and geolocalization for UAV search and rescue missions using color and thermal imagery. *Proceedings of the IEEE Aerospace Conference*. doi: 10.1109/AERO.2008.4526559.

[26] Razmjooy N., Ramezani M., and Estrela V. V. A solution for Dubins path problem with uncertainties using world cup optimization and Chebyshev polynomials. In: Iano Y., Arthur R., Saotome O., Vieira Estrela V., and Loschi H. (eds), *Proc. BTSym 2018. Smart Innovation*, Systems and Technologies, vol 140, 2019. Springer, Cham doi: 10.1007/978-3-030-16053-1_5, 2019.

[27] DeGarmo, M., and Nelson. G. 2004. Prospective unmanned aerial vehicle operations in the future national airspace system, In Proceedings of the AIAA 4th Aviation Technology, Integration and Operations (ATIO) Forum, Aviation Technology, Integration, and Operations (ATIO) Conferences, Chicago, USA doi: 10.2514/6.2004-6243.

[28] França, R.P., Peluso, M., Monteiro, A.C.B., Iano, Y., Arthur, R., and Estrela, V.V. Development of a kernel: a deeper look at the architecture of an operating system. In: Iano Y., Arthur R., Saotome O., Estrela V.V., and Loschi H. (eds) *Proc. BTSym 2018. Smart Innovation, Systems and Technologies*, vol 140. Springer, Cham, 2019.

[29] Park, C., Cho, N., Lee, K., and Kim, Y. 2015. Formation flight of multiple UAVs via onboard sensor information sharing. *Sensors*, 15, 17397–17419.

[30] Jain, R., Templin, F., and Yin, K.-S. 2011.Wireless datalink for unmanned aircraft systems: Requirements, challenges and design ideas. In Proceedings of the Infotech@Aerospace 2011, St. Louis, Missouri, American Institute of Aeronautics and Astronautics, doi: 10.2514/6.2011-1426.

[31] Sayyed, A., de Araújo, G.M., Bodanese, J.P., and Becker, L.B. 2015. Dual-stack single-radio communication architecture for UAV acting as a mobile node to collect data in WSNs. *Sensors*, 15, 23376–23401.

[32] Zhang, T., Li, W., Achtelik, M., Kuhnlenz, K., and Buss, M. 2009. Multisensory motion estimation and control of a mini-quadrotor in an air-ground multi-robot system. In Proceedings of the 2009 IEEE International Conference on Robotics and Biomimetics (ROBIO 2009), 45–50.

[33] Kretschmar, M., and Welsby, S. 2005. Capacitive and inductive displacement sensors, in Sensor Technology Handbook, J. Wilson (Ed.), Newnes: Burlington, MA.

[34] Grimes, C. A. Dickey, E. C., and Pishko, M. V. 2006. Encyclopedia of sensors (10-Volume Set), American Scientific Publishers. ISBN 1-58883-056-X.

[35] Bănică, F.-G. 2012. Chemical sensors and biosensors: Fundamentals and applications. Chichester, UK: John Wiley & Sons. ISBN 978-1-118-35423-0.

[36] Blaauw, F.J., Schenk, H.M., Jeronimus, B.F., *et al.* 2016. Let's get Physiqual – an intuitive and generic method to combine sensor technology with ecological momentary assessments. *Journal of Biomedical Informatics*, Vol. 63, 141–149.

[37] Nonami, K., Kendoul, F., Suzuki, S., Wang, W., and Daisuke Nakazawa, D. 2010. Autonomous flying robots, unmanned aerial vehicles and micro aerial vehicles, Springer, ISBN 978-4-431-53855-4.

[38] Doherty, P., and Rudol, P. 2007. A UAV search and rescue scenario with human body detection and geolocalization. In Proceedings of the 20th Australian joint conference on advances in artificial intelligence (AI'07), 1–13, Berlin, Heidelberg, Springer-Verlag. ISBN 3-540- 76926-9, 978-3-540-76926-2.

[39] Rudol, P., Wzorek, M., Conte, G., and Doherty, P. 2008. Micro unmanned aerial vehicle visual servoing for cooperative indoor exploration. In Proceedings of the IEEE Aerospace Conference, 1–10.

[40] Griswold, M. E. 2008. Spectrum management key to the future of unmanned aircraft systems? USAF, Air University Press Maxwell Air Force Base, Alabama.

[41] Shakernia, O., Koo, T. K. J., and Sastry, S. S. 1999. Landing an unmanned air vehicle: Vision based motion estimation and nonlinear control. *Asian Journal of Control*, Vol. 1, No. 3, 128–145. http://www2.eecs.berkeley.edu/Pubs/TechRpts/1999/3782.html

[42] Coelho, A. M., and Estrela, V. V. 2012. EM-based mixture models applied to video event detection, "Principal Component Analysis – Engineering Applications," Dr. Parinya Sanguansat (Ed.), 101–124, InTech. doi: 10.5772/38129.

[43] Coelho, A. M., and Estrela, V. V. 2013. State-of-the-art motion estimation in the context of 3D TV, "Multimedia Networking and Coding," Reuben A. Farrugia, and Carl J. Debono (Eds.). doi: 10.4018/978-1-4666-2660-7.ch006

[44] Werner, S., Dickmanns, D., Furst, S., and Dickmanns, E. D. 1996. A vision-based multi-sensor machine perception system for autonomous aircraft landing approach, In Proc. the SPIE – The Int. Soc. Optical Eng., Vol. 2736, Orlando, USA, 54–63.

[45] Schell, F. R., and Dickmanns, E. D. 1994. Autonomous landing of airplanes by dynamic machine vision. *Machine Vision and Applications*, Vol. 7, 127–134.

[46] Yang, Z. F., and Tsai, W. H. 1998. Using parallel line information for vision-based landmark location estimation and an application to automatic helicopter landing. *Robotics and Computer-Integrated Manufacturing*, Vol. 14, No. 4, 297–306.

[47] Espiau, B., Chaumette, F., and Rives, P. 1992. A new approach to visual servoing in robotics. *IEEE Transactions on Robotics and Automation*, Vol. 8, No. 3, 313–326.

[48] Kang, W., Yu, S., Ko, S., and Paik, J. 2015. Multisensor super resolution using directionally-adaptive regularization for UAV images. *Sensors*, 15, 12053–12079.

[49] Karpenko, S., Konovalenko, I., Miller, A., Miller, B., and Nikolaev, D. 2015. UAV control on the basis of 3D landmark bearing-only observations. *Sensors*, 15, 29802–29820.

[50] Yoon, I., Jeong, S., Jeong, J., Seo, D., and Paik, J. 2015. Wavelength-adaptive dehazing using histogram merging-based classification for UAV images. *Sensors*, 15, 6633–6651.

[51] Wen, M., and Kang, S. 2014. Augmented reality and unmanned aerial vehicle assist in construction management. Computing in Civil and Building Engineering. 1570–1577. doi: 10.1061/9780784413616.195 http://ascelibrary.org/doi/abs/10.1061/9780784413616.195.

[52] Li, H., Zhang, A., and Hu, S. A multispectral image creating method for a new airborne four-camera system with different bandpass filters. *Sensors* 2015, 15, 17453–17469.

[53] Aroma R. J., and Raimond K. A novel two-tier paradigm for labeling water bodies in supervised satellite image classification. 2017 International Conference on Signal Processing and Communication (ICSPC), 384–388, 2017.

[54] Aroma R. J., and Raimond K. A review on availability of remote sensing data. 2015 IEEE Technological Innovation in ICT for Agriculture and Rural Development (TIAR), 150–155, 2015.

[55] Chiang, K.-W., Tsai, M.-L., Naser, E.-S., Habib, A., and Chu, C.-H. 2015. New calibration method using low cost MEM IMUs to verify the performance of UAV-borne MMS payloads. *Sensors*, 15, 6560–6585.

[56] Roldan, J. J., Joossen, G., Sanz, D., del Cerro, J., and Barrientos, A. 2015. Mini-UAV based sensory system for measuring environmental variables in greenhouses. *Sensors*, 15, 3334–3350.

[57] Gonzalez, L. F., Montes, G. A., Puig, E., Johnson, S., Mengersen, K., and Gaston, K. J. 2016. Unmanned aerial vehicles (UAVs) and artificial intelligence revolutionizing wildlife monitoring and conservation. *Sensors*, 16, 97.

[58] Estrela, V. V., Monteiro, A. C. B., França, R. P., Iano, Y., Khelassi, A., and Razmjooy, N. Health 4.0: Applications, management, technologies and review. *Medical Technologies Journal*, 2(4) 262–276, 2019. doi: 10.26415/2572-004x-vol2iss1p262-27.

[59] Kehoe, B., Patil S., Abbeel P., and Goldberg, K. A survey of research on cloud robotics and automation, IEEE Transactions on Automation Science and Engineering, 2014.

Chapter 2

Computer vision and data storage in UAVs

Vania V. Estrela[1]*, Jude Hemanth*[2]*, Hermes J. Loschi*[3]*, Douglas A. Nascimento*[3]*, Yuzo Iano*[3] *and Navid Razmjooy*[4]

Unmanned aerial vehicles (UAVs) handle operations such as inspection, mapping, monitoring, and surveying, to cite a few, that assist image processing, aerial imaging, and computer vision (CV) algorithms. The UAVs with cameras collect a massive amount of images and videos for various research and commercial applications. Furthermore, UAVs have various sensors, e.g., thermal, magnetic, sound, light, and speed, to collect environment details for specific research as well as commercial usages. Thus, this chapter focuses on acquiring, storing, processing, and compressing images and videos. This chapter describes how CV software impacts tasks such as processing, communications, storage, and compression besides other applications specific to a UAV cyber-physical system (CPS). Additionally, Section 2.2 explains the general architecture of the cloud-based UAV-CPS, the challenges, and design goals. Section 2.3 discusses memory usage in UAV, specific requirements, limitations of onboard storage, and general solutions. Section 2.4 briefs the UAV data logging (DL), primary benefits, and protocol standardisation with examples. Section 2.5 grants a view of the different types of DL, requirements, and proposes solutions. Section 2.6 discusses future trends of data storage, data processing, control, the impact of big data, complexity, privacy barriers, infrastructure, and other challenges.

2.1 Introduction

The choice of onboard components takes into consideration the lowest possible weight, small size, low vibration sensibility for high recorded data quality and reliability, computational abilities, compatibility, power consumption, and interfacing ability. Computer vision (CV) algorithms tend to be computationally expensive, and an unmanned aerial system (UAV) project requires a compromise

[1]Universidade Federal Fluminense, RJ, Brazil
[2]Karunya University, Coimbatore, India
[3]LCV, FEEC, UNICAMP, Campinas, SP, Brazil
[4]Tafresh University, Tafresh, Iran

between computational resources and the other hardware requirements for a given maximum payload that can be better understood using the cyber-physical system (CPS) paradigm [1–8]. The term payload refers to the reduced physical space for transportation purposes in an aircraft.

The applications of UAVs involve restricted computational processing, and low energy resources of UAVs show a noteworthy test for real-time information processing, networking, and policymaking. To talk about the challenges enforced by the total bulk of acquired data, principally video, the irregular, and limited network means, and the limited UAV resources, a cloud-supported UAV application framework can be implemented. This structure can integrate video acquisition, file scheduling, information offloading/processing, and network state assessment to provide an efficient and scalable system. The framework includes a client-side set of mechanisms hosted on the UAV CPS, which selectively unloads the gathered data to a cloud-based server. Then, this server arranges for real-time processing and data feedback services to the associated ground station or control centre and client device/operator.

This work discourses the storage and software requirements for image acquisition, analysis, compression, and knowledge retrieval in UAV-CPS. Some issues worth discussion are:

(a) The system safely running from memory with the help of a boot method and remotely updated without difficulty;
(b) the operating system (OS) kernel;
(c) the file system with minimum utility programs for running and remotely accessing the UAV;
(d) an implementation methodology for the critical UAV software;
(e) a thorough review of all data logging methods; and
(f) the development of software simulators to model and test the hardware components and physics of the vehicles in a safe situation.

Developing UAV software can be an enormous task that involves constructing, incorporating, writing, simulating, and testing all the indispensable components for a functioning UAV platform. There are various problems anticipated while developing the full software from top to bottom for autonomous vehicles, thus offering a complete software framework for these vehicles, which can be applied in practice.

Designing a new product with minimum hardware is possible using simulation software. This reduction of hardware reduces the costs and the time. Further, it makes ease use of the product by turning it into an effortlessly upgradable, or even wholly modifiable, with a simple software update. These all add value to the avionics product. The heart and brain of a UAV-CPS software appear in Figure 2.1, which depicts its role along with its supporting hardware modules that are required to interface with the UAV ecosystem. The software block corresponds to software on either onboard and offboard units.

Near-all UAV categories follow the above architecture paradigm. Sensors perceive the environment and send their signals as inputs to the control hardware and software. Actuators implement alterations in the environment using the

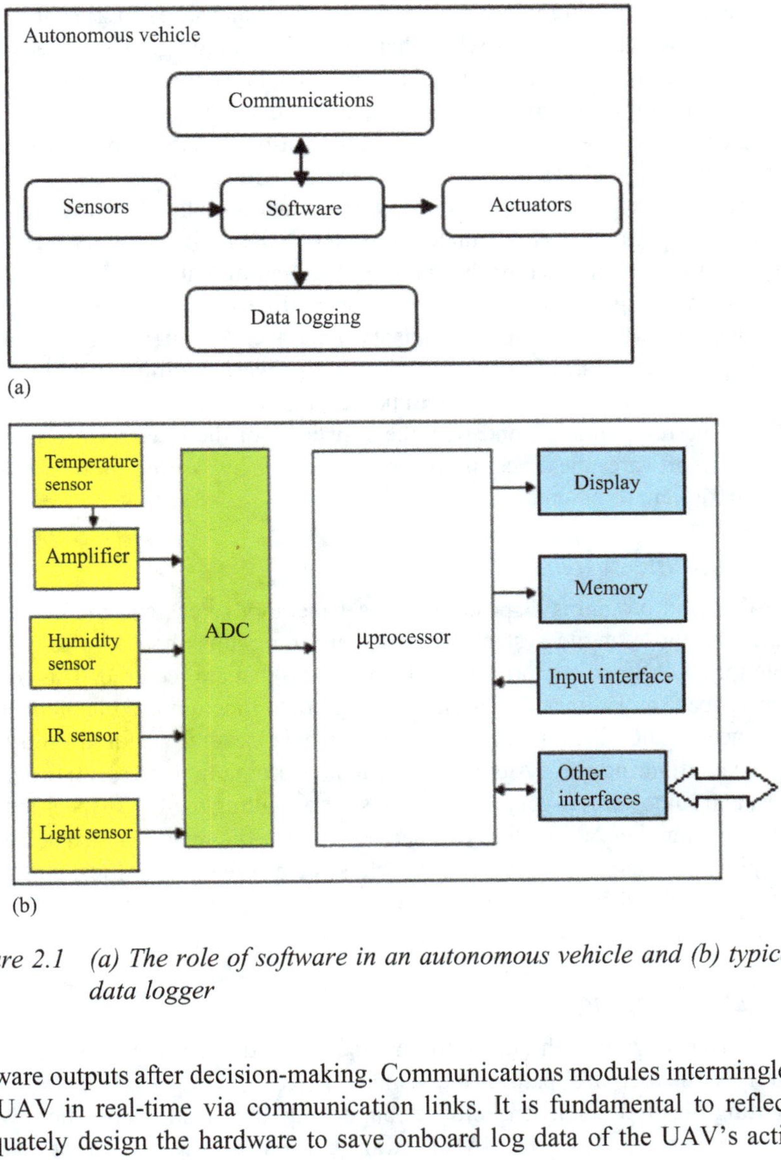

Figure 2.1 (a) The role of software in an autonomous vehicle and (b) typical data logger

software outputs after decision-making. Communications modules intermingle with the UAV in real-time via communication links. It is fundamental to reflect and adequately design the hardware to save onboard log data of the UAV's activities and findings.

2.1.1 Requirements

UAVs are planned for some high-level applications that cannot be translated into source code immediately. First, the development phase needs to state the operational necessities of the vehicles keeping the interface support and platform independence.

Since sensors and actuators exist with a huge variety of interfaces such as USB, Ethernet, RS232, I2C, RS422, RS485, SPI, Bluetooth, PWM, and so on, communications hardware will have to address these interfaces to keep the whole system working satisfactorily while handling data logging. The selection of UAV-CPS hardware and software platforms should support all currently and eventually necessary interfaces for a particular application.

Furthermore, the software and hardware platforms should be independent. Software portability allows exchanging computer hardware while using the same existing software without writing too many code amendments or, ideally, any more code. Software portability is still under work due to the multidisciplinary character of avionics and the existence of different civilian and military demands.

A free operating UNIX system kernel such as Linux meets these requirements since it has drivers for all the compulsory interfaces and it works on multiple hardware platforms, instead of using a specific hardware platform in addition to a non-portable software environment with partial interface support. Each UAV has a list of specific requirements, and with the assistance of the interfaces support and the portable software, these requirements can be met by writing high-level code, rather than drivers or assembly.

2.1.2 Root file system

All the files and programs dispose of the user memory space, and the kernel will save them in the root file system. The files and programs should be as small as possible for a UAV and comprise only the programs that are required. This reduces the storing space and allows for the smallest system to function and maintain easily. A UAV necessitates little memory, which can have several standard utility programs for configuring the system, a few daemon programs for essential network services, and some vehicle-specific programs for executing the UAV-specific tasks. However, the number of CV tasks that can be performed onboard is increasing, and similar programs demand more from the file system. Editors, libraries, compilers, debuggers, graphical user interfaces, and the like commonly do not fit in the aircraft file system.

2.1.3 Data logging

Although data logging is the most critical characteristic of a UAV, it should not disturb the critical control path of the vehicle.

Writing files is an indispensable task but potentially blocking operation, which may lead to crashing down the UAV by waiting for a file write operation to complete. UNIX-based operating systems have several data logging methods. Some data logging methods are suitable for specific applications, but some are not suitable. Hence, it is paramount to comprehend each DL method quantitatively for selecting a suitable method for each UAV-based applications.

The sensors of the UAV require data storage with a suitable compression technique which helps to store more information in the fixed amount of storage available in the UAV. Further, the data compression technology helps to transfer the UAV sensors data to the cloud or the ground station in minimum time with the limited available bandwidth.

2.1.4 Cloud support and virtualisation

Storage virtualisation stands for a technology that consists of making one collection of resources look like another group of resources, ideally with enhanced features. A logical representation of the set of resources must be independent of physical limitations and must conceal part of the complexity. Moreover, this logical representation must include and integrate new functions with existent services and can be nested or spread over multiple layers of a system [9].

The term virtualisation belongs to the broader concept of software-defined storage (SDS), which is a data storage method, which detaches the program controlling the tasks related to storage from the physical storage hardware. This concept permits SDS solutions to be located on any existing storage systems or, mostly, mounted on any commodity personal computer hardware and hypervisor. To change to an advanced level in the IT stack hierarchy leads to deeper integration and reaction to application needs regarding storage performance and abilities. SDS solutions afford a complete set of storage services (comparable to conventional hardware structures) with numerous persistent storing resources: internal disks, cloud, external storage systems, or object/cloud platforms [9].

This chapter is structured in seven sections. Section 2.2 investigates the architecture of the cloud-based UAV cyber-physical system. UAV needs versus memory use and architectures appear in Section 2.3. Section 2.4 discusses UAV data logging. Section 2.5 debates types of data logging in UAV. A discussion along with future trends appears in Section 2.6, followed by the conclusions in Section 2.7.

2.2 The architecture of the cloud-based UAV cyber-physical system

Figure 2.2 depicts a typical UAV CPS whose communication network environments are typically termed as disconnected, intermittent, and limited (DIL) and has impairments due to missing or corrupted information, intermittent or restricted connectivity.

Usually, cloud computing overcomes several UAV resource restrictions by handling some offload data. Nevertheless, to process data using offloading has a cost, e.g., enormous data volumes can severely lower the energy available for the UAV and require a considerable network bandwidth. Similarly to existing efforts on mobile cloud use for smartphone applications, customised algorithms to find the network infrastructure strength and readiness before offloading. These algorithms have inspired UAV-CPS frameworks relying on computing clouds as well as network components. The need for a network state measurement capability is graver in DIL environments than in mobile networks that use robust and redundant commercial structures and is a vital component of the UAV-CPS client–server configuration. The UAV hardware hosts the client, which gathers visual and contextual information and has a context-aware video scheduler unit that selectively offloads acquired info to the cloud-based on contextual data. The cloud infrastructure encloses the server and pays attention to incoming client traffic. After receiving

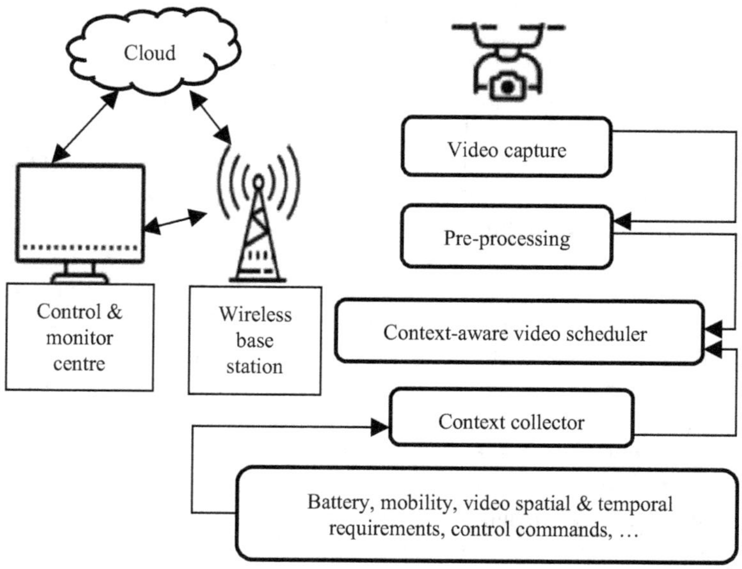

Figure 2.2 The UAV cyber-physical system architecture

meaningful data, the server offers valuable services like powerful processing capabilities for object identification, and scene comprehension that will result in choices that must be sent to advance processing services in the ground station or another control facility [10–12].

The deployment of UAVs for collecting critical data requires innovation, and it is a challenging task comprising some of the following problems [11,12]:

(a) The observed areas can change or mess up the UAV operation.
(b) UAVs battery has limitations that affect their flying time considerably, communication structure, and processing capacity.
(c) The UAV may connect over a DIL network environment where the network state sentience must be handled by an algorithm that decides when the information can be reassigned to the cloud server.
(d) The overall UAV sensing has severe requirements for correctness. For instance, in search and rescue scenarios, image or video evidence collected has to be processed appropriately.
(e) The UAV mobility pattern directly impacts the information acquisition and the execution of image processing processes. Particularly, the collected data quality may fluctuate during the flying time. Extra power and bandwidth will be consumed if all records go to the ground station.
(f) Object detection and other applications need a high-quality image and video data, which leads to the need to adjust UAV flight configurations to optimise the resolution, dimensionality, and quality of visual data.
(g) Intelligent video capture and intelligent video processing tools significantly better the quality of decision-making in disaster scenarios.

Since UAVs come across the same challenges as mobile communication networks, these systems use cloud computing and virtualisation. However, it is not always preferable to link to a cloud because the deployment of UAVs has considerable higher complexity than the use of smartphones in mobile networks. Other factors, such as time, task management, flight path, and information processing, may require specific ground-breaking solutions.

The design purposes of a UAV-CPS cloud framework rely on the following assumptions:

1. Efficient cost-efficient onboard video pre-processing procedures can be done by the previous framework to expand performance.
2. Frame-based video segmentation and communication with a convenient sampling rate can reduce the data volume offloaded to the cloud.
3. Sufficient UAV onboard storage to save all data captured during the entire deployment time with the assistance of current solid-state technologies.
4. Adequate wireless communications between the UAV, the ground station (GS), the cloud, and other control/management centres demand pre-processing programs on the UAV that can filter out superfluous frames; the remaining critical frames must be exchanged reliably among the cloud and the UAVs.
5. Cloud computing helps to deal with issues like the scalability and performance of UAV-CPSs with reasonable effort.

Figure 2.2 offers a higher-level view of the UAV-CPS client and server frameworks from the design perspective. The UAV-CPS is the client, consists of a context collector, a context-aware video scheduler, a camera for capturing video, and a video pre-processing unit. The cloud server stores video retrieves video, detects an object, and performs other types of data mining. Wireless networks connect the client and the server. The control centre accesses the processed data through the server, assisting automatic and human decision-making.

The client video recording and pre-processing units collect video data, store them in an onboard repository (e.g., a hard disk), in addition to running simple pre-processing. Further, it picks a frame at a pre-defined time instant and sends it to the pre-processing unit to attain material for the context-aware scheduler, which receives video data and information from a context collector.

The context collector instructs the onboard sensors and other systems to collect the UAV battery power level, real-time mobility patterns, control centre information, spatial coordinates, temporal data, and network state metrics concurrently. The context-aware scheduler processes the acquired material and the output image developed from the image pre-processing stage to later decide about the frame. For a more rigorous analysis, the frame is transmitted to the cloud.

A cloud supports virtualisation and hosts the server components, scalable resource allotment, and easy configuration characteristics like the way post-processing processes are applied to satisfy the power demanding and computationally intensive requirements. The server contains two focal modules: video retrieval units and data storage modules. The data storage systems support an assortment of data mining services and signal processing. The server retrieves video data with results and client information for wide-ranging processing (e.g., detected potential targets), which are directed to the control centre used for assessment. First, the scheduling process inspects the UAV status to be sure there is enough power to transfer data. If this holds, then it checks the UAV motion pattern to enforce the satisfactory quality of the acquired data and that they comprise interesting objects. In the end, a more sophisticated task, for instance, relates the histogram distribution of the frame undergoing processing with those obtained from previously obtained frames. If a significant change happens, then the network state is appraised to determine if the frame will be directed within a time proper for real-time evaluation.

If the network accessibility and capacity are sufficient, then the cloud will receive the frame for additional processing.

A virtual machine can run in a cloud which also hosts the server components. The video retrieval unit continually listens for arriving data, reconstructs the video data, and then post-processes it. The server also saves incoming video data for further treatment by other control services when extra knowledge is indispensable to support choices. The server is also responsible for a control centre interface [13].

Wireless and wired networks provide the basis for the communications among parts of the UAV-CPS: client, server, and ground station. A wireless connection connects UAVs and the ground station, which is connected to the cloud by wired and wireless connections, for exchanging data with the network connections. In each case, data may be transmitted over networks exhibiting DIL environment characteristics. The choice of communications model may suit a service-oriented environment where the communication reliability is achieved through a web services middleware.

There is a strong connection concerning network losses and the types of data storage/processing because these issues are essential for performance maximisation of the whole system. Decision-making relies significantly on network state metrics. If the wireless channels connecting the UAVs and ground station have high data rates, high definition videos, and different frame rates (depending on use and circumstances), then more records can be directed to the cloud. If the channel quality is strict, only critical frames and messages should flow in the network. Both pre- and post-processing algorithms hosted in the UAV and the cloud should be enhanced regarding efficiency, scalability and detection performance.

2.3 UAV needs versus memory use

When it comes to CV processing frameworks employed in UAVs, there are the teleoperation (TO) and the onboard vision processing (OVP) models depicted in Figures 2.3 and 2.4.

In the TO model, video and images acquired by onboard cameras undergo compression using a standard procedure such as JPEG, Wavelets, MPEG, or other and further wireless transferred to a GS. Contrarywise, image and video compression are computationally intensive tasks with an adverse effect on the system consumption

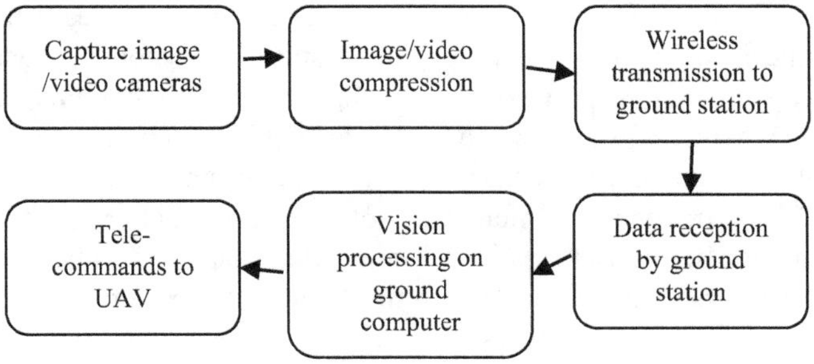

Figure 2.3 Teleoperation model

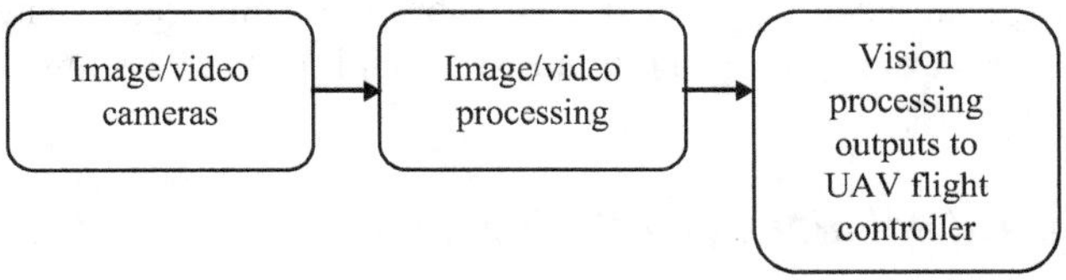

Figure 2.4 Onboard vision processing model

of power. Moreover, compressed image and video data use wireless transmission, which requires sufficient bandwidth that increases pressure on power system resources. Thus, image and video data arriving at the GS are generally noisy and delayed [14]. After ground computer processing, telecommands are sent back to UAVs via TO but can only assure near real-time operation at best. Likewise, human participation in TO is a significant downside that can both help and introduce some autonomy risks.

In contrast to TO, the OVP model encourages onboard processing as grasped from Figure 2.4. This model safeguards autonomy as well as real-time operation with scarce onboard computational resources. This model is only valid for simple image processing tasks. Even for simple CV operations, guarding system power consumption within reasonable bounds is a non-trivial undertaking. This chapter emphasizes implementation bottlenecks of onboard CV processing and possible solutions because autonomy is paramount in avionics research.

2.3.1 Limitations of OVP

CV programs are both computationally- and data-intensive. Even computers with multiple processors, deep pipelines, sophisticated memory hierarchies, and working at high power, when compared to UAVs, face challenges to run low-level CV algorithms like feature extraction for medium and high-resolution images. CV processing demands high-performance from embedded computers relying on commercial off-the-shelf (COTS).

With a simple, low-power onboard processor with limited computational capability, UAVs cannot always provide image/video data in real-time. Lack of computer architectures for processing image/video records in real-time is a significant hurdle in onboard UAV vision processing. Strict power constraints also imply low-power solutions to maximise the flight time while completing a mission in real-time. Dynamic power constraints also restrict the maximum operating clock frequency of the architecture because one way of increasing performance is to raise the system clock frequency, which it is not proper for battery-powered UAVs. Since UAVs have a compact area and low weight, that need to be accounted for when seeking a lightweight solution with the smallest form factor. Concisely, low-power CV processing architectures, functioning at low clock frequencies, lightweight and with small size for real-time operation, are fundamental for making UAVs autonomous.

To overcome the power constraints, UAVs can have solar panels to harvest energy for long endurance flights [15]. Laser power beaming can eventually deliver additional energy at night or when luminosity is minimal to allow long UAV flights without landing [16].

2.3.2 General solutions and their viability analysis

In general, UAV computational efficiency can be improved at subsequent levels:

Algorithmic: For the reasons talked over in the previous sections, CV algorithmic solutions would be the best possible way to deploy UAV applications. The development of new CV algorithms that reduces the computational load and data dimensionality can pave the way for real-time deployment on COTS low-power embedded computers.

Software: CV algorithms aiming at high execution speed like the speeded-up robust features [17] entail improvements to deliver real-time performance. As a result, a purely software-based approach is not always viable at this point because it can solve computational caveats, but a heavy reliance on computational capabilities also suffers impacts from the underlying hardware.

Hardware: Recent research [18–25] has shown that both hardware-driven and hybrid solutions can advance UAV-CPSs. Improvements in COTS will support all other levels without tailor-made architectures.

Hybrid solution: It refers to any conceivable combination of the three other solutions (e.g., hardware-software co-design or hardware resolution with some algorithmic optimisations to diminish computations). FPGAs can fulfill strict weight, size, t and power constrictions necessary for UAV-CPSs real-time performance owing to their low price and short design time. Some interesting field-programmable gate array-based solutions appear in [26,27].

2.4 UAV data logging

A data logger (DL) (aka data recorder) is as per Figure 2.1(b) a device that records evidence over time about the UAV parameters like pose either with built-in sensors or through external tools and sensors. As a rule, they are small, portable, battery-powered, and have at least a microprocessor, in-house memory for file storage, and sensors. Some DLs interface with other computers and use software that initiates the DL so that visualisation and analyses of the collected data can be done onboard or remotely.

DLs vary between multipurpose types for a variety of measurements to very particular instruments for gaging one environment or application kind only. General-purpose DLs are generally programmable. Although electronic DLs replaced some chart recorders, there are still static machines with a restricted number or unchangeable parameters.

DLs afford information collection on a 24-hour basis automatically, which is a chief benefit. DLs are characteristically set up and left unattended to sense and record material during the monitoring period upon activation. This fosters a comprehensive, accurate depiction of the environmental conditions experiencing surveillance, e.g., relative humidity and air temperature.

Standardisation of protocols as well as data formats posed a big problem but are now emerging in the industry. XML, JSON, and YAML, for instance, are little by little being embraced for data exchange. The increasing growth of the Semantic Web together with the Internet of Things will likely quicken this trend. Some standard protocols embrace a smart protocol called SDI-12 that allows the connection of some instruments to a variety of DLs. The SDI-12 also supports multi-drop instruments. Some DL companies also offer the MODBUS standard traditionally for industrial control, and there are countless industrial devices, which withstand this communication standard. One more multi-drop protocol widely used relies on the CAN Bus (ISO 11898). Some DLs employ a malleable scripting scheme to adapt to various non-standard protocols.

DLs typically have slow sample rates, and they are implicitly standalone devices. Characteristic data acquisition systems get data from connected computers. This standalone aspect of DLs implies onboard memory to save the acquired data. Ideally, this memory should be extended to house an extended.

Given the lengthy recording times of DLs, they normally feature a mechanism to associate a timestamp to guarantee that each recorded datum has the date and time of acquisition to create a sequence of events. By themselves, DLs usually employ integrated real-time clocks whose drift can be a paramount consideration.

DLs vary between simple single-channel input and complex multi-channel instruments. Overall, the simplest device results in the highest programming flexibility. Some more refined instruments tolerate cross-channel computations and alarms, given some predetermined conditions. The newest DLs can serve web pages with numerous people to observe a system remotely.

The automatic and remote nature of various DL applications requires them to work with a battery supplemented by solar power. These limitations turned the devices extremely power efficient since, in many cases, they operate in severe environmental circumstances where high reliability becomes an additional pre-requisite.

DLs must be exceptionally reliable because they work with little or no human involvement for long nonstop periods and often installed in harsh or inaccessible locations. Hence, they will not be unsuccessful in logging data for any cause provided they have power. As such, DLs are almost entirely immune to the problems affecting a general-purpose computer when a program crashes and eventual OS instability. Some examples of DLs follows:

(a) Specific data about the aircraft performance is acquired by the flight data recorder.
(b) A device called event data recorder is installed by a manufacturer to gather and save various data during the time immediately before as well as after a crash.
(c) In embedded systems and digital electronics design, specialised high-speed digital DL helps to bypass the restrictions of conventional instruments as a logic analyser. A DL can record very long flights to aid in the repairs of functional bugs that happen occasionally.
(d) Micro-UAV (MAV) can bring considerable improvements to the healthcare sector, and they require some particular concerns when it comes to DLs. An open-source electronic health DL can be fast built based on the Raspberry Pi [28–30]. A Holter monitor, for instance, is useful for continuously monitoring innumerable electrical activities of the cardiovascular system that can offer more flexibility to inpatient and outpatient care.

2.5 Types of data logging

The specifics on the airborne platform status are vital to accomplishing a safe flight. Ordinarily, the UAV receives control commands only from the operator and from control algorithms. Moreover, a secondary radio channel can send the necessary telemetry information.

This method works with a relatively low frequency to monitor mostly online status and the capacity to transmit high-level commands. Consequently, problems like the range and quality of wireless connections arise. During a flight, the UAV can surpass the radio signal coverage, which results in corruption or loss of the received data. So, the system needs an additional onboard logging unit to back up crucial status variables when there is an accident. This feature (aka black box) does not depend on the radio communication quality, and it permits to investigate the causes of system breakdowns. More sophisticated estimation and control algorithms entail an expressively higher data refresh rate and access to internal variables of the system. The verification process of algorithms monitors standard working conditions. A possible solution is also to log this material on the platform throughout the flight to process it offline. This subsystem should not obstruct the black-box scheme to preserve high reliability. As a final point, scientific platforms require data logging with low- and high-frequency rates. Both solutions demand extra memory storage for the main avionics elements. Figure 2.5 depicts the block diagram corresponding to the types of storage in a UAV-CPS.

Generally speaking, an avionics system contains three main modules: the mainboard (MB), the attitude and heading reference system (AHRS), and the power

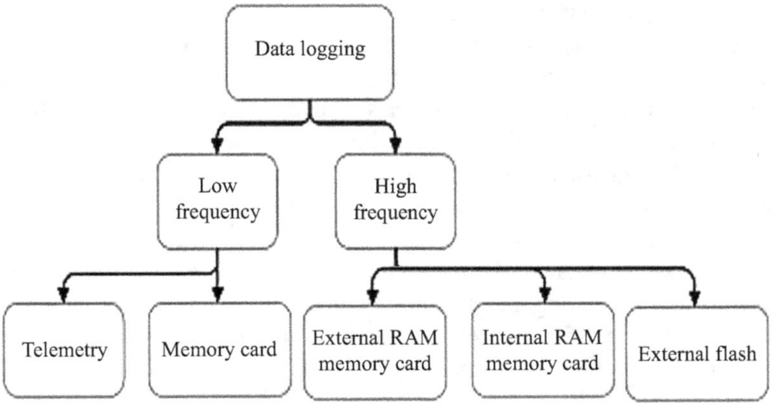

Figure 2.5 Block diagram for data logging procedures and storage forms

board [28,29]. The essential component is the MB since it executes programs and fuses all the information from the whole system. Thus, it should be the best location for extra logging memory. The MB can use a secure digital (SD) card, which is a non-volatile memory, and interfaced via direct memory access (DMA) and a serial peripheral interface (SPI).

The file system simplifies information uploading and downloading. During the initialisation phase, flight parameters come from the configuration file. At that point, the secondary file with a distinctive name is generated, which will store status variables. The writing process is sequential and corresponds to another line for each timestamp. The non-optimal written data amount and the need to convert floating-point variables to text limit the maximum write speed. It is worth noticing that the data logging task has the lowest priorities onboard. The alteration of the software procedure can enhance this subsystem performance. With the help of bigger buffers, data can be sent in sequential multi-block write with SD mode communication relying on a standard called secure digital input-output. As stated in [31], its data transfer works better than SPI with a memory card. This led the authors to try to an innovative communication approach (SD mode) constructed on hardware planned upgrades. This chapter will concentrate on modifications of current black box subsystems and on developing resolutions for a secondary logging method – the high-frequency logger. There are several open-source or corporate avionics modules on the market [32] equipped with specific procedures for data logging such as the Ardupilot [33].

Analysing some known avionics system reveals two options for data logging: (i) tlogs, where the telemetry log is sent to the ground control station (GCS) and saved by specialised software; and (ii) data flash logs relying on extra onboard flash memory. Online records can provide the necessary parameters.

Various commercial makers, such as Mikrokopter [34], distribute flight controllers and wide-ranging aerial systems, which is not very useful in high-frequency data logging. However, the onboard SD card limits its capabilities. The Pixhawk [35] is an advanced open-source system, which consists of the standard flight controller in addition to, in some versions, of another embedded computer. Besides, this avionics has the finest logging functionalities. Similar to previous designs, a telemetry link can transmit the data, and it has an additional SD card. High-frequency logging can be accomplished with an arbitrary rate for a given CPU load and memory speed. Regrettably, if the controller cannot catch the frequency, some packets may be missed, thus leading to unacceptable behaviour in many scientific research circumstances. All offered examples are open-source.

In consequence, a manufacturer can develop firmware to all-purpose hardware and develop customised logging routines. Professional UAVs found on the market can fully support scientific research include many desired functions. Unfortunately, such platforms are costly and do not provide open-source code.

2.5.1 Requirements and recommended solutions

All examples discussed previously have advantages and disadvantages. Those requirements are for low- and high-frequency logging: the minimum data frequency rate, the minimum quantity of saved variables, as well as the time of the data acquisition process last. Black-box logging tasks require a minimum refresh rate. This frequency helps to make a diagnosis of eventual system faults that happened during unforeseen behaviour. The lowest priorities of the data logging system call for pre-emption by the control and communication functions. So, this approach is superior to splitting written data into distinct vectors with a specific timestamp than accumulating them to save in big buffers. Besides being an arduous task, additionally, the generated file is at times closed and opened to preserve data during an emergency power cut-off. Declared parameters should be upheld or even improved with the drop of the processing time via modification in the software routine or SD mode usage during communication using the memory card. High-frequency logging needs the diagnostic of AHRS module estimation algorithms because it is a very demanding task. The high-frequency logging will record raw information from the inertial measurement unit sensor without missing samples. These activities are vital for the proper development of algorithms, for instance, to determine the variances of measurements and to debug control routines. Three main solutions described below arise from [36] and examples from the preceding Section.

2.5.2 Internal RAM with SD

The first solution relies on the already implemented mechanism utilising the memory card besides internal RAM for the MB microcontroller. This tactic needs modernisation because of the identified limitations. The table of measurements is declared, and filled with data in the idle RAM. The operator creates the sequence, and its duration depends strictly on some variables and sampling frequency. The frequency is diminished to save more variables within the same period. This solution does not involve any delay or strict timing requirements to be met when some given measurements are recorded, but another concern arises – in what manner to save these data to non-volatile memory. All records will disappear after the power goes off because of the RAM characteristics. Thus, a new file is formed in the SD card with a copy of the data right after the termination of the measuring phase. Regrettably, the standard frequency of the memory card logging will cause a considerable increase of time, which is a noteworthy disadvantage of the described methodology because there is no option to elevate the write frequency thru the flight to safeguard flight control tasks. Hence, a simple improvement is to raise a frequency only after the descent and with disarmed motors, which improved five times the highest data acquisition rate. Regrettably, the number of variables limitation persists.

2.5.3 External RAM with SD

The second strategy expands the first one by increasing the external RAM size interfaced to the controller with a flexible static memory controller. Sadly, as the number of variables raises, the longer becomes the time the data last into the SD card. This method diverges from the prior approach by the arbitrary address table and in the way external memory is handled.

2.5.4 External flash memory

The third solution uses an external non-volatile flash memory circuit. It disagrees with the other resolutions in some functional facets because of the memory cell structure [37]. The storage circuit can be divided into pages, blocks in addition to planes. This memory type abolishes the data transfer to the SD card and variables are written right to a non-volatile area. However, this approach has two noteworthy impediments: (i) data accessibility and (ii) writing speed. The implemented memory capacity encourages one to save intervals with multiple extents. This methodology improves data logging because it will employ a whole plane as an alternative to using only the first sectors with the proper recording process. Thus, it is necessarily a simplified system for recognising each sequence with the address range specification. This is realised in the first memory block. Each measurement has a distinct page, which saves these data along with their date and variable names.

2.6 Discussion and future trends

2.6.1 UAV-based data storage

Some UAV-CPS applications will dispatch any sensed data straight to the ground station (GS) or other units. While others save these data in UAVs due to (i) the perpetual absence of high communication bandwidth; (ii) the transmission of acquired information is not always compulsory and straightway, but when they are obligatory, and (iii) the presence of onboard storage.

Another improvement is to move the sensed information to the ground station only after treating the data and promptly accumulated during the manoeuvre. Both the storage and information gathering abilities of drones can be alike or heterogeneous since UAVs gather equal or different information amounts depending on the application. If UAVs gather dissimilar volumes of data or if the UAVs have different storages, then a collaborative knowledge storage mechanism is desirable for the swarm to accumulate the collected information ably.

2.6.2 UAV-based data processing

Applications such as high-resolution image/video processing, pattern recognition, data mining for the streaming period, and online task preparation may demand cooperation amongst UAVs and high-performance computation. High-performance information processing tasks can use one computation unit in a UAV or several processing units in multiple UAVs. When there are multiple UAVs, reliable distributed processing methods require implementation. Furthermore, all processors in the sky must be accessible and efficiently operated. This is crucial if the UAVs are working in regions far from the GSs and when instantaneous outcomes are needed to trigger an appropriate action. For example, a UAV may need to recognise a particular variety of pests in crops within proximity of some responsive units. In this case, CV and pattern recognition are indispensable to discern the object of interest to remediate the problem immediately. Sometimes, the procedure cannot wait for reliable records from a distant GS and the corresponding feedback and has to act on the site straight away. Hence, the UAVs in the zone can finish the analysis together and respond accordingly.

2.6.3 Distributed versus centralised control

Secure and effective deployment of multiple UAVs requires unique distributed and collaborative real-time controls. The coordination of numerous UAVs helps to accomplish a specific task, use their resources effectively, arrange for safe manoeuvres, and sustain the fault tolerance mechanism. Still, the previous requirements demand different control mechanisms, which may be challenging to perform with a centralised approach. This occurs owing to three motives: (1) A centralised control system may have a single point failure problem; (2) Not all the drones will always be connected to the GCS, since control signals may not reach the UAV; and (3) A centralised control can cause communication and security a bottlenecks.

2.6.4 Impact of big data in UAV-CPSs

Some caveats in UAV-CPSs caused by big data (BD) include outdated IT infrastructure, the intrinsic complexity in addition to disorderliness of big (and raw) data, lack of data science knowledge within organisations, privacy concerns, and organisational cultures that do not facilitate data-driven operations and decision-making [30]. These challenges are discussed below.

2.6.4.1 Infrastructure readiness

The development of IT infrastructure for BD analytics requires noteworthy software and hardware analyse massive records in real-time while looking at the requirements of BD analytics that are on the rise continuously. Cloud and heterogeneous computing frameworks can offer solutions for BD, but these technologies habitually fail, either from a technical or cost-effective standpoint, when a large information volume is treated. Thus, these data processing bottlenecks must be addressed for superior use of BD.

2.6.4.2 Complexity

To leverage the BD power cause problems linked to data complexity and the inherent difficulties from untreated records (raw data). Habitually, data are saved using different formats, comprising unstructured databases and discrete text files [38]. Moreover, the data volume increases on a daily basis, which messes up handling data from several sources and using different formats more puzzling [39]. The specific managerial and technological aspects that may suffer from problems caused by data complexity are discussed underneath.

Most organisations are not ready to solve the problem of having a large rate of data growth at reasonable cost satisfactorily. Many groups merely choose to delete old data instead of trying to put up with data growth.

The obstacle of merging different data modalities arises from the distinct data sources results in semantic conflicts deriving from and between the numerous types of information sources. For example, terms like assets, earnings, and health may vary from one data to another. Information can be frequently stored in different formats with both (i) unstructured data that include text documents, SMS, emails, images, videos, audio files, and transaction files; and (ii) structured data which organised typically in relational databases. The complication of managing and analysing differing data formats is frequently outside the capability of many organisations.

2.6.4.3 Privacy

Privacy concerns often hinder the adoption and use of BD analytics within companies. Oftentimes, BD analytics employs people with data collected for an utterly different purpose [38]. People's evidence, along with other data sources, can constitute legal and ethical problems, like to leak reserved information about the person (e.g., fiscal situation, medical records, domestic relationships, and awkward behaviour). What makes the impasse even more onerous is that many companies are not upfront about the use of their customer data [40].

2.6.4.4 Barriers to BD processing in UAV-CPSs

Companies need to address the barriers related to people, technology, and corporate domains to leverage BD as a vehicle for improving organisational performance.

Building a novel and independent platform for BD analytics is the best alternative for new businesses but not practical for legacy IT systems. In general, a solution relying on both modern BD platforms and existing legacy systems [30] is the best alternative.

Luckily, BD platforms can be designed and developed using low-cost commercial hardware (that can typically comprise legacy IT systems). BD infrastructure should store and process big data volumes in real-time and demand protection against service disruption or failure [40,45]. The usage of a large number of commercial servers to save and process information simultaneously can fulfil these requirements. The commodity servers are categorised either as slave nodes or as master nodes. The master nodes provide information management while the slave nodes store and process data. Commercial servers or nodes can be building blocks for BD storage units. This cost-efficient scalable architecture allows for an increase in the number of nodes to expand the processing power and storage with relative ease [30].

While developing a BD infrastructure using commercial servers as well as storage systems, some caution is needed when connecting via Ethernet or fibre networks. Network bandwidth requires attention to data interchange across servers. Thus, the network infrastructure must support the high throughput and bandwidth associated with high bulks of data moving through servers.

It is paramount to assimilate some BD technologies and platforms (e.g., Hadoop, NoSQL, MapReduce, In-Memory DB, etc.). Some secondary skills are advanced knowledge in maths, statistics, predictive analytics, machine learning, decision-making models, and data visualisation, to name a few [40–45].

2.6.5 Challenges related to privacy and the protection of personal information

The rapid advancements in UAV-CPS and the associated BD technologies impact personal privacy and human rights. As personal and sensitive, the rapid advancements in UAV-CPS and the associated BD technologies have been closely linked to personal privacy and human rights. As personal and sensitive material becomes more visible, the need to protect people's privacy becomes essential to circumvent legal or ethical polemics and to ensure customer buy-in of BD initiatives. Although technology faults may lead to confidentiality or safety breaches, it is often the behavioural side that causes more problems. In a way, as long as individuals are in charge of their data, it is irrelevant how resilient or vanguards the technical dimension of security can be.

Currently, several behavioural solutions exist to enable individuals to preserve control over their data and avoid security and privacy incidents. For example, individuals should possess the right to delete their past data, to stipulate data expiration dates, and to own info about their social connections. The creation and adoption of regulations that validate these ideas [9,13] are necessary to strengthen these practices. These laws and guidelines will not only defend consumer privacy but also inspire individuals to share their info in a way that assists both the consumers and the organisations to upsurge their performance relying on this data.

2.6.6 Organisational and cultural barriers

Organisational and cultural barriers related to BD are deemed significant and thought-provoking issues to overcome.

Therefore, a business must first revise its culture so that it backs fact-based decision-making to take full advantage of BD opportunities. Positive cultural modifications result from documenting, realising, and communicating a clear organisational vision about BD, ensuring top management pledge to this vision and handling the drivers impelling the executive culture rather than altering the culture itself [30]. A clear vision of how BD fits in with the whole mission strategy should hurry and solidify the BD acceptance within the organisation. The framed idea has to be transformed into specific business processes and significant initiatives that rely on BD to improve organisational performance.

The original standalone DL model is shifting to a scheme that brings together data while also establishing wireless communications to set alarms of events, to report data automatically and to permit remote control. DLs can serve web pages for readings in progress, e-mail their alarms, to allow the use of FTP their results into databases or to direct to the users. A popular trend is to use free and open-source hardware and software approaches in lieu of brand-name products. The small Raspberry Pi computer is a widespread single-board platform to host real-time Linux or pre-emptive Linux kernel operating systems with many interfaces, e.g., SPI, I2C, or UART permitting the direct connection of sensors and computers, and unlimited configurations to display measurements over the Internet in real-time, process records, plot charts/diagrams, and so on [46–50].

There are open-source initiatives with increasingly community-developed projects for data acquisition and DL.

2.7 Conclusions

Logging methods appear commonly in many avionics designs as a telemetry connection or onboard memory storage (aka black box). The data from a DL can be used to investigate failure analysis, repetitive divergences, and accident studies. As stated earlier, range limits telemetry, throughput and leads to faulty data. So, the solution founded on the SD card is better matched for this task. Common tools found in DLs are:

(a) Processing unit;
(b) digital and analogue input modules;
(c) interfacing and communication hardware;
(d) digital and analogue (D/A) inputs; and
(e) IC's memory.

The network management of data loggers entail:

(a) Front-end application software at a central control place;
(b) status of all the DLs in the network;
(c) online status of D/A inputs;
(d) data discriminating viewing;
(e) online graphical displays mimicking all GS layouts; and
(f) auto-backup on exceeding the configured size of the database.

Recently, the SD mode communication standard with DMA replaced SPI implementations of data transfer to the memory card and reduced the processing time. This pool can be used to implement new algorithms or rise the data recording sampling frequency. All mentioned approaches for high-frequency logging present advantages and disadvantages. Hence, it is problematic to relate to them effortlessly.

The first solution applies to any avionics system using a memory card as well as a microcontroller with plenty of RAM. This concept is stress-free to implement and supports researchers to validate their algorithms and sensors outputs [1–8, 46–48] using both onboard and cloud resources.

The second method involves software modifications besides hardware modernisation of avionics. Even when there are external SDRAM memory units, other existing storage, and enough measurement time, it prolongs the transmission time to the SD card. Deplorably, at times there are no conditions to run the platform for a long time.

The third solution relies on external flash memory, and it also requires much firmware and hardware adjustments. Extra PC software is equally necessary, and it makes room for the prospect to improve a reconfigurable logging framework. It also has the best performance. This tactic is a superior suite for wholly enclosed avionics modules spending output data buses and abolishes mechanical interface with avionics (introducing memory card). From current development plans, future work must address investigation on different procedures for data buffering beforehand saving these records to the memory card.

Taking on superior buffers should enhance the achieved data rate in addition to optimise the processing time. Moreover, memory cards with higher speeds can be tested.

In summary, the DLs present the following advantages:

(i) A DL helps to monitor typical failures like intermittent failures;
(ii) It helps in analysing the cause of accidents;
(iii) It helps to perceive the human failure;
(iv) DLs allow for Drivers Passing Signal at Danger (SPAD);
(v) Operational mistakes done by operating staff Signal and Telecom interferences in safety circuits;
(vi) A DL can spot interferences and/or failures;
(vii) A DL can be a diagnostic tool for preventive maintenance with signalling mechanisms;
(viii) DLs can be connected inserted into networks to observe the PI/RRI/EI remotely; and
(ix) Networked DLs can produce failure reports remotely to track events online and offline whenever possible.

References

[1] Razmjooy N, Mousavi BS, Khalilpour M, and Hosseini H. Automatic selection and fusion of color spaces for image thresholding. *Signal, Image and Video Processing*, 2014;8(4):603–614.

[2] Hemanth DJ, and Estrela VV. Deep learning for image processing applications. Adv. Par. Comp. IOS Press. ISBN978-1-61499-821-1 (print) 978-1-61499-822-8 (online) 2017.

[3] Mousavi BS, Soleymani F, and Razmjooy N. Color image segmentation using neuro-fuzzy system in a novel optimized color space. *Neural Computing and Applications*, 2013;23(5):1513–1520.

[4] Estrela VV, Magalhaes HA, and Saotome O. Total variation applications in computer vision. In Handbook of Research on Emerging Perspectives in Intelligent Pattern Recognition, Analysis, and Image Processing, pp. 41–64. IGI Global, 2016.

[5] Moallem P, Razmjooy N, and Mousavi BS. Robust potato color image segmentation using adaptive fuzzy inference system. *Iranian Journal of Fuzzy Systems*, 2014;11(6):47–65, 2014.

[6] Mousavi B, Somayeh F, and Soleymani F. Semantic image classification by genetic algorithm using optimised fuzzy system based on Zernike moments. *Signal, Image and Video Processing*, 2014;8(5):831–842.

[7] Razmjooy N, Estrela VV, and Loschi HJ. A survey of potatoes image segmentation based on machine vision. In: Applications of Image Processing and Soft Computing Systems in Agriculture, 2019:1–38.

[8] Estrela VV, and Coelho AM. State-of-the-art motion estimation in the context of 3D TV. In: Multimedia Networking and Coding. IGI Global, 2013: 148–173. doi:10.4018/978-1-4666-2660-7.ch006.

[9] Bernasconi A, Goodall E, Shea J, *et al.* Implementation guide for IBM spectrum virtualize for public cloud. IBM Technical Report. 2017. http://www.redbooks.ibm.com/redpapers/pdfs/redp5466.pdf.

[10] Luo C, Nightingale J, Asemota E, and Grecos C. A UAV-cloud system for disaster sensing applications. In IEEE 81st Vehicular Technology Conference (VTC Spring), 2015:1–5, 2015. doi: 10.1109/VTCSpring.2015.7145656.

[11] Kosta S., Aucinas A., Hui P., Mortier R., and Zhang X., ThinkAir: Dynamic resource allocation and parallel execution in the cloud for mobile code offloading. In Proceedings of IEEE INFOCOM, 2012:945–953.

[12] Namboodiri V, and Ghose T. To cloud or not to cloud: A mobile device perspective on energy consumption of applications. In Proceedings of the 2012 IEEE International Symposium on a World of Wireless, Mobile and Multimedia Networks (WoWMoM).

[13] Muthurajkumar S, Vijayalakshmi M, and Kannan A. Resource allocation between temporal cloud database and user using access control. In Proceedings of the International Conference on Informatics and Analytics (ICIA-16). ACM, New York, NY, 2016. doi: https://doi.org/10.1145/2980258.2980338.

[14] Ehsan S, and McDonald-Maier, KD. On-board vision processing for small UAVs: Time to rethink strategy. In Proceedings NASA/ESA Conference on Adaptive Hardware and Systems, 2009:75–81.

[15] Sun Y., Ng DWK, Xu D, Dai L, and Schober R. Resource allocation for solar powered UAV communication systems, arXiv preprint arXiv: 1801.07188, 2018.
[16] Sheet NAF. Beamed laser power for UAVs, NASA–2014.http://www.nasa.gov/centers/armstrong/news/FactSheets/FS087-DFRC.html.
[17] Bay H, Tuytelaars T. Gool, T, and Luc V. SURF: Speeded Up Robust Features, Proceedings 2006 ECCV, 2006.
[18] Saeed A, Neishaboori A, Mohamed A, and Harras KA. Up and Away: A visually-controlled easy-to-deploy wireless UAV cyber-physical testbed. In Proceedings of the IEEE 10th International Conference on Wireless and Mobile Computing, Networking and Communications (WiMob). 2014:578–584.
[19] Fok C, Petz A, Stovall D, Paine N, Julien C, and Vishwanath S. Pharos: A testbed for mobile cyber-physical systems. Univ. of Texas at Austin, Tech. Rep. TR-ARiSE-2011-001, 2011.
[20] Jung D, Levy E, Zhou D, *et al.* "Design and development of a low-cost testbed for undergraduate education in UAVs. In Proceedings of 44th IEEE Conference on Decision and Control 2005 and 2005 European Control Conference, CDC-ECC'05. 2005:2739–2744.
[21] Lupashin S, Schollig A, Sherback M, and D'Andrea R. A simple learning strategy for high-speed quadrocopter multi-flips. In Proceedings of 2010 IEEE International Conference on Robotics and Automation (ICRA). IEEE, 2010:1642–1648.
[22] Michael N, Mellinger D, Lindsey Q, and Kumar V. The grasp multiple micro-UAV testbed. *IEEE Robotics & Automation Magazine*, 2010:17(3)56–65, 2010.
[23] Brown TX, Doshi S, Jadhav S, and Himmelstein J. Testbed for a wireless network on small UAVs. Proceedings of AIAA 3rd Unmanned Unlimited Technical Conference, Chicago, IL, 2004:20–23.
[24] Engel J, Sturm J, and Cremers D. Camera-based navigation of a low-cost quadrocopter. Proceedings of 2012 IEEE/RSJ International Conference on Intelligent Robots and Systems (IROS). 2012:2815–2821.
[25] Crenshaw TL, and Beyer S. Upbot: A testbed for cyber-physical systems. In Proceedings of 3rd International Conference on Cybersecurity experimentation and test. USENIX Association. 2010:1–8.
[26] Fowers SG. Stabilization and Control of a Quad-Rotor Micro-UAV using Vision Sensors, Master of Science Thesis, Brigham Young University, USA, 2008.
[27] Edwards B, Archibald J, Fife W., and Lee DJ. A vision system for precision MAV targeted landing, Proceedings of 2007 IEEE International Symposium. on Computational Intelligence in Robotics and Automation, Jacksonville, FL, USA, 2007.
[28] Bondyra A, Gardecki S, and Gąsior P. Distributed control system for multirotor aerial platforms. *Measurement Automation Monitoring*, 2015:61(7):343–346.
[29] Bondyra A, Gardecki S, Gąsior P, and Kasiński A. Falcon: A compact multirotor flying platform with high load capability. *Advances in Intelligent Systems and Computing*, 2015;351:35–44.

[30] Gąsior P, Bondyra A, and Gardecki S. *Measurement Automation Monitoring,* 2017:63(5), ISSN 2450–2855.

[31] TOSHIBA SD Card Specification, 2006.

[32] Lim H, Park J, Lee D, and Kim HJ. Build your own quadrotor. Open source projects on unmanned aerial vehicles. *IEEE Robotics & Automation Magazine*, 2012:19:33–45.

[33] http://www.ardupilot.co.uk/ [Accessed 2018 Aug 10].

[34] http://www.mikrokopter.de/en/home [Accessed 2018 Aug 10].

[35] https://pixhawk.org/ [Accessed 2018 Aug 10].

[36] Suzdalenko A. Guidelines for autonomous data logger design. IEEE International Symposium on Industrial Electronics (ISIE), 2011:1426–1429.

[37] Micron Technology Inc.: NAND Flash 101: An Introduction to NAND Flash and how to design it in to your next product. Micron Technology Inc. Tech. Note. 2006.

[38] Douglas M. Big data raises big questions. *Government Technology*. 2013: 26(4):12–16.

[39] Johnson JE. Big Data + Big Analytics = Big Opportunity, *Financial Executive*, 2012:28(6):50–53.

[40] Van Rijmenam M. Think bigger: Developing a successful big data strategy for your business. New York: AMACOM, 2014.

[41] Alharthi, A., Krotov, V., and Bowman, M., Addressing barriers to big data, *Business Horizons*, 2017:60(3):285–292.

[42] McAfee A, and Brynjolfsson E. Big data: The management revolution, *Harvard Business Review*, 2012:90(10):60–68.

[43] Miller S. Collaborative approaches needed to close the big data skills gap. *Journal of Organization Design*, 2014:3(1):26–30.

[44] Schadt E. E., The changing privacy landscape in the era of big data. *Molecular Systems Biology*, 2012:8(1):1–3.

[45] Schouten P. Big data in health care solving provider revenue leakage with advanced analytics. *Healthcare Financial Management*, 2013:67(2):40–42.

[46] Estrela, VV, Monteiro ACB, França RP, Iano Y, Khelassi A., and Razmjooy N. Health 4.0: Applications, management, technologies and review. *Med Tech Journal*, 2019;2(4):262–276, http://medtech.ichsmt.org/index.php/MTJ/article/view/205.

[47] Gupta S, Girshick RB, Arbeláez PA, and Malik J. Learning rich features from RGB-D images for object detection and segmentation. Proceedings of 2014 ECCV, 2014.

[48] Zhou Y, Li H, and Kneip L. Canny-VO: Visual odometry with RGB-D cameras based on geometric 3-D–2-D edge alignment. *IEEE Transactions on Robotics*, 2019;35:184–199.

[49] Penson W, Fazackerley S, and Lawrence R. TEFS: A flash file system for use on memory constrained devices. Proceedings of IEEE Canadian Conference on Electrical and Computer Engineering. 2016:1–5.

[50] http://www.asctec.de/en/ [Accessed 2018 Aug 10].

Chapter 3

Integrated optical flow for situation awareness, detection and avoidance systems in UAV systems

William Sanchez Farfan[1], Osamu Saotome[1], Vania V. Estrela[2] and Navid Razmjooy[3]

Optical flow (OF) plays a decisive role in visual situation awareness, detection and obstacle avoidance systems for unmanned aerial vehicles (UAVs), which are cyber-physical systems (CPSs) that interact with the environment through sensors and actuators. The use of cameras allows the integration of computer vision (CV) algorithms with the inertial navigation systems (INS). The movement of characteristics of the image fused with the dynamic of the UAVs allows us to improve the process of remoting sense, avoid obstacles or estimate the position and velocity of the UAV. In the literature, there are various algorithms to locate characteristics points between two consecutive images. However, the computation time and consumption of physical resources such as memory features are due to embedded systems. This chapter shows (i) how to integrate the movement of the pixel textures (OF) in the image with INS data, (ii) compares different algorithms to match points between consecutive images, (iii) implements a process to encounter points between consecutive images and (iv) implements a computationally less expensive and with less memory consumption algorithm. A case study about using the field-programmable gate array (FPGA) as part of the visual servoing is discussed showing how to integrate results into the CV hardware system of a UAV and addressing the need to handle issues such as multi-resolution.

3.1 Introduction

The unmanned aerial vehicles (UAVs) known as drones or aircraft remotely piloted (ARP) are systems employed in operations such as supervision, intelligence and reconnaissance, mapping, search and rescue, among others. However, since early in the XX century, the ARPs have been used mainly in military operations. From 1979

[1]DCTA-ITA-IEEA, Sao Jose dos Campos, SP, Brazil
[2]Universidade Federal Fluminense, RJ, Brazil
[3]Trafesh University, Trafesh, Iran

on, it was incorporated in the aerial photogrammetry science [1] by Przybilla and Wester-Ebbinghaus. Since early, it has been improved due to the apparition and incorporation of microelectronic systems as the inertial measurement unit (IMU), which is composed of an accelerometer and a gyrometer mainly. Nowadays, we can see another sensor that can be integrated on the ARP such as LIDAR, SAR, optical and acoustic sensors, among others.

The cameras are light sensors that catch the waves of the objects in the visible band of the spectrum; it passively provides information about its environment. The low price and the reduced size impact the payload of any mission, in a variety of civil and military applications giving the ARP a little autonomy, incorporating characteristics as the ability to detect and avoid obstacles, which are essential to the integration of the ARPs in the aerial traffic space. Additionally, the integration of the image acquisition (i.e., camera) with the navigation system has permitted us to estimate most confidently the position and velocity of the UAV with the basic navigation system (INS)/global navigation satellite system (GNSS).

The CV algorithms and the navigation system augment the UAVs perception of their environment to make them more autonomous and improved its situational awareness to make a decision and carry out the mission without the necessity of the operator's supervision. Although the camera and other sensors can provide it characteristics of autonomy to the ARP nowadays, unfortunately, the regulations of the aeronautical authorities do not permit that ARPs share the aerial space with the commercial aircraft. Now the challenge is implementing inside of the ARPs a system reliable to detect and avoid any collision act, improving its perception and comprehension of the objects nearby its environment. Recent investigations address the possibility of using cameras to implement this characteristic [2,3].

Currently, the navigation consists of the fusion of the INS, the IMU and GNSS data. However, there are situations where the signal is corrupted or is not available because the ARP is located on an urban or forest area or the signal has been altered by an attack of spoofing or hacking [4,5]. The CV algorithms are good candidates to maintain the estimate of the position and velocity when the ARP loses the GNSS signal. To fuse the data from the CV algorithms and the navigation system, Bonin-Font *et al.* [3] use two strategies: (i) to map previously established locations concerning its localisation and (ii) without maps using only the characteristics of the environment.

The utilisation of CV algorithms with the INS based data on the pre-established map to localise its position requires sufficient memory to store the maps. Another possibility the Ground Control Station stores the images of the maps online assuming a robust connection and significant bandwidth to send it. Some works [6–13] are related to the integration of geo-referenced maps. Others, for example [14], made a study over a navigation system that recognises marks on the ground and extract the longitude and latitude of the geo-referenced image using artificial neural networks to identify marks of interest (landmarks) on the drone image. For this last case, a similitude is established that indicates the geographic coordinates of the ARP. To flight at night, Da Silva [15] has taken thermal images

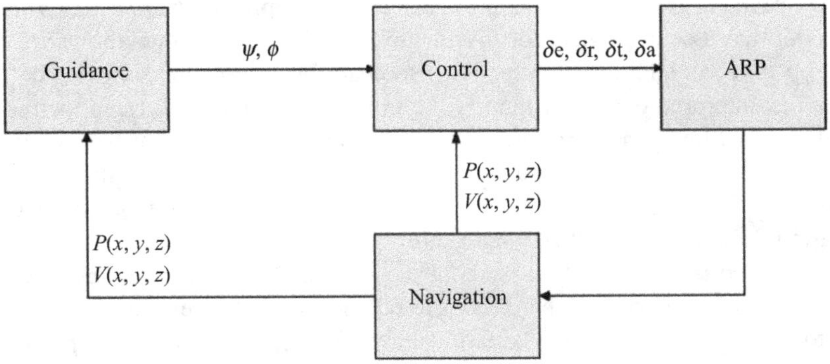

Figure 3.1 Basic systems inside of the ARP

by UAV and compared them with geo-referenced map previously established if the images are very similar, and the geographic coordinates of the map are the same geographic coordinates of the ARP (Figure 3.1). On the other hand, the navigation systems without maps are based on the environmental characteristics to self-localisation, for example, SLAM [16]. This navigation system detects and avoids obstacles in addition to tracking a target.

Aerial photogrammetric is a sub-area of remoting sense that uses ARPs for mapping terrain, and the output information can be used for interpretation and classification of images, orthophotos, digital elevation maps, digital terrain maps, digital surface maps, and 3D modelling of buildings and terrain.

There is a variety of applications relying on a camera embedded in the payload. Although the navigation system is an essential part of the ARP, it is not unique. Figure 3.1 shows the different systems that are inside the ARP and its integration.

The planning system sends new coordinates to the guidance system depend on the information perceived by its surroundings using an optical sensor and the type of mission to achieve, for example, tracking of a target or send information about a possible obstacle. It is also possible to use the optical sensor to provide redundant information to estimate the velocity and position most confident.

3.2 Computer vision

The images are a vital source of passive knowledge that provides a significant quantity of information about the environment that is filming. It is a numeric representation of an object illuminated with a radiant source. Mathematically, the image becomes a matrix of $n \times n$ cells where each cell represents a pixel of the image. A cell comprises the union of three colours red, green and blue and the number in each cell represent the value of images intensity in that one pixel.

CV extracts knowledge from an image to be interpreted or modified. The data extraction can occur bitwise, locally or globally. The binary operations are done pixel by pixel as addition, subtraction, Boolean operations and others. The local operations interpret, extract or modify the information of a small region within the image and a filter generally determines the characteristics of the region such as contours, corners, blobs and others. Finally, the global operations modify the whole image, creating warp, panoramic images or another image version via a global operation such as the RANSAC algorithm.

One important information is about the movement of the pixels between two consecutive images, so the change in the position of the pixels can be integrated into the navigation systems assisted without maps. This change of position is known as OF, and it requires the presence of some textures.

3.2.1 Optical Flow

The movement of the pixels is the primary CV cue used to integrate motion data into the navigation systems, projecting the movement of the scene in 3D on the plane of the 2D image. The OF aids image segmentation, calculation of the time of the collision, and the estimation of the direction of movement among many other tasks. There are a variety of techniques that determine the OF, such as methods based in matching region, based in energy, based in phase or technical based in the differential or based in the gradient [17], but the methods most known are those based on the calculation of the gradient.

3.2.1.1 Methods based on the brightness gradient

The methods based on the gradient start as of supposition that the intensity of the image does not vary; that is to say, in an interval of time $(t, t + \Delta t)$, the intensity of the pixels is constant or in mathematical terms:

$$f(x, y, t) = f(x + \Delta x, y + \Delta y, t + \Delta t) \tag{3.1}$$

where $f(x, y, t)$ is the intensity of the image in instant t and Δx and Δy are the change of the pixel position in instant $t + \Delta t$. Applying the Taylor series yields:

$$f(x + \Delta x, y + \Delta y, t + \Delta t) = f(x, y, t) + \frac{\partial f}{\partial x}\Delta x + \frac{\partial f}{\partial y}\Delta y + \frac{\partial f}{\partial t}\Delta t + O^2 \tag{3.2}$$

Removing the high terms of order O^2 and inserting (3.1) into (3.2), the result is an equation that describes the condition of the OF as follows:

$$I_x \Delta x + I_y \Delta y + I_t \Delta t = 0 \tag{3.3}$$

According to (3.3), there is one equation to determine the value of three unknown variables, and this problem is known as the aperture problem that demands some constraints to determinate these unknown variables. There are two methods to address the solution of (3.3) either using global restrictions and to calculate the movement of the whole image or localising restrictions inside of the image to find its projection in the next image.

Horn and Schunk method

The Horn–Schunk (HS) method determines the OF in the whole image based on the restriction on the equation of the smoothing involving small changes in the intensity of the pixel with a few discontinuities (smoothness constraint).

The implementation of this restriction imposes a mathematical equation defined as:

$$Ec = \frac{\partial u^2}{\partial x} + \frac{\partial u^2}{\partial y} + \frac{\partial v^2}{\partial x} + \frac{\partial v^2}{\partial y} \tag{3.4}$$

According to the OF and the smoothness constraint, it is necessary to minimise the sum of (3.3) and (3.4):

$$\iint \left((I_x u + I_y v + I_t) + \alpha^2 \left(\frac{\partial u^2}{\partial x} + \frac{\partial u^2}{\partial y} + \frac{\partial v^2}{\partial x} + \frac{\partial v^2}{\partial y} \right) \right) dxdy \tag{3.5}$$

where $\Delta x = u$, $\Delta y = v$, $\Delta t = 1$ and α is a weighting factor associated with the error. To determine the values of u and v that minimise (3.5), the HS introduces an estimation of the Laplacian:

$$(\alpha^2 + I_x^2)u + I_x I_y v = (\alpha^2 \overline{u} - I_x I_t) \quad \text{and} \quad \left(\alpha^2 + I_y^2\right)v + I_x I_y u = (\alpha^2 \overline{v} - I_y I_t), \tag{3.6}$$

where $\overline{u}$ and $\overline{v}$ are the average of the Laplacian, so the solution for u and v is:

$$\left(\alpha^2 + I_x^2 + I_y^2\right)u = \left(\alpha^2 + I_y^2\right)\overline{u} - I_x I_y \overline{v} - I_x I_t \quad \text{and}$$
$$\left(\alpha^2 + I_x^2 + I_y^2\right)v = (\alpha^2 + I_x^2)\overline{v} - I_x I_y \overline{u} - I_x I_t. \tag{3.7}$$

A direct solution for (3.7) is costly computationally speaking. Still, Horn–Schunk proposed an interactive solution where u^{n+1} and v^{n+1} are new solutions based on the velocity average previous u_n and v_n, so the iterative solution is:

$$u^{n+1} = \overline{u}^n - \frac{I_x\left(I_x \overline{u}^n + I_y \overline{v}^n + I_t\right)}{\alpha^2 + I_x^2 + I_y^2} \quad \text{and} \quad v^{n+1} = \overline{v}^n - \frac{I_y\left(I_x \overline{u}^n + I_y \overline{v}^n + I_t\right)}{\alpha^2 + I_x^2 + I_y^2} \tag{3.8}$$

Lucas and Kanade method

In contrast to the HS approach, the Lucas and Kanade (LK) technique calculates the movement of the pixels around a specific characteristic of the image. This method determines the OF in a small neighbourhood of $n \times n$ pixels, calculating the movement of the pixels according to (3.3) around a specific point centred on a window of $n \times n$ and assuming that the OF is constant, in the $n \times n$ window leads to:

$$\begin{aligned} I_{x_1}u + I_{y_1}v &= -I_{t_1} \\ I_{x_2}u + I_{y_2}v &= -I_{t_2} \\ \vdots \quad \vdots \quad &= \ \vdots \\ I_{x_n}u + I_{y_n}v &= -I_{t_n} \end{aligned} \tag{3.9}$$

The overdetermined system from (3.9) becomes more tractable as follows:

$$\begin{bmatrix} I_{x_1} & I_{y_1} \\ I_{x_2} & I_{y_2} \\ \vdots & \vdots \\ I_{x_n} & I_{y_n} \end{bmatrix} \begin{bmatrix} u \\ v \end{bmatrix} = \begin{bmatrix} I_{t_1} \\ I_{t_2} \\ \vdots \\ I_{t_n} \end{bmatrix} \tag{3.10}$$

Equation (3.10) can be solved by minimising the mean square error related to the solution:

$$v = (\boldsymbol{A}\boldsymbol{A}^T)^{-1}\boldsymbol{A}^T(-b) \tag{3.11}$$

However, it gives more importance to the pixel in the centre of the window by using a Gaussian convolution filter, such that the final solution becomes:

$$v = (\boldsymbol{A}\boldsymbol{W}\boldsymbol{A}^T)^{-1}\boldsymbol{A}^T\boldsymbol{W}(-b) \tag{3.12}$$

$$\begin{bmatrix} u \\ v \end{bmatrix} = \begin{bmatrix} \sum WI_x^2 & \sum WI_xI_y \\ \sum WI_xI_y & \sum WI_y^2 \end{bmatrix}^{-1} \begin{bmatrix} -\sum WI_xI_t \\ -\sum WI_yI_t \end{bmatrix} \tag{3.13}$$

The LK method found is not invariant to the scale, yves Bouguet [18] proposes a method based on pyramids; in this way, the OF calculation will be invariant to scale.

The HS method calculates over the whole image the movement of the pixels, and it is computationally expensive and consumes much memory, but it produces a dense OF over the whole image. Now, the LK is a method with few OF, based in characteristic own of the image, and it is less computationally expensive.

3.2.1.2 Feature extractor algorithm

Features are regions that present elements belonging to one object, this element can be a corner, edge, blob, a new colour and other types of features. The extraction of the feature points is determined based on the filters of convolution. The LK uses corners inside the image to determine what points to track. Some algorithms used to extract feature points include the SIFT, SURF, ORB and Harris.

Scale-invariant feature transform (SIFT)

The SIFT is an algorithm patented by Lowe [19] that localise and match key-points through two steps. The first step detects the key-point, and the second step described and matched it with the next image. For this chapter, only the first step is exposed.

To detect key-points, SIFT uses Gaussian filters over the whole image at different scales, which results in smoother images. In Figure 3.2, the smoothness of the images is represented as the red box. The determination of the key-points is made employing the differences of Gaussians (DoGs). The maximum or minimum DoGs are considered a key-point in Figure 3.2 is represented in the green box.

Seeded-up robust features

The seeded-up robust features (SURF) is an algorithm patented by [20], which is almost the same as the SIFT. It has two phases: detect and describe, but, in this chapter, only we going to explicate the detect stage.

The use of integral images on the part of SURF and box filter to determinate feature point make that it is less expensive. An integral image is a window where each cell is the result of summing the previous pixels, and the box filters are approximations of Hessian filters. In Figure 3.3, it is possible to see the Hessian filter and its corresponding box filters.

The determination of feature points is based on the Hessian matrix determinant:

$$Det(Hessian) = \boldsymbol{D}_{xx}\boldsymbol{D}_{yy} - w\boldsymbol{D}_{xy} \tag{3.14}$$

where $\boldsymbol{D}_{xx}$, $\boldsymbol{D}_{yy}$ and $\boldsymbol{D}_{xy}$ are the approximations of the box filters for the Hessian filters in direction x, y and xy and w is a factor of weighting to correct the approximation between the Gaussian kernels.

The SURF uses a pyramid to determine features point on different scales, and in each level of the pyramid, SURF calculates the Hessian determinant over the

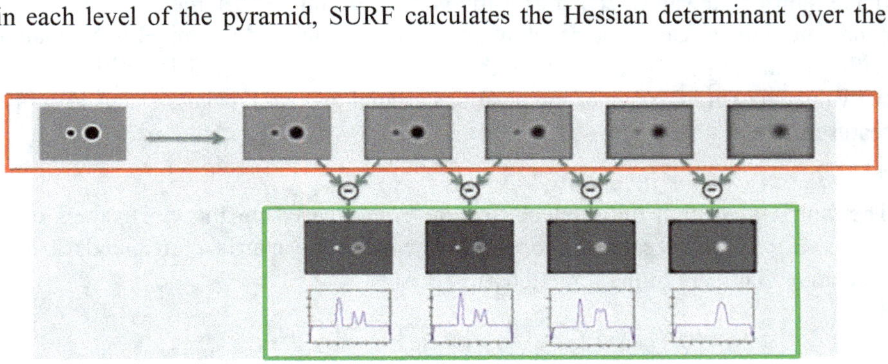

Figure 3.2 Smoothing an image using the Gaussian filter

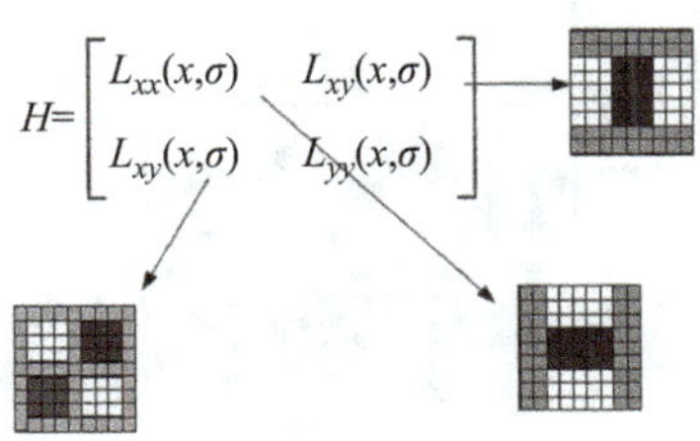

Figure 3.3 Hessian matrix and its respective box filter

whole image so that the determinants within an area of 3×3 is compared. Again, the result is compared with the scale-up and the scale down, and the final value is considered a feature point.

Oriented fast and rotated BRIEF (ORB)
The ORB is an algorithm that arose as an alternative to the SIFT and SURF [21]. Similar to the above algorithms, this method detects and matches key-points. However, ORB is not invariant to the scale.

The detection of key-point hinges on the comparison of the central neighbourhood pixel with the surrounding pixels, as shown in Figure 3.4.

The first comparison employs the pixels located in the 1st, 5th, 9th and 13th positions. The comparison is made according to the next equations:

$$f(n) = \begin{cases} I_i \leq I_p - th \text{ dark} \\ I_p \leq I_i \leq I_p - th \text{ similar an,} \\ I_p + th \leq I_i \text{ light} \end{cases}$$

$$f(I_i) = \begin{cases} 1 & f(n) \text{ is dark or light} \\ 0 & f(n) \text{ is similar} \end{cases}, \tag{3.15}$$

where I_p is the value of the central pixel, I_i is the value of the pixel in each one of the position I, and th is a threshold. According to (3.15), if the pixel at the position I is dark or light, $f(I_i)$ has a value of one; otherwise, the value is zero. Consequently, if the sum of all values $f(I_i)$ in positions 1, 5, 9 and 13 is larger than 3, it is a candidate to be a feature point. Now, the central neighbourhood pixel is compared with surrounding pixels and realised the comparison again with (3.15), if the sum of the values $f(I_i)$ at each one of the positions around the pixel, is larger than 13 it is a feature point.

The Harris Corner detector
The Harris algorithm only detects feature points based on the eigenvalues of a matrix 2×2 where each one of the elements of the matrix corresponds to the derivative Gaussian in the directions x and y.

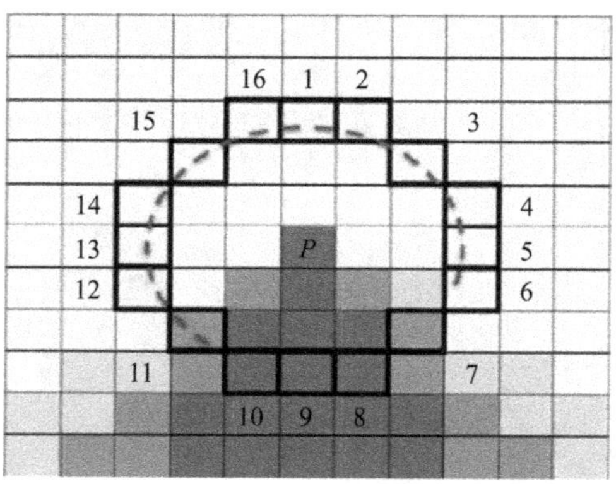

Figure 3.4 Circle around the central pixel

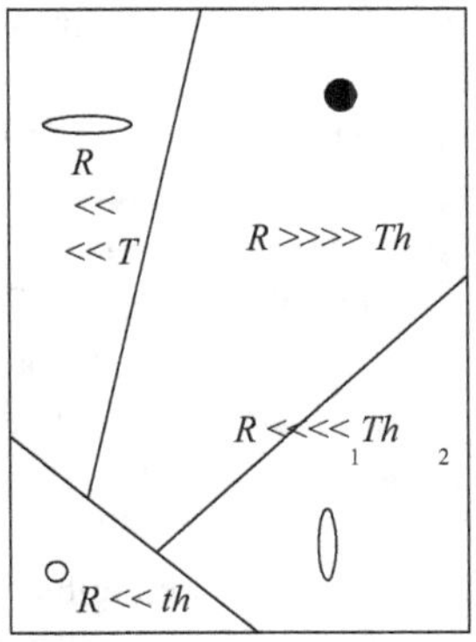

Figure 3.5 Relation between the eigenvalues of the Harris matrix with the equation alternative (3.17)

If one of the eigenvalues is greater than the other eigenvalue, it is a line. If both eigenvalues are small, then it is a blob. If both eigenvalues are greater than a certain threshold, then it is a corner. Therefore, a feature point A can be associated to its form to determinate a corner without calculating the eigenvalues applying this equivalence:

$$R = Det(Harris) - \alpha * trace^2(Harris) > th \tag{3.17}$$

If R is less to threshold *th,* then it is an outline or blob. If R is greater than *th* it is a corner and, therefore, it is considered a feature point, as shown in Figure 3.5.

This method is invariant to the rotation without scaling.

3.3 Optical flow and remote sensing

Remote sensing (RS) is the process of detecting and monitoring the physical characteristics of an area by measuring its reflected and emitted radiation at a distance from the targeted area. Special cameras collect remotely sensed images of the Earth, which help researchers sense things about the Earth [22]. To get this information can be realised by means of satellites or cameras embedded inside aerial vehicles a sub-area of the remote sensing is the aerial photogrammetry.

The aerial photogrammetry is the science of extracting and interpreting information over objects and its environment using images and pattern registered by the electromagnetic spectrum, this science is divided by two areas [23], the interpretative and metric photogrammetry. The interpretative image has as the main goal the identification and reconnaissance of objects over determinate area based on its qualities as size, pattern, texture, contours, bright, contrast and others. On the other hand, the metric photogrammetry extracts an exact measure on the image based on the information of the sensor to measure the relative positions among points on the image as distances, angles, volumes, size and others, and it is widely used in planimetry and altimetry. The information extracted or interpreted of the images can be applied to topography, urban planning, archaeology and other sciences.

In metric aerial photogrammetry, it is necessary to know the orientation of the image with respect to one reference point, [24] established two references, one internal, where the pixels of the image are related to the camera's coordinates and one external reference, where a relation between the objects represented in the image and the objects located in the terrain is established. The orientation of the image can be realised directly either by means of the GNSS/IMU inside of the navigation system or by means of control ground, located on the terrain where the exact geographic coordinates x, y and z are known. The process to determinate the orientation of the image is known as aerial triangulation (AT).

The tendency is that the indirect method will be eliminated because of its high cost and consumption of time and it will be replaced with the direct method in future [25], and the integration of the OF improves the estimation of the image orientation.

3.3.1 Aerial Triangulation

Aerial Triangulation (AT) helps to determine the orientation that is necessary to have a good exactitude of the information provided by the sensors. However, it is not right, especially in the sensors of low cost. Improvement in the orientation correctness is possible by incorporating the information from the cameras. Next, tie points can be extracted and tracked for further fusion with the navigation system INS/GNSS.

The use of the OF in the tracking of tie points is demonstrated by [26], and the collinearity equations represent the projection of the objects in the terrain in relation to the image coordinates. Figure 3.6 shows the relation of the object represented in the image and the objects on the terrain.

One tie point is a feature of the scene represented in two images consecutive, and these points have no geographies' coordinates so that the collinearity equations can determine these coordinates in the instant t:

$$\begin{bmatrix} X_P \\ Y_P \\ Z_P \end{bmatrix} = \lambda_1 \begin{bmatrix} r_1 - r_0 \\ s_1 - s_0 \\ -f \end{bmatrix} + \begin{bmatrix} X_{c1} \\ Y_{c1} \\ Z_{c1} \end{bmatrix}, \tag{3.18}$$

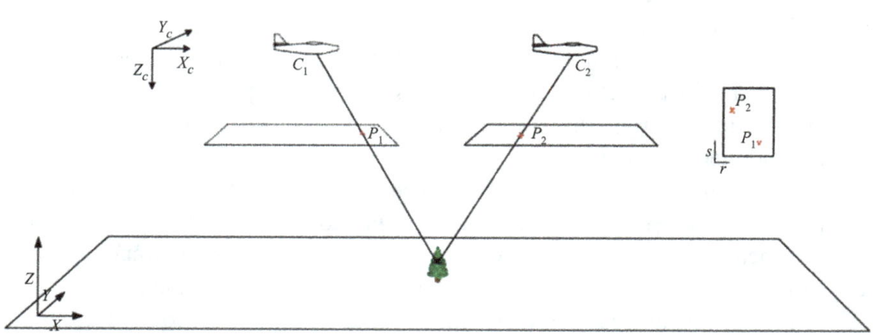

Figure 3.6 Relation of tie points involving two consecutive images

where X_p, Y_p and Z_p are the geographic coordinates of a point in the scene, λ is a factor of scale, R is the matrix of rotation formed by the measure of the IMU, r and s are the coordinates in the plane of image, f is the focal length and X_c, Y_c and Z_c are the geographic coordinates from the GPS.

Now known the tie point in geographic coordinates, the projection of the tie points in the coordinates of the next image at the instant of time $t + \Delta t$ is calculated by means of the next equation:

$$\begin{bmatrix} r_2 - r_0 \\ s_2 - s_0 \\ -f \end{bmatrix} = \lambda_2 R_2 \begin{bmatrix} X_P - X_{c1} \\ Y_P - Y_{c1} \\ Z_P - Z_{c1} \end{bmatrix}. \tag{3.19}$$

From the previous statements, one can see that the aerial photogrammetry uses the collinearity equation with the information provided by the GNSS and the INS, as implemented in [26] based on the tracking of tie points using the movements of feature points (OF).

On image using the tracking of tie points with the method KLT and the tie point represented as $P_1 = (r, s)$ in Figure 3.6 is determined at point $P_2 = (r + dx, s + dy)$ on the next image, where dx and dy is calculated using the LK algorithm (3.3).

The integration of the tie point in the image according to (3.3) is:

$$f(x, y, \Delta t) = f(r_2 + \Delta x, s_2 + \Delta y) \tag{3.20}$$

where r_2 and s_2 are the tie points calculate for the collinearity equation (3.19).

So, to enhance the tracking of tie points [26], the integration of the OF (3.3) with the calculate of the tie points based in (3.19) is proposed according to:

$$f(r, s, \Delta t = f(x + (r_2 - x) + \Delta x, y + (s_2 - y) + \Delta y)) \tag{3.21}$$

3.4 Optical flow and situational awareness

Nowadays, the UAVs are operating from a long place of control ground station, being so the environment perceived by the operator is different to the environment perceived by the ARP. To have success in the mission, the operator of the ARP may perceive the ambient as if he-self was inside of it. According to [27], it is necessary to have in mind various actors that intervene in the operation of the system and the mission, and they are the human or operator, the UAV and the objective of the mission; each one of these actors has a relation between themselves.

- Human-UAV:
  What the operator requires knowing about the UAV, for example monitor the systems of the UAV (battery, autopilot, sensors, position, velocity and others), auto-landing, return to home.
- UAV-Human:
  What the UAV requires knowing about the operator, for example detection and avoid obstacles, tracking of targets, reconnaissance, and situational awareness.

- Mission:
  The type of mission that will be developed, for example, mapping, tracking, recreational, search and rescue.

According to the above, there is a high grade of dependence of the UAV with respect to the operator and the operator with respect to the UAV. The ARP sends information constantly to the operator to monitor its systems and the environment around it. The accumulation of tasks and information can cause stress and fatigue in the operator. So, these factors could affect the success of the mission.

However, there are various tasks that the ARP can make by itself without the need of the overload of the operator, for example, detect and avoid obstacles or tracking of targets, so the operator can make decisions most complexes.

The situation awareness is defined as "The perception of the elements in the environment within a volume of time and space, the comprehension of their meaning and the projection of their status in the near future" [28]. In this definition, we can find three main aspects that can be applied in the unmanned systems.

Perception is the knowledge of all objects situated around it.

Comprehension is the capacity of establishing when an object becomes a possible threat.

Projection is the information that it should consider to evade the threat.

The situational awareness permits one to know what is happening in the surrounding, why it is happening, what would happen now and what one can be made to make decisions and maintain the control according to the situation.

3.4.1 Detect and avoidance system

One of the challenges that ARP faces when flying in urban or forest areas and that occupy more time of attention of the operator is the avoid obstacles, affecting the situational awareness, these characteristics can be operated automatically. According to [29], there are three stages that an ARP carries out to avoid the collision with an object: the first stage is sensing, the second stage is detecting and the last state is the avoidance. These same concepts are specifics according to the description of situational awareness, perception, comprehension and projection.

3.4.1.1 Perception

The monitoring of the elements around the ARP is made by means of the information of the sensors as LIDAR, acoustic sensors, optical and infra-red sensors. However, the optical sensors have a lot advantages as its low cost, small size and low weight.

In the perception stage it is necessary to know some characteristics own of the objects that are being filmed, this characteristics can be lines, contours, corner or blob areas, to determinate lines there are algorithms as the Hough transformation, to determinate corners we can use the SIFT, SURF ORB or Harris algorithms. With these feature points it is possible to determinate the OF and with it, the objects inside of the image can be isolated.

3.4.1.2 Comprehension

Of the all objects represented by the image, some of it would be a threat and others not. Determination of the direction of movement of the objects in the scene in relation to the ARP is necessary to determinate which of it is possible a threat. The determination of the direction of the movement is known as focus of expansion (FOE).

Focus of expansion

The focus of expansion (FOE) is the point of congruence of the vectors that form the movement of the objects in the whole image, and it lets one determine the direction of the movement of the ARP about each one of them; the movement can be rotational, transnational or a combination of both (Figure 3.7).

The focus of expansion can be calculated through the OF around the whole image or on feature points, so [30] propose a solution based on the movement on features points:

$$FOE = (A^T A)^{-1} A^T b \tag{3.22}$$

with

$$A = \begin{bmatrix} a_{00} & a_{01} \\ \vdots & \vdots \\ a_{n0} & a_{n1} \end{bmatrix} \quad \text{and} \quad b = \begin{bmatrix} b_0 \\ \vdots \\ b_n \end{bmatrix}$$

where each feature point $p(x, y)$ and its corresponding vector flow $V = (u, v)$ is $a_{i,0} = u_i \; a_{i,1} = v_i$ and b is defined as $b_i = xu_i - yv_i$.

Time of collision (TOC)

There are various ways to calculate the time of collision [31–33], one is based on the expansion of the feature points, and the other is by means of the OF. TOC is calculated by the following relation:

$$\frac{y}{f} = \frac{Y}{Z} \;\rightarrow\; y = f\frac{Y}{Z} \tag{3.23}$$

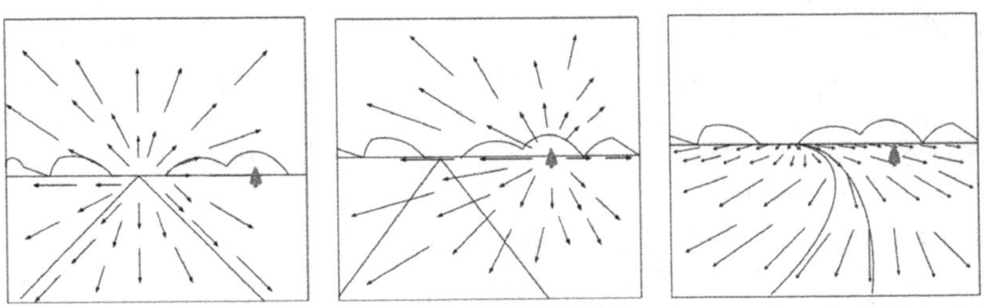

(a) Translational (b) Translational (c) Rotational

Figure 3.7 FOE in different movements

where Y is the position of a point in ground coordinates, Z is the distance between the FOE and the position of camera f is the length focal and y is a projection of point Y in the image.

Differencing (3.23) with respect to the time:

$$\frac{\partial y}{\partial t} = f\left(\frac{\frac{\partial Y}{\partial t}}{Z}\right) - fY\left(\frac{\frac{\partial Z}{\partial t}}{Z^2}\right). \tag{3.24}$$

As Y is a point fixed in the scene, the differentiation of $\partial y/\partial t = 0$ and according to the definition of y in the (3.23) yields:

$$\frac{\partial y}{\partial t} = -y\frac{V}{Z}$$

Rearranging the previous expression leads to:

$$\frac{y}{\frac{\partial y}{\partial t}} = -\frac{Z}{V} = \tau \tag{3.25}$$

where $V = \partial Z/\partial t$ is the velocity of the ARP in the direction of the FOE and τ is the TOC.

3.4.1.3 Projection

The projection is a stage that search determinate if the obstacle that is in front really is a threat, if the object really is a threat, it generate a new path to avoid the obstacle.

3.5 Optical flow and navigation by images

From the Second World War, the INS has been a part important in the calculate of the position and velocity of a lot of aircraft, boat and underwater [32]. It is constituted essentially by a gyroscope and an accelerometer embedded in an IMU. However, this sensor is inaccurate due to drift, bias and other errors. All these errors indicate that this sensor is not viable to calculate the position and velocity. To correct this problem, it is necessary to integrate the system with an additional sensor and calibrate the errors generated in the estimation of the states. The GNSS is one of the sensors most utilised in navigation systems that are integrated with the INS using generally extended Kalman filters [30,34].

In the navigation, the sensors can be classified as internal and external reference sensors [27]. The IMU sensor reads the acceleration and angular velocity by itself, in contrast to the GNSS sensor that receives its signal of reference of an external source. Therefore, for, this condition, the GNSS is most susceptible to suffer interferences or interceptions, as occurred in 2011 when an ARP was captured by Iranian forces [33]. It indicates the vulnerability of the navigation systems to be intercepted, hacked or suffer from a spoof. Another cause of losing the GNSS signal is when it flies among a forest or in the middle of the builds.

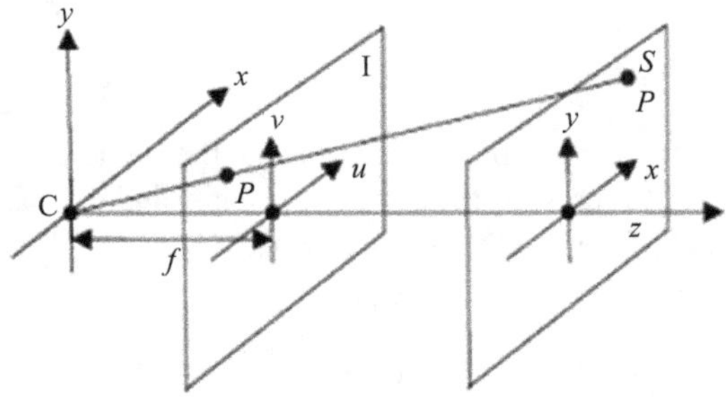

Figure 3.8 Pinhole model for the camera

The estimation of the movement based on the fusion of different sensors with a vision system is made through the projection of the movement of the object in a scene that is recorded in the OF [35]. The process of calculating the movement of the ARP in the plane of the image is known as egomotion, and the estimation of this movement is conditioned to the model of the camera; in this chapter, we are going to applicate the pinhole model. In Figure 3.8, the model of the camera is presented.

In Figure 3.8, C is the camera placed at the origin of the camera coordinate system *xyz*, *S* is the scene that is being filmed, and *f* is the focal distance from C until the origin of plane I (*P* lies on I). Since the camera provides information about the structures being recorded, it is necessary to map these captured objects on the two-dimensional plane XY, and the relation between *u* and *Y*, and the one between *v* and *X* become, respectively,

$$\frac{u}{f} = \frac{r}{Z} \Rightarrow u = f\frac{r}{Z} \quad \text{and} \quad \frac{v}{f} = \frac{s}{Z} \Rightarrow v = f\frac{s}{Z} \tag{3.26}$$

Resolving the equation before results in the matrix form

$$\begin{bmatrix} u \\ v \end{bmatrix} = \frac{1}{Z}\begin{bmatrix} 0 & f \\ f & 0 \end{bmatrix}\begin{bmatrix} r \\ s \end{bmatrix}. \tag{3.27}$$

3.5.1 Egomotion

Egomotion is the estimation of the camera movement relative to the scene that is being filmed, using feature points to the scene represented in the image and transforming from the image plane until the origin of the camera we can find the system movement with relation to one point in the ground. To estimate the movement, it is necessary to consider that the *Z* axis of the camera is aligned with the *Z* axis of the ARP's body, and the localisation of the camera may coincide with the IMU's position; additionally, the earth may be considered plane and the altitude and attitude are known.

The next equations were derivative based on the pinhole model camera.

The velocity of the camera with respect to the a point is determined by:

$$R^c = -R^e - w^{ce}XR \tag{3.28}$$

where R is the geographic coordinates of a point in the scene $[X\ Y\ Z]^T$. By defining $R^e = V_{CAM} = [V_x\ V_y\ V_z]^T$ represents the camera's velocity and $w^{ce} = w_{CAM} = [w_x\ w_y\ w_z]$ represents the measure of the IMU sensor, it follows:

$$\begin{bmatrix} \dot{X} \\ \dot{Y} \\ \dot{Z} \end{bmatrix} = - \begin{bmatrix} V_x \\ V_y \\ V_z \end{bmatrix} - \begin{bmatrix} w_y Z - Y w_z \\ w_z X - Z w_x \\ w_x Y - X w_y \end{bmatrix} \tag{3.29}$$

Deriving the equations of the pinhole model (3.26) with respect to the time yields the system:

$$\dot{X} = \dot{r}\frac{Z}{f} + \dot{Z}\frac{r}{f} \quad \text{and} \quad \dot{Y} = \dot{s}\frac{Z}{f} + \dot{Z}\frac{s}{f} \tag{3.30}$$

Combining (3.29) with (3.30) gives:

$$\begin{aligned} -V_x - w_y Z + Y w_y &= \dot{r}\frac{Z}{f} + \frac{r}{f}\left(-V_z - w_x Y + X w_y\right) \quad \text{and} \\ -V_y - w_z X + Z w_x &= \dot{s}\frac{Z}{f} + \frac{s}{f}\left(-V_z - w_x Y + X w_y\right) \end{aligned} \tag{3.31}$$

Arranging the above equations, the movement of the pixels in the coordinates of the image becomes:

$$\begin{aligned} \dot{r} &= \frac{f}{Z}\left[-V_x - w_y Z + w_z Y\frac{Z}{f} - \frac{X}{f}\left(-v_z - w_x Y\frac{Z}{f} + w_y Z\frac{Z}{f}\right)\right] \text{ and} \\ \dot{s} &= \frac{f}{Z}\left[-V_y - w_x Z + w_z X\frac{Z}{f} - \frac{Y}{f}\left(-v_z - w_x Y\frac{Z}{f} + w_y X\frac{Z}{f}\right)\right] \end{aligned} \tag{3.32}$$

For one feature point, the matrix is defined as:

$$\begin{bmatrix} \dot{r} \\ \dot{s} \end{bmatrix} = \begin{bmatrix} -f & 0 & r & \frac{rs}{f} & -\left(f + \frac{r^2}{f}\right) & s \\ 0 & f & s & \left(f - \frac{s^2}{f}\right) & -\frac{rs}{f} & r \end{bmatrix} \begin{bmatrix} V_x/Z \\ V_y/Z \\ V_z/Z \\ w_x \\ w_y \\ w_z \end{bmatrix} \tag{3.33}$$

Now, defining

$$M(f, r_n, s_n) = \begin{bmatrix} -f & 0 & r & \frac{rs}{f} & -\left(f + \frac{r^2}{f}\right) & s \\ 0 & f & s & \left(f - \frac{s^2}{f}\right) & -\frac{rs}{f} & r \end{bmatrix} \tag{3.34}$$

for a set of feature points results in:

$$\begin{bmatrix} \dot{r}_1 \\ \dot{s}_1 \\ \dot{r}_2 \\ \dot{s}_2 \\ \vdots \\ \dot{r}_n \\ \dot{s}_n \end{bmatrix} = \begin{bmatrix} M(f, r_1, s_1) \\ M(f, r_2, s_2) \\ \vdots \\ M(f, r_n, s_n) \end{bmatrix} \begin{bmatrix} V_{x,y,z} \\ w_{x,y,z} \end{bmatrix} \tag{3.35}$$

Now, the systems of the form $Ax = b$ can be resolved by a minimum mean square, which leads to the estimation of the linear and angular velocities:

$$\begin{bmatrix} V_{x,y,z} \\ w_{x,y,z} \end{bmatrix} = \left(M^T M\right)^{-1} M^T (\Delta r, s) \tag{3.36}$$

To resolve (3.36), it is necessary to have a minimum of three image coordinates of the feature points.

3.6 Case study: INS using FPGA

According to the sections above, the determination of OF is the concept that joins the computer vision algorithms with the navigation systems. In this case of study, an architecture to fuse the OF with the navigation system INS/GPS is proposed.

The computation of the vision algorithms is computationally costly, and it consumes much memory to process and store the images; the process has characteristics of real-time to respond most rapid to the stimulus of the environment or the operator [36]. To differentiate from processors, FPGAs are devices with circuits projected by the system designer. The developer decides which task is the input, how processing stages are interconnected, and which tasks can be executed in parallel form. This characteristic is essentials to work on computer vision.

The initial point to merge the OF with the INS/GNSS is to detect the image feature points in need of tracking via various algorithms to extract them, e.g., SIFT, SURF, ORB and Harris (please, refer to Table 3.1 for computing times).

Table 3.1 Calculate of time in seconds of the feature points in different sizes of images

	Resolution			
Method	**320 × 240**	**584 × 388**	**640 × 320**	**1,024 × 768**
SIFT	0.590584	2.187848	3.295912	11.315503
SURF	0.691454	2.591132	3.659905	12.27442
ORB	0.048693	0.097650	0.116491	0.241904
Harris	0.039246	0.06228	0.076807	0.15

According to Table 3.1, it is possible to see that SIFT and SURF determine the feature points in a time very similar, so the SIFT and SURF need 0.59 and 0.69 s, respectively, to determine feature points on one image of 320 × 420 pixels. Now, the time of computation to determinate the feature points on an image of 1,024 × 768 pixels is about 11.315 s to SIFT and 12.274 s to SURF, so in general, the time increases exponentially when the size of the image increases. This elevation time is due to the prominent features points that are processed as is observed in the figure.

In contrast, the LK and the ORB have very similar times of computation. However, the corresponding times of calculation are smaller than the time for the SIFT and SURF techniques (Figure 3.9). The calculation of the feature points on one image of 320 × 420 pixels, the ORB needs 0.048 s and the LK 0.039 s. If the resolution of the image increases to 1,024 × 768, then the time of calculation becomes 0.2419 s for the ORB and 0.15 s for the LK. The quantity of feature points is minimal when compared to the SIFT and SURF. Hence, the quantity of feature points that the ORB and the LK can estimate OF is computed in less time.

The embedded systems are physical units with a few resources, and the characteristics of real-time are necessary, for it, two candidates to implement the algorithm in the FPGA are the ORB and the Harris (Figure 3.10). However, the ORB is not invariant to rotation so that the Harris algorithm is the candidate.

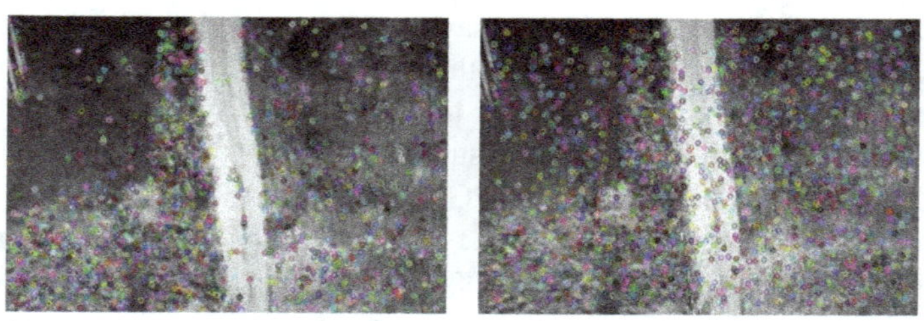

(a) (b)

Figure 3.9 Feature points: (a) SIFT and (b) SURF

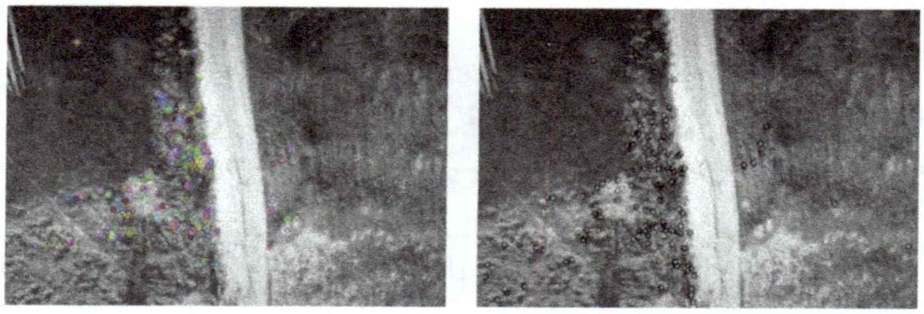

(a) (b)

Figure 3.10 Feature points: (a) ORB and (b) Harris

3.6.1 Architectural proposals

The algorithms Lucas–Kanade, SIFT, SURF and ORB include two steps to calculate feature points and matching it or know the position in the next image, so the proposed architecture has two units to calculate feature points and other units to match it or calculate its movement (Figure 3.11).

3.6.1.1 Control unit (CU)

It is the main component inside of the architectural, it receives the image from the input component and send the information to the interface with the memory physical to store it in the SDRAM memory, also send the pixels to the feature detector component and receive the information of it to store in the RAM, another task is sending the feature points from the RAM to the matching component and receive the OF to send to the output component.

3.6.1.2 Generation of time

The architecture works with different periods, so the input component handles 112500 bits per second. The SDRAM memory works at 133 MHz and the rest of the architecture with another frequency. Hence, the control unit needs to generate different periods for each one of the components.

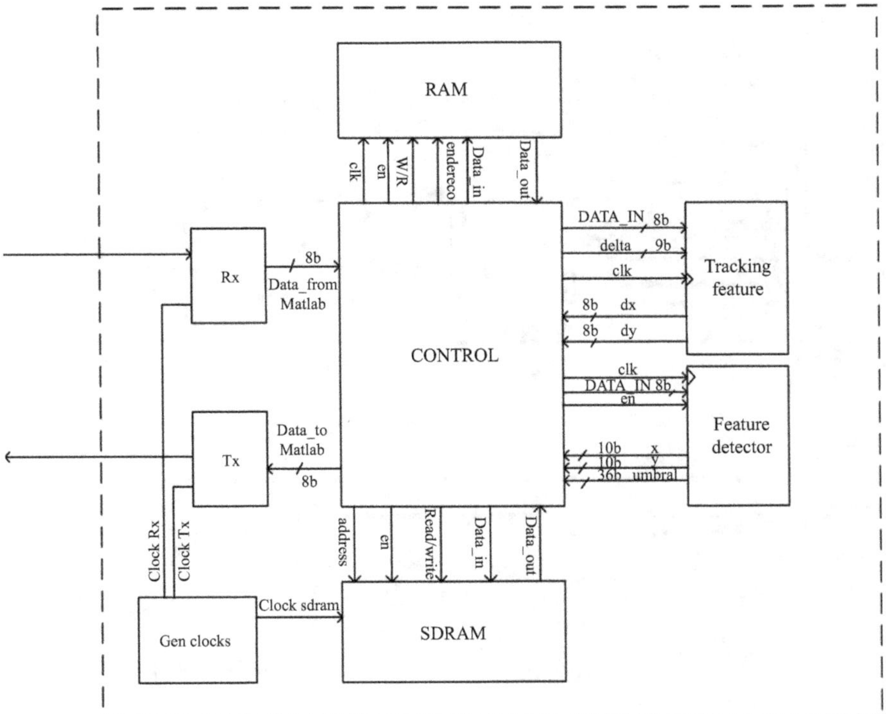

Figure 3.11 Proposed architecture

3.6.1.3 Feature points detector

In this component is the Harris code, internally this component has five subcomponents: the first component makes a matrix of 3 × 3 according to the received information, the second component calculates the derivative of the points in the x and y directions, the third component calculates the derivatives according to the Harris matrix, the fourth component sums up the result of multiplication of the derivatives and the fifth component determines the feature point according to (3.16); additionally, the whole process is controlled for a unit that determines the position of each pixel (Figure 3.12).

3.6.1.4 OF calculation

The calculation of the displacement of the feature points is very similar to the Harris features detector component since it hinges on the calculation of the Harris matrix, whose first four components have the same function as the component above. The difference with the feature points elements is the fifth component. The displacement of the feature points (Figure 3.13) (d_x, d_y) results from:

$$d_x = \frac{-\sum I_t I_x \sum I_y^2 + \sum I_x I_y \sum I_t I_y}{\sum I_y^2 \sum I_x^2 - \sum I_x I_y^2} \tag{3.37}$$

$$d_y = \frac{-\sum I_t I_y \sum I_x^2 + \sum I_x I_y \sum I_t I_x}{\sum I_y^2 \sum I_x^2 - \sum I_x I_y^2} \tag{3.38}$$

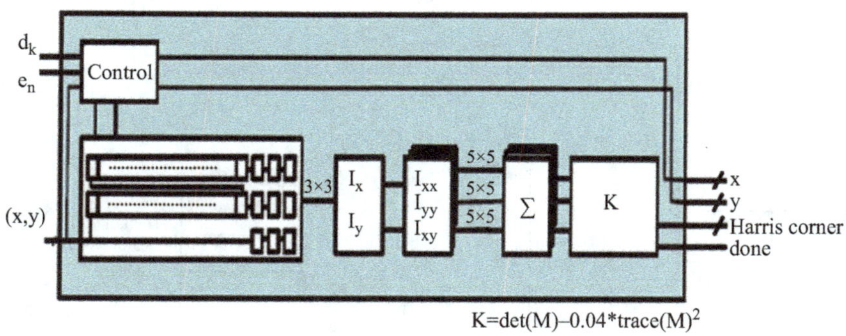

Figure 3.12 Unit selector for feature points

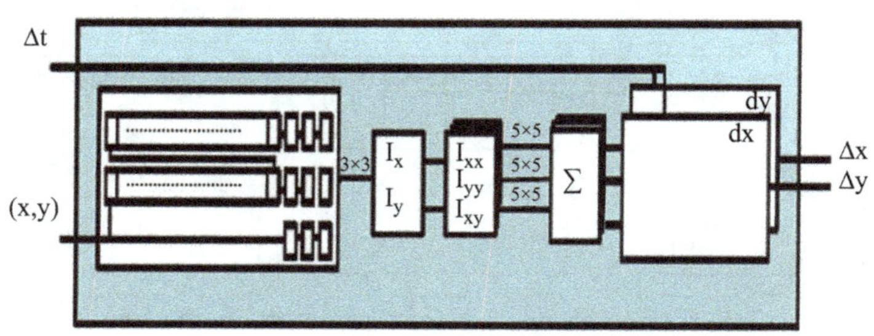

Figure 3.13 Feature points' unit tracking

3.6.1.5 Input and output component

The function of the input component is to receive the information about the camera in the form of bits and convert it into an array of 8 bits; in contrast, the output sends the information in bits, and for last the interface SDRAM component sends the information to the SDRAM memory to store the image.

3.6.2 Integration INS/GPS/OF using a Kalman filter

There are many ways to integrate the OF with the navigation system using the Kalman filter [30,34]. However, the deduction that here is presented was made by Ho *et al.* [37].

To integrate the INS/GPS, the model ψ-angle that consists of a model where the perturbations are made in the local lever northing-pointing coordinate system can be used. The equations of this model are defined in the following equations:

$$\begin{aligned} \delta\dot{v} &= -(w_{ie} + w_{in})x\delta V - \delta\psi x f + \delta g + \nabla \\ \delta\dot{r} &= -w_{en}x\delta r + \delta V \\ \delta\dot{\psi} &= -w_{in}x\delta\psi + \varepsilon \end{aligned} \tag{3.39}$$

where, 24 states to implement the Kalman filter were defined, it is described below:

$$\begin{aligned} X_{NAV} &= [\delta r_N, \delta r_E, \delta r_D, \delta V_N, \delta V_E, \delta V_D, \delta\psi_N, \delta\psi_E, \delta\psi_D]^T \\ X_{Acc} &= [\nabla bx, \nabla by, |\nabla bz, \nabla fx, \nabla fy, \nabla fz]^T \\ X_{Gyro} &= [\varepsilon_{bx}, \varepsilon_{by}, \varepsilon_{bz}]^T \\ X_{Grav} &= [\delta_{bx}, \delta g_{by}, \delta g_{bz}]^T \\ X_{Ant} &= [\delta L_{bx}, \delta L_{by}, \delta L_{bz}]^T \end{aligned}$$

where δv, δr and $\delta\psi$ are velocity, position and attitude error, respectively, w_{ie} is the earth rate vector, w_{in} is the angular rate vector of the true coordinate system with respect to the inertial frame, w_{en} is the angular rate vector of the true coordinate system concerning the earth, ∇ is the accelerometer error vector, δg is the error in the computed gravity vector as is described by Ho *et al.* [37].

So, the vector of observation is the difference between the position vector calculated by the INS and the position provided by the GPS, being as in the equation of observation:

$$Z_k = P_{INS,k} - P_{GPS,k} \tag{3.40}$$

To add the equations of the OF inside of the Kalman filter [37] assume that the ground is flat and the OF is the sum of the translational and angular velocity, and the ARP is flying in a constant height, and there are not high dynamic. Therefore, the simplified OF equation is:

$$Q_x = \frac{V_x}{H} + w_y \quad \text{and} \quad Q_x = \frac{V_y}{H} + w_x \tag{3.41}$$

where Q_x and Q_y are the OF vectors, V_x and V_y are the velocity components along the camera body x and y and w_x and w_y are the rotational angular rates of the UAV camera, and H is the ground height, which is measured from the camera to the ground surface.

When the OF measures are perturbed, the error of the model can be written as below:

$$\delta Q_x = \frac{\delta V_x}{H} + \delta w_y - \frac{V_x}{H^2}\delta H \quad \text{and} \quad \delta Q_y = \frac{\delta V_y}{H} - \delta w_x - \frac{V_y}{H^2}\delta H \tag{3.42}$$

As the camera and the IMU are fixed in the body, the longitudinal and lateral velocity of the camera are defined employing the velocity in the coordinate system NED as:

$$\begin{bmatrix} V_x \\ V_y \end{bmatrix} = \begin{bmatrix} \cos\psi & \sin\psi \\ -\sin\psi & \cos\psi \end{bmatrix} \begin{bmatrix} V_N \\ V_E \end{bmatrix} \tag{3.43}$$

Now, the height is defined as:

$$H = H_0 + H_s + H_d \tag{3.44}$$

where H_0 is the initial ground height, H_s is the ground height of ground surface during UAV flight and H_d is the height variation due to UAV dynamics, differencing (3.44)

$$\dot{H} = \dot{H}_s - V_d \tag{3.45}$$

where V_d is the velocity of the component vertical in the NED coordinate.

Now, substituting (3.44) and (3.43) on the perturbed model equation, we have the state augmented to the Kalman Filter (3.46), so the observer vector to the Kalman filter according to [37] is:

$$Z_k = \begin{bmatrix} P_{INS,k} - P_{GPS,k} \\ Z \end{bmatrix} \tag{3.46}$$

3.7 Future trends and discussion

Besides the issues related to OF and navigation mentioned before, other topics deserve further investigation and are addressed below.

3.7.1 3D optical flow

A photonic mixer device (PMD) camera is a TOF camera working with the TOF principles and resembling a LIDAR scanner. TOF cameras rely on a modulated light pulse sent by the illumination source. The target distance results from the estimation of the time the pulse takes to reflect from the target and back to the reception unit. PMD cameras can produce the range data, which is nearly independent of the illumination conditions, optical appearance and the intensities of the image pixels, similar to visible light cameras. It acquires high-resolution range data fast [37,38].

As the PMD range camera delivers sufficient evidence about the obstacles, it can estimate the moving obstacles trajectories. These TOF cameras provide a 3D point cloud consisting of a set of surface points in a 3D coordinate system (X, Y, Z), for all entities in the field of view of the camera.

3D OF quantifies the motion of each volume voxel between adjacent volumes. The moving patterns correspond to temporal variations of the image intensity. Generally, determining the OF employs the smoothness constraint equation. The spatiotemporal derivatives of the image intensity are used in differential techniques to obtain the OF.

Differential methods can be local or global. Local techniques encompass the optimisation of a local functional, like in the LK scheme. The global procedures find the flow vector through minimisation of a global functional, as in Horn and Schunck. Local methods are robust to noise but do not deliver dense OF. Global techniques produce extremely dense OF fields, with much larger noise sensitivity.

Any OF is the scene flow projection onto the image plane of a camera. In a completely non-rigid world, the motions of the scene points may altogether be independent of each other. Hence, the 3D scene motion corresponds to a dense 3D vector field defined for all points on every single surface in the scene. This type of OF requires the ability to handle a heavy computational load in a distributed cyber-physical system [8–13,35,36,39–41].

3.7.2 Multispectral and hyperspectral images

Multispectral (MSI) and hyperspectral (HSI) images contain many continuous light wavelength spectral bands that can be indexed. These bands afford profuse data on the spatial-spectral distribution of the target object materials and their physical, chemical, as well as geometrical properties. Compared to RGB images, MSI and HSI data have also revealed great potential in 3D CV, above all for situations requiring fine examination of the object spectral responses.

Despite its full CV potential, the use of 3D reconstruction from MSI and HSI imaging in navigation and collision avoidance is still very limited. In most works that have incorporated MSI and HSI data into 3D models, the 3D shapes were generated by range camera rather than from MSI and HSI images directly. Then, the spectral information can be mapped to the 3D shape.

MSI and HSI imagers can acquire high spatial resolution band images from near-ultraviolet to near-infrared range to be integrated into a 3D scanning system to capture the spectral reflectance and fluorescence of objects.

A 3D model can also be generated based on depth data captured by a laser scanner, with MSI and HSI images mapped to this 3D model. The 3D models reconstructed from different bands have demonstrated different properties, which complement each other.

Various reasons cause the differences between band-wise 3D models. The first reason is that the object surface has different reflectance due to the variation of composition materials. Henceforth, some facts visible in one band may turn out to be invisible in another band.

The second reason is that the point of focus varies due to wavelength because the focal length of a lens is associated with its refractive index, which declines with the rise of wavelength. Hence, most available MSI and HSI cameras can only be well focussed on several bands and become unfocussed in other bands. So, different feature points are extracted from each band, which produces different band-wise 3D models with varying scales and point sets preventing direct registration to form a complete 3D model. Each 3D model from a set of MSI and HSI images can be the result of merging all key-points from several bands for 3D reconstruction. Nevertheless, the fine structural and spectral data from the band images are lost.

The use of MSI and HSI images demands a lot from the CPS concerning distributed computing [38,42–45] and fast communication links [45,46]. Since dimensionality is quite a burden, computational intelligence [47–50] is called for to handle the computational load both via software, hardware and hybrid frameworks.

A complete MSI and HSI 3D model combines all band-level 3D models into a single one with some structural descriptor developed to characterise the spatial relationship among points belonging to the same set, which can then match points between two 3D models created at different bands.

Complete 3D models obtained from MSI and HSI images to help a UAV detect and avoid obstacles by generating band-wise 3D models with excellent structural, spectral object data integrated into the fully reconstructed model. This exceeds traditional 3D registration and reconstruction from RGB images because this brings in accurate and efficient band-wise merging models to aid navigation, which facilitates the analysis about the way structural data from different band models affect the whole scenery.

3.8 Conclusion

Throughout the chapter, the importance of CV algorithms to estimate the movement of pixels in a pair of consecutive images to obtain several cues about the UAV outer environment was analysed. The OF can be integrated into the navigation system to help to improve the orientation and the tracking of tie points. We also saw how the OF could help prevent acts of collision. Finally, a case study based on the calculation of the OF through a hardware architecture relying on FPGA is examined and to show how OF can be integrated into the navigation system using an extended Kalman filter.

The use of the camera is a fundamental piece in the development of any mission. However, the cameras also have suffered a significant step forward in recent times. Today's cameras have high resolution but in the long term, these are going to be fully integrated into the navigation system with even better imaging characteristics. These and other future advancements will require systems with faster processing units that with enough memory to store the imageries and execute more cutting-edge algorithms compatible with them.

This chapter talked about a few algorithms to determine the OF and gave a comprehensive literature survey. Nonetheless, many other algorithms can detect feature points even faster than ORB or Harris, as SUSAN, AKAZE and others, which can be integrated into the OF calculation. To improve the real-time performance is crucial to know which of them is/are adequate to the application in mind and, in case of necessity, for a better FPGA implementation.

It should be pointed out that information fusion from multimodality imaging sources may require extensive distributed processing through the cloud with the need to balance communication, onboard and offboard processing [8–13,35,36,39–41].

References

[1] Eisenbeiss H, *et al.* A mini unmanned aerial vehicle (UAV): system overview and image acquisition. International Archives of Photogrammetry Remote Sensing and Spatial Information Sciences. 2004;36(5/W1):1–7.

[2] Cigla C, Brockers R, and Matthies L. Image-based visual perception and representation for collision avoidance. In: IEEE International Conference on Computer Vision and Pattern Recognition, Embedded Vision Workshop; 2017.

[3] Bonin-Font F, Ortiz A, and Oliver G. Visual navigation for mobile robots: a survey. Journal of Intelligent & Robotic Systems. 2008 Nov; 53(3):263–296. Available from: http://dx.doi.org/10.1007/s10846-008-9235-4.

[4] Schmidt GT. Navigation sensors and systems in GNSS degraded and denied environments. Chinese Journal of Aeronautics. 2015;28(1):1–10.

[5] Kerns AJ, Shepard DP, Bhatti JA, *et al.* Unmanned aircraft capture and control via GPS spoofing. Journal of Field Robotics. 2014;31(4):617–636.

[6] Mota RLM, Ramos AC, and Shiguemori EH. Application of self-organizing maps at change detection in Amazon Forest. In: 2014 11th International Conference on Information Technology: New Generations (ITNG). IEEE; 2014. p. 371–376.

[7] Martins MP, Medeiros FL, Monteiro MV, *et al.* Navegacao Aerea Autonoma por Imagens. In: VI WAI – Workshop Anual de Pesquisa e Desenvolvimento do Instituto de Estudos Avancados, Sao Jose dos Campos; 2006.

[8] Razmjooy N, Mousavi BS, Khalilpour M, and Hosseini H, Automatic selection and fusion of color spaces for image thresholding. Signal, Image and Video Processing. 2014;8(4):603–614.

[9] Mousavi BS, Soleymani F, and Razmjooy N. Color image segmentation using neuro-fuzzy system in a novel optimized color space. Neural Computing and Applications. 2013;23(5):1513–1520.

[10] Estrela VV, Magalhaes HA, and Saotome O. Total variation applications in computer vision. In Handbook of Research on Emerging Perspectives in Intelligent Pattern Recognition, Analysis, and Image Processing, pp. 41–64. IGI Global, 2016.

[11] Mousavi B, Somayeh F, Razmjooy, and Soleymani F. Semantic image classification by genetic algorithm using optimised fuzzy system based on Zernike moments. Signal, Image and Video Processing. 2014;8(5):831–842.

[12] Razmjooy N, Estrela VV, and Loschi HJ. A survey of potatoes image segmentation based on machine vision. In Applications of Image Processing and Soft Computing Systems in Agriculture, pp. 1–38. IGI Global, 2019.

[13] Estrela VV, and Coelho AM. State-of-the art motion estimation in the context of 3D TV. In: Multimedia Networking and Coding. IGI Global, 2013. 148–173. doi:10.4018/978-1-4666-2660-7.ch006.

[14] Silva Filho P. Automatic landmark recognition in aerial images for the autonomous navigation system of unmanned aerial vehicles [dissertation]. ITA. Sao Jose dos Campos, (SP); 2016.

[15] Da Silva W. Navegacao autonoma de vant em periodo noturno com images infravermelho termal [dissertation]. INPE. Sao Jose dos Campos, (SP); 2016.

[16] Fabresse F, Caballero F, Merino L, *et al.* Active perception for 3D range-only simultaneous localization and mapping with UAVs. In: Unmanned Aircraft Systems (ICUAS), 2016 International Conference on. IEEE; 2016. p. 394–398.

[17] Barron JL, Fleet DJ, and Beauchemin SS. Performance of optical flow techniques. International Journal of Computer Vision. 1994;12(1):43–77.

[18] Yves Bouguet J. Pyramidal implementation of the Lucas-Kanade feature tracker. Intel Corporation, Microprocessor Research Labs. 2000.

[19] Lowe DG, inventor; University of British Columbia, assignee. Method and apparatus for identifying scale-invariant features in an image and use of same for locating an object in an image, 1999-03-08.

[20] Funayamam R, inventor; Katholieke Universiteit Leuven, assignee. Method and apparatus for identifying scale invariant features in an image and use of same for locating an object in an image; 2006-04-28.

[21] Rublee E, Rabaud V, Konolige K, *et al.* ORB: An efficient alternative to SIFT or SURF. In: 2011 IEEE international conference on Computer Vision (ICCV). IEEE; 2011. p. 2564–2571.

[22] Science for a changing world [homepage on the Internet]. The Association; c1995-2002. AMA Office of Group Practice Liaison. Available from: https://www.usgs.gov/faqs/what-remote-sensing-and-what-it-used-0?qt-news science products=7#qt-news science products.

[23] Zomrawi N, Hussien MA, and Mohamed H. Accuracy evaluation of digital aerial triangulation. International Journal of Engineering and Innovative Technology. 2011;2(10):7–11.

[24] Cheng L, Chen SX, Liu X, *et al.* Registration of laser scanning point clouds: A review. Sensors. 2018;18:1641.

[25] Munoz PM. Apoyo aereo cinematico y aerotriangulacion digital frente a los sistemas de navegacion inercial; D.Sc. dissertation, 2004.

[26] Tanathong S, and Lee I. Using GPS/INS data to enhance image matching for real-time aerial triangulation. Computers & Geosciences. 2014;72:244–254.

[27] Hartmann K, and Steup C. The vulnerability of UAVs to cyber attacks – an approach to the risk assessment. In: 2013 5th International Conference on Cyber Conflict (CyCon). IEEE; 2013. p. 1–23.

[28] Endsley MR. Design and evaluation for situation awareness enhancement. In: Proceedings of the Human Factors Society Annual Meeting, vol. 32. SAGE: Los Angeles, CA; 1988. p. 97–101.

[29] Chand BN, Mahalakshmi P, and Naidu V. Sense and avoid technology in unmanned aerial vehicles: a review. In: 2017 International Conference on Electrical, Electronics, Communication, Computer, and Optimization Techniques (ICEECCOT), IEEE; 2017. p. 512–517.

[30] Sabatini AM. (2006). Quaternion-based extended Kalman filter for determining orientation by inertial and magnetic sensing. IEEE Transactions on Biomedical Engineering, 53, 1346–1356.

[31] Zsedrovits T, Zarándy Á, Vanek B, Peni T, Bokor J, and Roska T. (2011). Collision avoidance for UAV using visual detection. 2011 IEEE International Symposium of Circuits and Systems (ISCAS), 2173–2176.

[32] Chao H, Gu Y, and Napolitano MR. (2013). A survey of optical flow techniques for UAV navigation applications. 2013 International Conference on Unmanned Aircraft Systems (ICUAS), 710–716.

[33] Gageik N, Benz P, and Montenegro S. (2015). Obstacle detection and collision avoidance for a UAV with complementary low-cost sensors. IEEE Access, 3, 599–609.

[34] Wang C, Ji T, Nguyen T, and Xie L. (2018). Correlation flow: robust optical flow using kernel cross-correlators. 2018 IEEE International Conference on Robotics and Automation (ICRA), 836–841.

[35] Estrela VV, Monteiro ACB, França RP, Iano Y, Khelassi A, and Razmjooy N Health 4.0: Applications, Management, Technologies and Review. Med Tech J, 2019;2(4):262–276, http://medtech.ichsmt.org/index.php/MTJ/article/view/205.

[36] Gupta S, Girshick RB, Arbeláez PA, and Malik J Learning rich features from RGB-D images for object detection and segmentation. Proceedings of 2014 ECCV. 2014.

[37] Ho HW, De Wagter C, Remes BDW, and de Croon GCHE, Optical-flow based self-supervised learning of obstacle appearance applied to MAV landing, Robotics and Autonomous Systems, Vol. 100, 2018, pp. 78–94, ISSN 0921-8890, https://doi.org/10.1016/j.robot.2017.10.004.

[38] Kaldestad KB, Hovland G, and Anisi DA, 3D sensor-based obstacle detection comparing octrees and point clouds using CUDA, Modeling, Identification and Control, Vol. 33, No. 4, 2012, pp. 123–130, ISSN 1890–1328 doi:10.4173/mic.2012.4.1

[39] Zhou Y, Li H, and Kneip L Canny-VO: visual odometry with RGB-D cameras based on geometric 3-D–2-D edge alignment. IEEE Transactions on Robotics, 2019;35:184–199.

[40] Peng HX, Liang L, Shen X, and Li GY Vehicular communications: a network layer perspective. IEEE Transactions on Vehicular Technology, 68, 2018, 1064–1078.

[41] Brahmbhatt S, Amor HB, and Christensen HI Occlusion-aware object localization, segmentation and pose estimation. Proceedings 2015 BMVC, 2015.

[42] Aroma RJ, and Raimond K. (2019) Intelligent land cover detection in multi-sensor satellite images. In: Ane B, Cakravastia A, and Diawati L (eds) Proceedings of the 18th Online World Conference on Soft Computing in Industrial Applications (WSC18). WSC 2014. Advances in Intelligent Systems and Computing, vol. 864. Springer, Cham.

[43] Li C, Yang SX, Yang Y, *et al.* Hyperspectral remote sensing image classification based on maximum overlap pooling convolutional neural network. Sensors. 2018;18:3587

[44] Zhao H, Wang Z, Jia G, Li X, and Zhang Y. Field imaging system for hyperspectral data, 3D structural data and panchromatic image data measurement based on acousto-optic tunable filter. Optics Express. 2018;26(13):17717–17730.

[45] Estrela VV, Saotome O, Loschi HJ, *et al.* Emergency response cyber-physical framework for landslide avoidance with sustainable electronics. Technologies. 2018;6:42. doi:10.3390/technologies6020042.

[46] Coppola M, McGuire KN, Scheper KY, and Croon GC. On-board communication-based relative localization for collision avoidance in micro air vehicle teams. Autonomous Robots. 2018;42:1787–1805.

[47] Hemanth DJ, and Estrela VV. (2017). Deep learning for image processing applications. Advances in Parallel Computing Series, vol. 31, IOS Press, ISBN 978-1-61499-821-1 (print), ISBN 978-1-61499-822-8 (online).

[48] de Jesus MA, Estrela VV, Saotome O, and Stutz D. (2018) Super-resolution via particle swarm optimization variants. In: Hemanth J, and Balas V (eds) Biologically Rationalized Computing Techniques For Image Processing Applications. Lecture Notes in Computational Vision and Biomechanics, vol. 25. Springer, Cham.

[49] Jia Z, and Sahmoudi M. A type of collective detection scheme with improved pigeon-inspired optimization. International Journal of Intelligent Computing and Cybernetics. 2016;9:105–123.

[50] Duan H, and Luo Q. New progresses in swarm intelligence-based computation. IjBIC. 2015;7:26–35.

Chapter 4

Introduction to navigation and intelligence for UAVs relying on computer vision

Suraj Bijjahalli[1] and Roberto Sabatini[1]

Vision-based sensors (VBSs) provide several advantages to unmanned aircraft systems (UASs) primarily due to a large amount of data they are able to capture, and their reduced size, weight, power, and cost compared to other state-of-the-art sensors. A number of vision-based navigation (VBN) methods have emerged recently, which aim to maximise state-estimation performance and reduce reliance on the global navigation satellite system. This chapter identifies and describes some of the most popular visual navigation strategies for the UAS to acquaint the reader with this important field of study. VBN methods presented here include visual servoing, optical flow-based state estimation, visual odometry and terrain referenced visual navigation. Reference system architectures and relevant mathematical models for these methods are presented to facilitate a more in-depth understanding. A review of these methods and their applications to various UAS use-cases is conducted, focussing primarily on seminal work in this domain. The limitations of this sensing modality are also presented, along with a discussion of future trends including multi-spectral imaging and biomimetic systems to inform the reader of key gaps and research avenues in this field.

4.1 Introduction

An overall reduction in manufacturing costs of cameras and reductions in the form factor of processors has led to the advent of a large market for commercial-off-the-shelf (COTS) VBN systems for UAS operations. A wide range of VBSs is increasingly employed in UAS operations for sensing the surrounding environment, localising the host platform and tracking nearby intruder platforms for functions such as separation assurance and collision avoidance. Scanning LIDARs are an alternative sensing modality capable of perceiving the surrounding environment at a high level of detail and are typically employed in surveying and mapping applications. Although advances have been made in lowering the form factor of LIDAR

[1]RMIT University, Bundoora, Australia

systems to support implementation on small UAS platforms, VBSs are better suited to this application owing to their relatively low size, weight and power, the passive nature of their operation and their low cost. A typical UAS system architecture is shown in Figure 4.1. The UAS comprises both the unmanned aerial vehicle (UAV) and the ground control station (GCS). Outputs from the navigation system including vehicle state estimates and/or low-level cues from sensors are used to drive guidance strategies and control. The navigation system outputs are also transmitted to the GCS via a telemetry downlink. Uplink from the GCS facilitates transmission of manual control commands by the remote pilot.

Within the navigation system, VBN sub-systems are typically employed in tandem with a suite of sensors. Common sensor systems are the global navigation satellite system (GNSS) and the inertial navigation system (INS) to enhance UAS state estimation and to increase navigation system redundancy in the event of individual sensor failure or loss. For example, VBSs are used for vehicle state estimation in environments such as urban canyons where GNSS is completely or intermittently unavailable.

Raw visual data is acquired and processed to compute vehicle states in a standalone configuration. However, a more common practice is to integrate the visual sub-system with other sensors via a suitable data-fusion algorithm. Depending on the functionality demanded of the visual sub-system, processing modules of varying complexity can be implemented, ranging from simple visual servoing (VS) routines to computationally intensive visual odometry (VO) techniques. For example, attitude estimation during the landing phase of a fixed-wing UAV can be implemented using a simple image processing module that employs edge detection methods and optical flow (OF) techniques to extract the horizon and calculate attitude based on the apparent motion observed [75, 76].

Image processing is a branch of computer knowledge that is concerned with processing a digital signal representing images taken with a digital camera or scanned by a scanner [1–9].

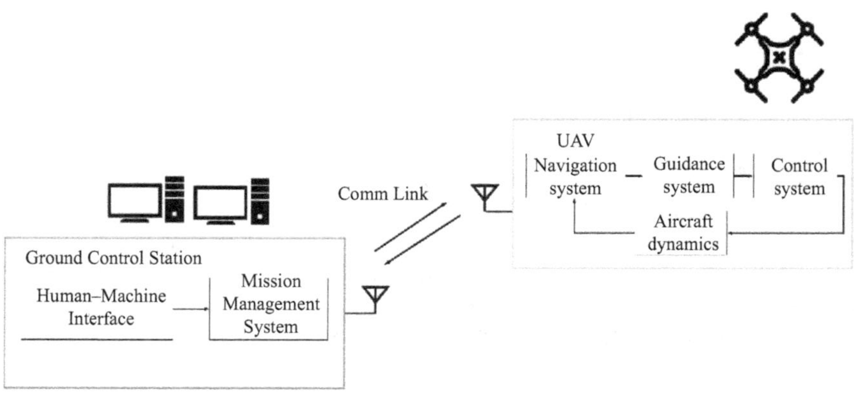

Figure 4.1 UAS system architecture

In contrast, VO provides the full navigation state vector but requires more computation owing to the requirement to extract visual features and track them across consecutive frames.

The following section presents a brief overview of commonly employed methods for VBN and introduces basic terminology associated with them.

4.2 Basic terminology

A large body of knowledge covers the integration of a low-cost VBS in UAS navigation system architectures. Distinctions between different VBN algorithms can be made based on the type of sensors employed (hardware) or on the types of features extracted and the implemented processing algorithms (software). Despite differences between VBN methods in the literature, a certain amount of overlap or similarity can be observed in different implementations, making strict classification difficult. Nevertheless, it is useful to define broad categories based on the employed navigation method.

Visual servoing (VS) refers to a number of methods in which inputs derived from raw visual data are fed to the UAS control system to navigate relative to observed features in the environment. The 3D pose (position and attitude) of the platform relative to the observed features can be explicitly derived. However, in most instances, state estimation is not explicitly performed relative to a global reference frame, and the focus is on deriving control inputs to achieve the desired flight pattern, viz., maintain altitude, maintain clearance from static objects and track mobile targets (Figure 4.2). These methods differ from each other in the extracted visual features.

Terrain-referenced navigation relies on *a priori* storage of maps of the environment. State estimation is performed through comparison of observed features with a stored database of static environmental features. Referring to Figure 4.3, the key

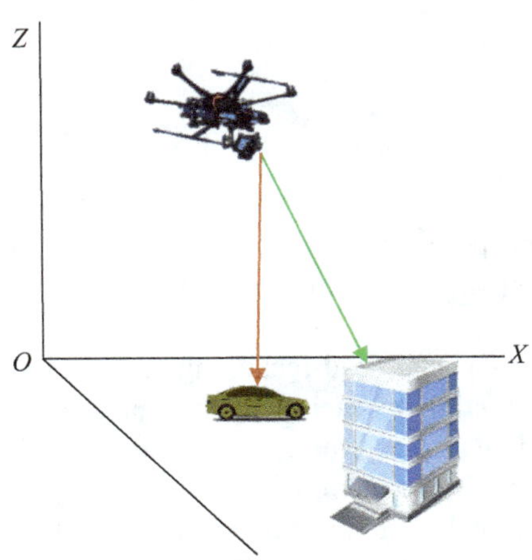

Figure 4.2 Visual servoing overview

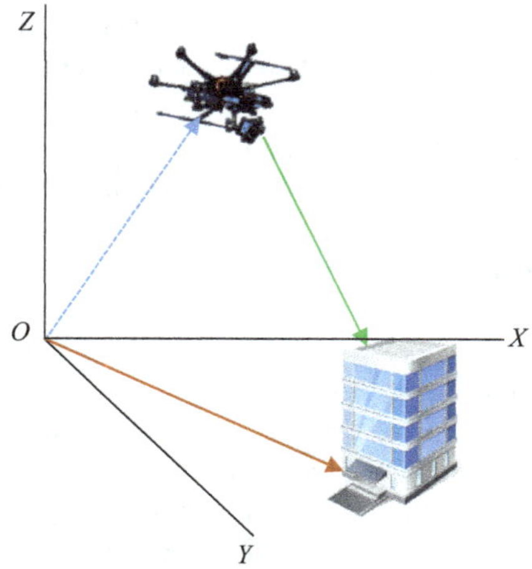

Figure 4.3 Terrain-referenced navigation overview

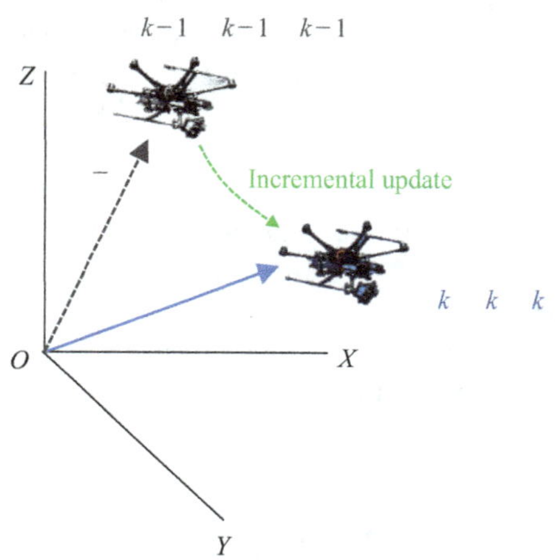

Figure 4.4 VO overview

task is to estimate vehicle pose R relative to a global coordinate system given *a priori* locations s of environmental features and online observations r of said environmental features.

Visual odometry is a visual dead-reckoning technique in which vehicle states R_k are estimated incrementally at each discrete time-step given the vehicle states R_{k-1} in the previous time step and the apparent motion of the observed scene. VO is a subset of the larger structure-from-motion (SFM) problem and is an intermediate step in any visual-simultaneous localisation and mapping (V-SLAM) algorithm (Figure 4.4).

Each of these methods will be expanded in greater detail, followed by an exploration of likely future work in this field of study.

4.2.1 Visual servoing

VS can be concisely defined as the use of data acquired by the VBS for the feedback control of the pose/motion of an agent (a UAS in this context). The method is well established in manufacturing for the control of robotic actuators. A position-based visual servoing (PBVS) approach is illustrated in Figure 4.5.

Features extracted from images is used to reconstruct the pose (position and attitude) of the camera (and by extension, the aerial platform). The estimated pose is compared with a reference (desired) pose to generate a Cartesian pose error signal, which drives the aircraft to the desired position and attitude. 3D reconstruction is typically the most computationally demanding phase of the cycle. Furthermore, the method is highly sensitive to camera calibration parameters. An alternative approach is image-based visual servoing (IBVS). Contrary to PBVS, the 3D pose is not reconstructed. Extracted image features are compared with reference images to generate a driving error signal (Figure 4.6).

The control law, therefore, essential functions in the image plane. The method is more insensitive to intrinsic and extrinsic camera calibration parameters than PBVS. Since the 3D pose does not need to be explicitly determined, hardware requirements from a computational point of view are typically low, making it readily suitable for small UAS. IBVS is alternately referred to in the literature as

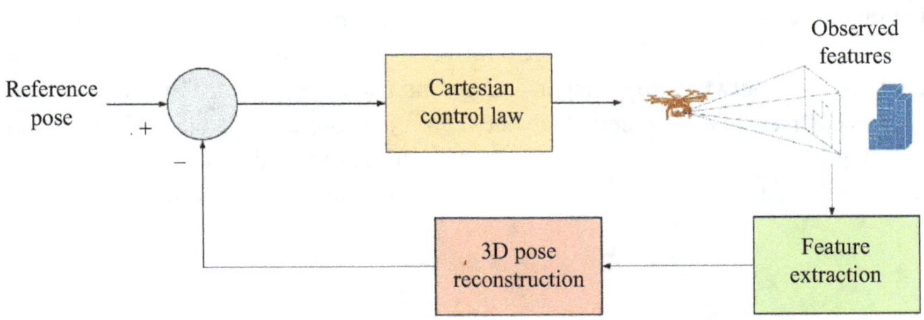

Figure 4.5 PBVS concept

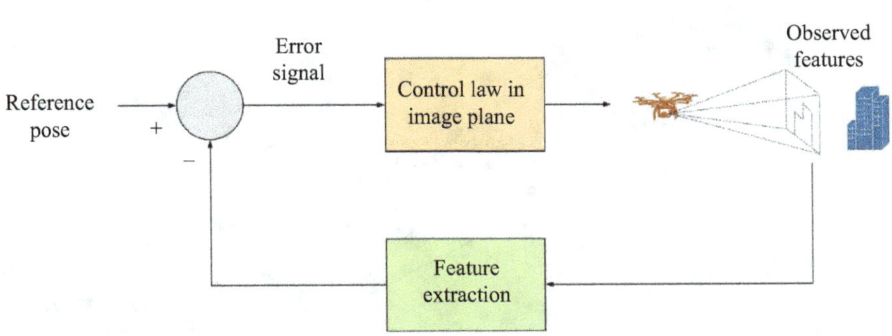

Figure 4.6 IBVS concept

the appearance-based approach. An application of the concept to a fixed-wing UAS navigation during the landing phase can be found in [10].

The environment is represented as a sequence of key images taken at various points, describing the path to be followed by the platform. The procedure consists of three phases:

- Learning stage: a sequence of frames is captured during a flight connecting the initial and target locations to serve as a visual reference route. The keyframes are sampled, geo-tagged and stored in the onboard database for guidance during the autonomous flight.
- Initialisation: A subset of images at the start of the visual route is compared with the current view, and the key image with the least matching error indicates the initial position for the autonomous run. The correlation between the reference image and the current view is typically used as a metric in localising the platform.
- Autonomous run: The key images of the visual route serve as waypoints for the aircraft to traverse in sensor space. The current view is compared to the key images to generate error signals to drive the control system.

A major drawback of this method is its limited applicability. The key images need to be recaptured each time the platform operates in a new workspace, making the approach suitable for routine operations in the same environment. Figure 4.7 illustrates the VBN system. Key image 2 has been identified as the current key image during the localisation stage and is used for VS. The system tracks the matching error between key images 2, 3 and its current view until the error between 2 and the current view exceeds that between 3 and the current view. At this stage, key image 3 is used to control the aircraft and the matching errors between key images 3, 4 and the current view are monitored. This process is repeated until the

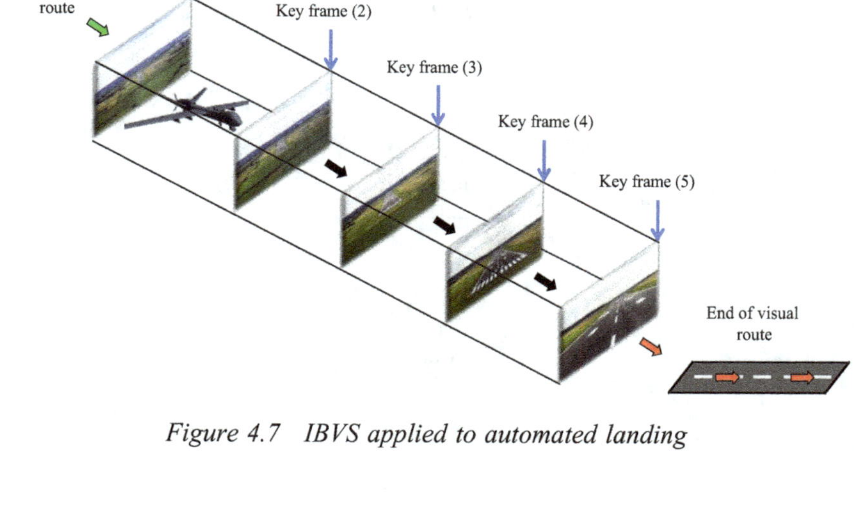

Figure 4.7 IBVS applied to automated landing

aircraft reaches its destination. Similar to this image-matching approach, OF is used in several instances as visual cues that drive navigation and control relative to objects in the environment. OF is formally defined as the 2D velocity field describing the apparent motion or the relative 3D motion between the camera and the environment. Stated differently, it is the 2D projection onto the camera image plane of the 3D apparent motion of objects in the field of view. This apparent motion can be represented through motion field models that will be described later on. It has been discovered that OF is used by insects and birds for navigation tasks [11,12]. The appeal of using OF for navigation lies in the rich source of information it provides and its relatively simple computation. In insects, OF, rather than complex stereo-vision mechanisms, is a source of low-level cues that guides seemingly complex behaviour and tasks, e.g. honeybees use OF to perform tasks such as grazing landing, travel distance estimation, obstacle avoidance, and flight speed regulation.

Figure 4.8 illustrates OF as sensed by a dragonfly. The OF field is a result of both the ego-motion (motion of the observing platform) and the object motion (motion of the object being observed).

OF can be computed by tracking feature patterns over consecutive images. It is important to note the key assumptions made in most of the computation algorithms:

- The constancy of brightness: Given a region in an image, over consecutive frames, the intensity or brightness is constant for incremental deviations from the region of interest.
- Spatial smoothness: Motion is assumed to be uniform over a small neighbourhood of pixels.

The intensity value I of a pixel in a given image is expressed as a function of the pixel location (x,y) and the time t at which the observation is made. The brightness constancy constraint can be expressed as:

$$I(x,y,t) = I(x+\Delta x, y+\Delta y, t+\Delta t) \tag{4.1}$$

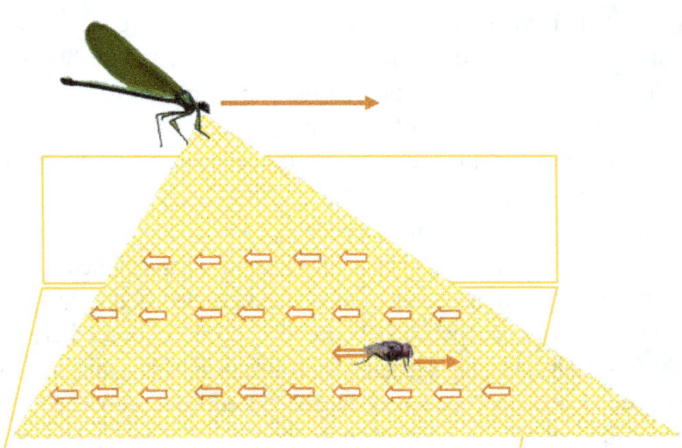

Figure 4.8 OF to estimate relative motion

Equation (4.1) expresses the assumption that intensity I does not change over incremental shifts in pixel coordinates and over time.

A Taylor series expansion applied to the right-hand side of (4.1) yields:

$$I(x + \Delta x, y + \Delta y, t + \Delta t) = I(x, y, t) + \frac{\partial I}{\partial x}\Delta x + \frac{\partial I}{\partial y}\Delta y + \frac{\partial I}{\partial z}\Delta z \tag{4.2}$$

By comparison of (4.1) and (4.2), (4.3) is written as:

$$\frac{\partial I}{\partial x}\frac{\Delta x}{\Delta t} + \frac{\partial I}{\partial y}\frac{\Delta y}{\Delta t} + \frac{\partial I}{\partial t}\frac{\Delta t}{\Delta t} = 0 \tag{4.3}$$

Or more concisely,

$$\frac{\partial I}{\partial x}v_x + \frac{\partial I}{\partial y}v_y + \frac{\partial I}{\partial t} = 0 \tag{4.4}$$

Equation (4.4) is an equation in two unknowns termed the aperture problem, and additional constraints must be applied in order to solve it. Different methods specify different constraints in including differential methods, region-based matching methods and energy-based methods, to mention a few. A description of each method is beyond the scope of this book. A review of OF methods for robotic navigation is found in [13]. The oldest known (and hence, most widely implemented) algorithms are the classical differential techniques such as the Lucas–Kanade method. Differential methods compute the spatial and temporal derivatives of image intensity to obtain OF. The method is generally recognized as being more robust to image noise than another classical differential technique, the Horn–Schunck method, which optimizes energy globally, to obtain a dense flow field. A relevant method for the target application in this review is the image interpolation method intended for small UAV's [14], which has more recently, been shown to be capable of a maximum rate of 20,000 OF vectors per second [15]. A bio-inspired approach based on the centring response of bees has been implemented in [16,17], wherein OF derived from images captured by side-facing cameras and an omni-directional camera was used to determine the relative distance between the aerial platform and the buildings in an urban canyon. The phenomenon is analogous to observing the environment through the window of a moving car – nearby features are perceived as moving quickly, whereas distant features are perceived as moving more slowly. The flow field is, therefore, a function of the distance to the observed object. The concept is illustrated in Figure 4.9.

The estimated lateral distance was used as an input to the heading control loop to produce yaw rate commands and centre the vehicle in the canyon. A similar approach can be used for fixed-wing UAS landing applications wherein the OF will increase with proximity to the ground. A limitation with this approach is that absolute range measurements are not provided through OF alone. The emphasis, in this case, is in guiding the vehicle relative to objects in the environment, rather than obtaining an absolute position estimate. However, the problem can be addressed by jointly implementing a range sensing modality along with OF estimation in a single

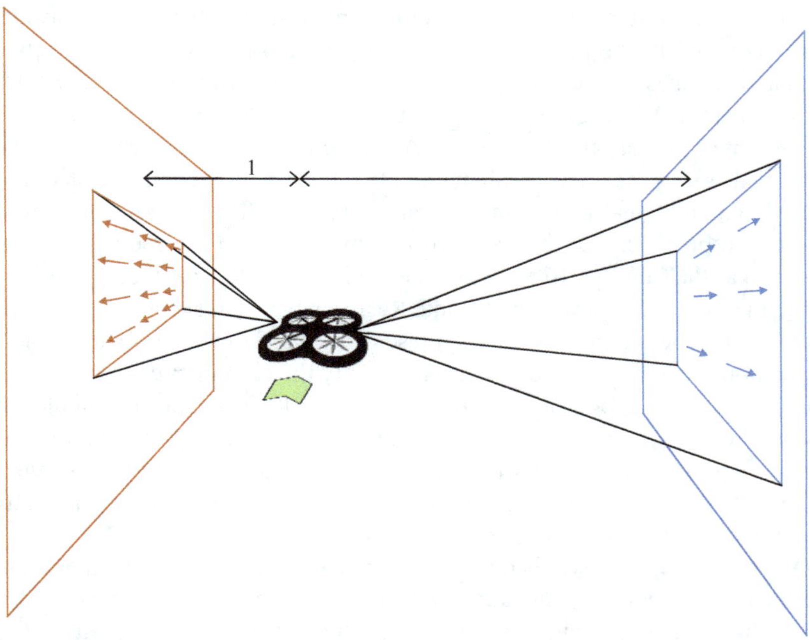

Figure 4.9 Use of OF to centre a UAV between two planar walls. The measured optic flow is a function of the lateral distance to each wall

platform. For example, the UAS architecture implemented in [18] combines estimation and stereo-vision in a single platform to safely traverse urban canyon scenarios. One drawback with OF is its reliance on the translational motion of the camera to generate flow fields. Additionally, unless the motion of the UAV is precisely known, this technique cannot produce absolute range measurements. It is therefore not suitable for situations in which a rotorcraft UAV is required to move slowly or hover in place near obstacles, such as for surveillance or structure inspections. A key problem observed in the research in [18] occurred when negotiating 90° bends or T-junctions in urban canyons. Under these conditions, the plane vertical sidewall assumption no longer applied, leading to an inability to control using OF alone. The perception and control algorithms were designed, assuming perfectly planar canyon walls. However, real-world operations are likely to violate this assumption, with no published tests in these conditions. Other prominent examples of OF used to drive aircraft control directly include terrain following and landing operations in static and dynamic environments [19,20]. Apart from VS applications which may or may not explicitly output UAS PVA, state estimation using OF has been investigated in the literature. The performance of OF-based motion estimation (specifically platform rotation rate) was compared against integrated GPS-INS systems in [21]. Besides OF, vision-based control of the aircraft relative to a perceived object can also be accomplished through feature extraction and tracking over an image sequence. The error signal for velocity control of a helicopter was obtained from Kalman-filter-based feature tracking in [22]. It is noted in the paper that the employed algorithm could be sensitive to substantial changes in ambient light, brightness and shadows. A dedicated study of the robustness of VS to these factors remains to be carried out.

A point of primary importance in UAS is the limited payload capability which makes real-time processing challenging. Pixel count is directly related to the required computational power onboard the platform. The development of sensors dedicated to providing OF has resulted in commercial-off-the-shelf (COTS) products such as the CentEye 16×32 pix chip which can achieve OF computation at a maximum rate of 545 frames per second. Research that stands out in this vast body of knowledge pertaining to OF is presented in [23], which focuses on low computation processing capability, as is typically the case in micro-aerial vehicles (MAVs) [24]. The image interpolation algorithm (I^2A) is typically employed in UAV and MAV systems to compute OF owing to the fact that it is non-iterative, does not require identification or tracking of individual features in the image and does not require the calculation of high-order spatial or temporal derivatives, making it robust to noise and quick to execute.

A research gap that stands out in implementations of UAS navigation utilising OF is the lack of rigorous quantitative evaluation of standalone OF systems, and even more so for integrated navigation systems. Such an analysis requires detailed error models that cover various failure modes of OF, and more importantly the sensitivity of data-fusion algorithms to these errors. Preliminary investigation of OF error models in the context of integrated GPS-INS-VBS systems has been carried out in [25], and error statistics of attitude estimation using OF are presented. A more comprehensive and flight trial-based evaluation was performed in [26]. A detailed examination of the modelling and implementation of VS can be found in [27].

A summary of VS methods in the literature is presented in Table 4.1.

4.2.2 Visual odometry

Several implementations of VO have been carried out for UAVs to compensate for the loss or denial of GPS availability. VO, as the name implies, analyses a sequence of captured images to compute vehicle position and attitude in incremental steps, similar to how the position is computed from integrating successive accelerometer measurements in an INS. As mentioned previously, VO is a preliminary step for resolving SFM in robotics. SFM is concerned with the 3D reconstruction of both the surrounding environment and the sequence of camera poses from sequential image sets, with an offline optimisation typically being performed to refine the reconstructed structure. The overall process is computationally intensive, and the time required is proportional (as would be intuitively expected) to the number of images captured, making real-time implementation challenging. VO, on the other hand, is tactical in its implementation and is concerned with estimating the 3D trajectory of the camera sequentially as frames are captured in real-time. Real-time SFM methods lie in the domain of SLAM, where the goal is the generation and maintenance of a global, consistent map along with tracking the trajectory of the

Table 4.1 Summary of VS methods in the literature

Method	Literature	Description	Similar work
Visual servoing	VBN through urban canyons [18]	Visual servoing to centre UAV in an urban canyon. Use of side-facing cameras for bio-inspired OF computation and a forward-facing stereo-vision camera for frontal obstacle avoidance	[16,17,19,28,29]
	Vision-based terrain following for an unmanned rotorcraft [30]	Experimentally demonstrated terrain following of an unmanned rotorcraft using a measure of GPS-derived ground speed in conjunction with OF to compute height. An additional contribution was an extension of the I^2A to accommodate large pixel flow rates	
	Implementation of widefield integration of optic flow for autonomous quadrotor navigation [23]	Gradient-based method to extract OF from images captured by an omnidirectional camera. Online bio-inspired OF-based attitude control for safe indoor corridor guidance	[19,26,31]
	Visual servoing of an autonomous helicopter in urban areas using feature tracking [22]	Thresholding, feature extraction and template matching, followed by Kalman-filter-based tracking of feature coordinates over the image sequence. The filter output drives the control law to move the platform relative to observed features (VS)	

robot/aircraft as opposed to VO, where the focus is solely on accurate estimation of the local trajectory. VO is an intermediate step for a full V-SLAM implementation, and, for many use-cases in UAS, is sufficient to provide or augment vehicle state estimation in urban environments where maps and databases can be assumed to be available *a priori*. The exception is for applications where map generation is an explicit requirement as in surveying. The decision to implement VO or V-SLAM ultimately depends on the trade-off between accuracy and real-time performance. V-SLAM is potentially more precise than VO, owing to more constraints applied on the generated trajectory, but is not necessarily more robust [32] with errors and outliers early in the process cycle (e.g. image correspondences, loop closure outliers) leading to large map inconsistencies. Additionally, far fewer instances of real-time UAS SLAM are found in the literature owing to computational limitations. The rest of this section focuses on reviewing VO implementations for UAS navigation, beginning with the fundamentals of the algorithm. The formulation of the VO problem is illustrated in Figure 4.10.

For simplicity of explanation, the camera coordinate frame C is also assumed to be the body reference frame of the aerial vehicle. Images I_{k-1}, I_k and I_{k+1} are captured at time-steps $k-1$, k and $k+1$, respectively, as the UAS moves in the environment. Features in the 3D environment (edges, corners, etc.) are projected onto the image plane of the camera and must be tracked across the successive frames. Consider time-steps $k-1$ and k: Given the presence of features that are tracked across two successive frames $k-1$ and k, VO is essentially the computation of the rigid body transformation $T_{k-1|k}$ comprising translation and rotation between the poses C_{k-1} and C_k. Once the transformation is computed, it is concatenated with C_{k-1} to output the current pose C_k. Without loss of generalisation, the same explanation describes VO when a stereo-camera is employed to acquire images rather than a monocular camera. In this case, image-pairs are acquired at each time-step instead of a single image. Additionally, the depth to the observed features can be estimated with each captured image-pair, which is a significant advantage over monocular vision where scale cannot be directly estimated. Regardless of whether monocular or stereo-vision is employed, the steps in each cycle are shown in Figure 4.11.

Feature detection is performed on each set of images captured at each given time-step, followed by matching the same set of features over successive frames (correspondence). Optimization routines estimate motion between frames with the tracked features forming constraints. Each type of optimisation routine depends on the type of feature correspondences performed between frames and has distinct advantages and limitations. The speed and accuracy of feature correspondences are typically boosted through the fusion of visual and inertial sensor measurements in order to exploit their complementary nature. In [33], the tight coupling of the two

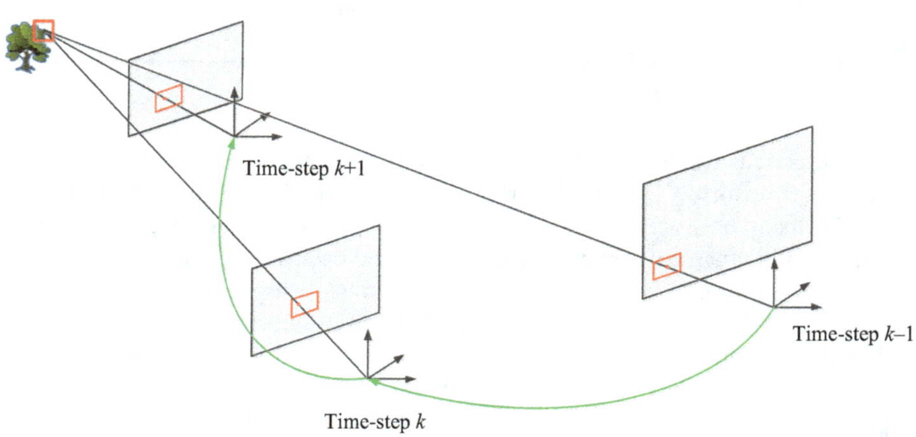

Figure 4.10 VO for UAS operations

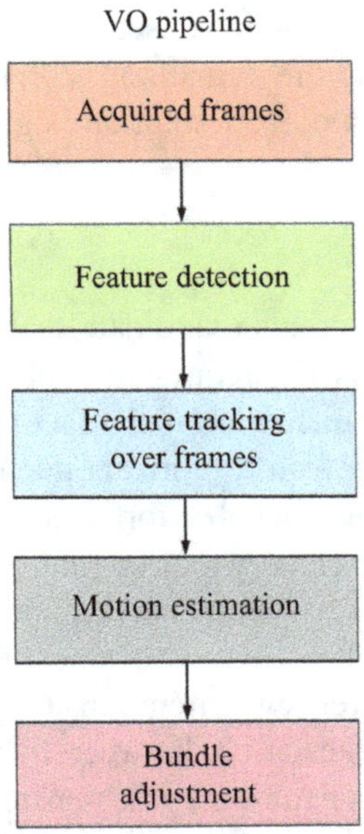

Figure 4.11 VO pipeline

sensors is achieved through inertial aiding of visual feature correspondence across frames. A similar aiding scheme that reduces the feature tracking search space for ground vehicles using downward-facing cameras and an EKF for predicting image transformation is found in [34]. A detailed description and review of visual-inertial fusion are covered in Chapter 2.

A comprehensive review that covers VO fundamentals, evolution and performance analysis is found in [32,35], which provides insight into optimal choices of feature detection and correspondence algorithms for specific applications. As would be expected, trade-offs between accuracy, robustness and efficiency exist. The scale-invariant feature transform (SIFT) descriptor has been demonstrated to be stable against large changes in illumination, rotation and scale, which makes it a good choice for dynamic UAS use-cases where large changes in camera orientation can be expected. However, it is noted that SIFT performance when operating on corners (prolific in urban environments) is sub-optimal when compared to blobs which lie in textured regions of the image. A crucial sub-task in the VO pipeline is outlier removal, i.e. the removal of wrong data associations or feature mismatches between frames. Image noise, occlusions, blur and changes in perspective and scene illumination typically result in matched feature sets being contaminated by outliers, which must be removed to refine motion estimation. This process is potentially a primary bottleneck for VO in small UAS operations in urban environments, where robust and high-integrity navigation is required to be performed in real-time by processing payloads that have limited computational ability. Random Sample Consensus (RANSAC) has emerged as the standard algorithm of choice for outlier removal in robotics. Given feature correspondences between frames, a motion model is fitted to a random subset of correspondences and iteratively checked for consistency with the remaining correspondences. The number of iterations necessary to ensure a correct solution is given by [35]:

$$N = \frac{\log(1-p)}{\log\left[1-(1-\varepsilon)^{s}\right]} \tag{4.5}$$

where s is the number of data points (candidate feature correspondences), ε is the percentage of outliers and p is the requested probability of success. A point of importance for real-time implementation is the fact that the number of iterations is exponential in the number of feature correspondences required to estimate the motion model. As a result, significant effort has gone into developing minimal model parameterisations. If the camera motion (and by extension the UAV motion) is unconstrained, a minimum of five-point correspondences is required for solving a six degree of freedom motion between two frames. Five-point RANSAC is therefore commonly used as it requires fewer iterations (and hence, less time) to arrive at a motion model (rotation and translation) between frames than six-, seven- or eight-point RANSAC. For real-time operations, an extension of RANSAC known as pre-emptive RANSAC has emerged as being particularly useful, in that it allows for the number of iterations to be fixed *a priori*. As with any dead-reckoning process, uncertainty in state estimation due to noise and errors in image processing propagates over each estimation cycle, resulting in a drift that renders the standalone VO estimate unusable after a certain time interval. The time interval over which a VO estimate can be considered valid depends on the uncertainty of the position used to initialise the process, and the uncertainty of estimation of each subsequent transformation (motion estimate) as illustrated in Figure 4.12.

The uncertainty propagation in Figure 4.12 applies to any dead-reckoning navigation algorithm. A periodic update or re-initialization of the state is required from an external sensor to 'reset' the dead-reckoning sensor. This is typically

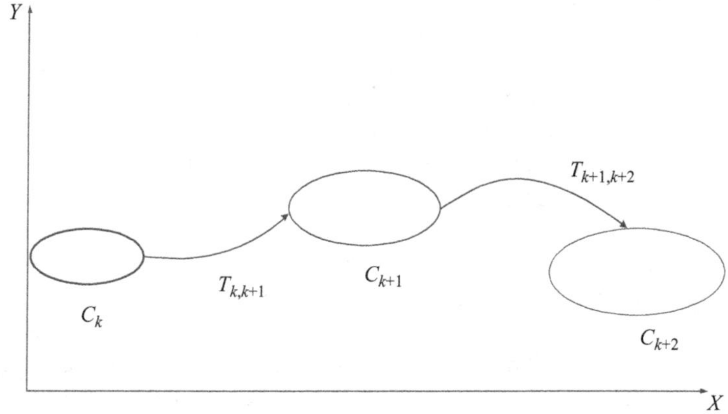

Figure 4.12 Uncertainty propagation in VO

accomplished through integration via a data-fusion algorithm of dead reckoning with an absolute positioning system, such as GNSS. Although both inertial navigation and VO are dead-reckoning methods subject to drift, they are frequently integrated to perform state estimation in UAS use-cases. The inertial measurement unit (IMU) provides guaranteed short-term continuity independent of environment or trajectory and is capable of capturing highly dynamic manoeuvres [73, 74]. Inertial data can compute the absolute scale information from the observation of acceleration due to gravity that is unobservable from monocular vision alone. Image sequences provide multi-pose constraints, which reduce drift from IMU integration and increase IMU bias observability. However, the accuracy, quantity and rank of these constraints are dependent on the environmental features observed in the scene. A fused vision and inertial navigation system combines the advantages of both approaches and accumulates error at a lower rate than either standalone technique [36]. Examples of vision-aided inertial navigation in urban environments attempt to exploit visual regularities in the scene. In [37], line segments in the image are used to detect vanishing points and hence infer the occluded horizon, enabling correction of gyroscope errors in the IMU. A recent and relevant example of real-time vision aided inertial navigation of micro aerial vehicles can be found in [38] where the IMU measurements are used to compute the rotation between frames and to estimate the reprojection error in order to aid inlier detection. In general, stereo-vision systems are preferable to monocular–vision systems for VO computation, since the depth (distance) to the observed scene from a monocular camera cannot be recovered from the images alone, making it necessary to have *a priori* knowledge of the object(s) being observed or to capture multiple images of the same scene from different vantage points that are separated by a wide baseline. Depth-perception, however, is possible using two cameras in a stereo-rig. Stereo-vision is analogous to the concept of stereopsis in humans wherein perception of depth and 3D structure is possible by visual information obtained from two 'sensors' – eyes.

The concept is illustrated in Figure 4.13. Two cameras are mounted in a stereorig and observe the same scene. A visual feature that is observable in both the left and right cameras will have different pixel coordinates in each image. Referring to Figure 4.13, the corner of a building that is observed in both cameras is offset by different distances d_L and d_R. The difference in offset is known as the disparity, which is a function of the depth of the observed feature. The disparity of each image point is inversely proportional to the depth of the scene at the corresponding pixel location. The relationship between disparity d and scene depth Z is given by:

$$Z = Tf/d, \tag{4.6}$$

where T is the distance or baseline between the left and right cameras and f is the focal length of each camera. From (4.6), it can be seen that depth-perception in stereo-vision is dependent on two intrinsic parameters of the stereo-vision camera: the focal length and the baseline.

Stereo-vision has been employed for altitude estimation and subsequent regulation in autonomous helicopters [39]. A similar approach was utilised for altitude

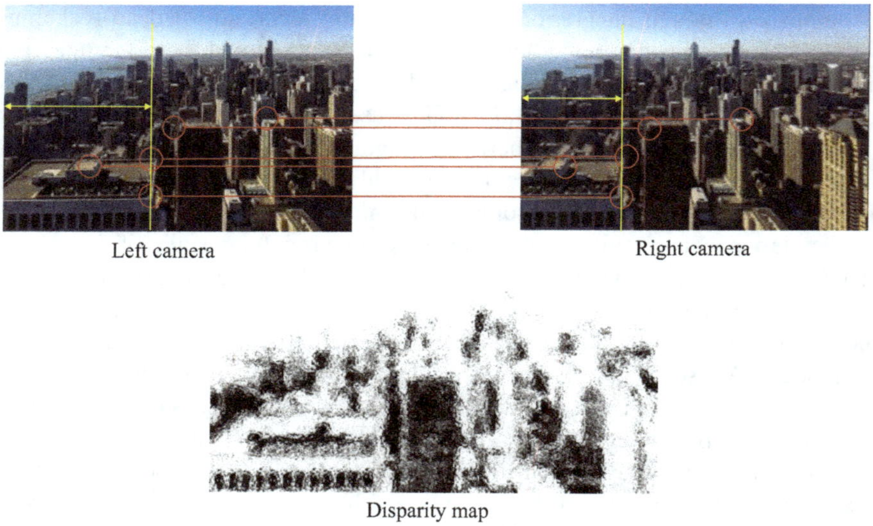

Figure 4.13 *Stereo-vision concept. The same scene is observed from two cameras. Features that are common to both cameras have different pixel coordinates in either image – the disparity between the coordinates is a function of the depth of the observed feature*

estimation in [40] although the results could not be experimentally validated in real-time. Maintaining an accurate scale, therefore, becomes increasingly difficult as the range to the perceived object grows, effectively reducing utility to that of a monocular setup. Additionally, vibration due to platform dynamics can cause small deformations in the stereo-rig, potentially resulting in large triangulation errors. Some research aimed at addressing these gaps in stereo-VO has begun to emerge recently [41,42].

4.2.3 Terrain-referenced visual navigation

Structured environments can be used to assist in the navigation of unmanned platforms. Visual targets with an *a priori* known location and structure have long been used in aerial robotic applications; that is, the visual targets that are employed are man-made landmarks with known attributes (size, colour and pattern) which allows the pose of the aerial platform relative to the landmark to be retrieved. The fact that the landmark attributes are known allows for ambiguity in scale to be removed (when using monocular vision). Early examples of this type of work can be found in [43,44] where a vision-aided inertial system was used to navigate relative to a known target. Similar tight coupling of inertial and visual sensors for navigation relative to known landmarks has been performed for planetary explorers as well [45]. This differs from a typical VS system in which visual data is directly mapped to control without explicit state estimation as an intermediate step. Known characteristics of the runway region and boundary have been used for autonomous runway recognition during the UAV landing phase [46].

Terrain-aiding is commonly achieved through image registration – comparison of images captured in-flight with a stored database of images of the operational area. In some regard, this is similar to the IBVS method described earlier. The key difference is that the stored images in this instance are geo-tagged, i.e. the location of the visual features in the stored database is known *a priori*, thereby enabling the aircraft position to be explicitly computed, whereas the absolute coordinates of the visual features are not known *a priori* (or required) in IBVS. The images can be captured using a standard downward-facing camera. Infra-red (IR) capability allows operations to be conducted at night and is less sensitive to adverse weather conditions. The captured and stored images are typically correlated after extracting boundaries or distinct features in order to reduce the database storage requirement and correlation load. More importantly, the correlation of boundaries or features renders the process more robust to variation in lighting [47]. Illustration of the decomposition of an image into boundaries or distinct regions is shown in Figure 4.14.

Image registration accuracy is dependent on the terrain, as distinct features must be observable for matching. Sources of error in this method arise from the low fidelity of the employed database, camera lens distortions and uncompensated scaling variations. Noise-like errors arise from focussing errors, and the limited resolution of the camera, database and image-matching algorithms. These errors must be modelled and included in the image registration algorithm to ensure robust operation. The image comparison is essentially a correlation of the captured image with the images stored in the database to ascertain the highest correlation value (closest match). Figure 4.15 shows a successful match between a captured image and a stored image in a database. Figure 4.16 illustrates the corresponding spike in cross-correlation when the captured image is compared against the stored image.

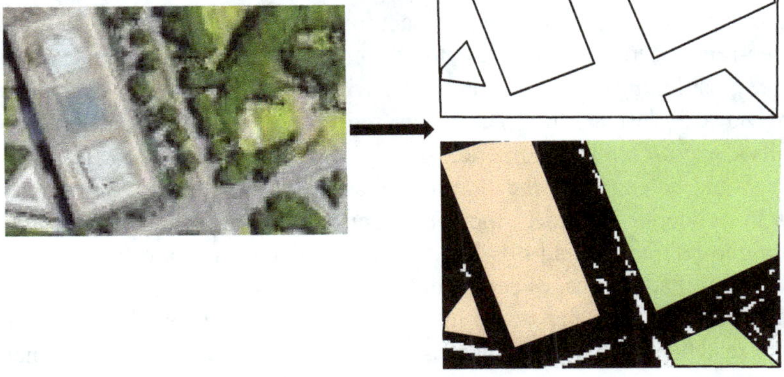

Figure 4.14 Extraction of boundaries/features before image matching

Figure 4.15 Image matching between captured and stored image

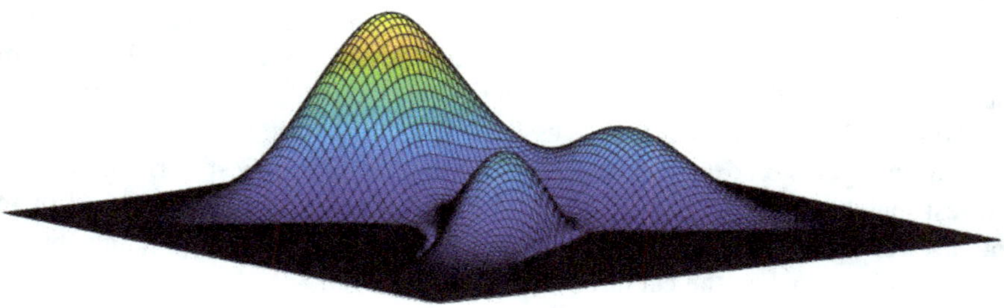

Figure 4.16 Cross-correlation between the captured image and the stored image

In [48], a combination of terrain-aided visual navigation and VO was employed for unmanned helicopter navigation. The terrain-aiding was through image registration: matching images captured online during the UAV trajectory with geo-referenced aerial images stored onboard. The image matching provides drift-free positioning, which is integrated with a Kalman-filter-based positioning system via a Bayesian filter. The image registration was performed through the normalised cross-correlation metric.

As would be expected, the algorithm was found to be sensitive to sudden variations in helicopter attitude. Another example of experimentally validated work utilising geo-referencing is [49]. The highlights of this work are (i) environmental classification of aerial images captured during the mission and (ii) rotation-invariant template matching with pre-stored maps. The use of a probabilistic framework for template matching supported robustness in the presence of error sources including noise and image occlusion. As would be expected, the performance is related to the presence of distinctive features observed in the images. Regions that consist of a

Table 4.2 Summary of Terrain-Referenced Navigation Methods in the Literature

Method	Literature	Description	Similar Work
Terrain-aided visual navigation	Vision-only approach and landing [43]	Vision-aided inertial navigation is relative to a target with known attributes. The use of a known target allows for ambiguity in scale to be resolved	[44,45,48,49]
	Geo-referencing for UAV navigation using environmental classification [49]	Environment classification and image registration with an *a priori* stored map of the operational area	

number of environmental classes (asphalt, grass and houses) can be effectively used to provide geo-referenced measurements, whereas uniform regions with only a single environmental class have a low likelihood of achieving a template match (Table 4.2) [78].

4.3 Future trends and discussion

VBSs are increasingly deployed on UAS owing to the need to support robust navigation in environments where GNSS is either unavailable or degraded. Apart from localisation, modern imaging sensors on the market boast high-resolution and frame-rates advantageous for many UAS navigation functions such as obstacle detection and forward collision avoidance and warning. Despite progressive improvements in miniaturisation and onboard computing performance, VBSs still exhibit vulnerability to factors such as low scene illumination, challenging high dynamic range scenes, blur and sparse, repetitive and high-frequency image texture [50,51]. As such, soft computing methods that lend the ability to learn and to adapt provide greater robustness to unpredictable factors [24]. In this regard, biomimetic methods inspired by biological nervous systems provide a significant advantage. Artificial neural networks and genetic algorithms support online learning and adaptation and as such, represent areas of significant research interest [1,52–55]. 3D time-of-flight imaging chips (resembling the ones available in flash LIDARs) are gradually emerging as an inexpensive, reliable means of capturing images and range measurements [56]. Thermal cameras promise advantages in several photogrammetry applications, and the expense will likely reduce sufficiently for aerial applications in the near- to mid-term [57,58,77].

Mid-to-far IR (MFIR) imaging sensors operating in the millimeter-wave (mmW) range offer enhanced performance in reduced visibility conditions (such as fog, haze and smoke) and the dark. Airborne IR imaging for object recognition and tracking has been so far been employed primarily for military applications. However, advances in semiconductor lasers and IR detector technology are likely to lead to cost-effective MFIR sensors for UAS navigation systems in the future [59–61]. Similarly, the development of cost-effective hardware is likely to lead to greater uptake of multi- and hyper-spectral imaging for UAS applications [61–65] in the future. The methods presented in this chapter can be extended to aid multi-UAV coordination and control scenarios, which remains an area of significant research interest [66–71].

4.4 Conclusions

This chapter has identified and presented several visual cues and methods commonly applied in visual navigation for UAS. Methods employed in the literature have been reviewed and concisely presented to introduce the reader to the field of study. VBSs present several advantages for UAS navigation, primarily by the amount of information they can capture and their small size, weight and power requirements. Despite this, however, several limitations exist that suggest a potential direction for future research – sensitivity to environmental conditions such as occlusion and variation of ambient light can hamper the availability and accuracy of this sensing modality. Artificial intelligence-based methods are likely to be beneficial by supporting the development of visual systems that learn and adapt to a number of operating conditions. One of the primary future requirements for UAS access to unsegregated airspace is the need to certify that a given navigation sensor can meet a stipulated level of performance with the desired probability. This will require rigorous quantitative evaluation of visual sensors that are implemented in standalone and integrated systems and even more so for integrated navigation systems. Along with the development of detailed error models that cover various failure modes and more importantly the sensitivity of data-fusion algorithms to these errors.

References

[1] C. Hwang (2015). Image processing and applications based on visualizing navigation service. ICDIP.

[2] N. Razmjooy, B.S. Mousavi, F. Soleymani, and M.H. Khotbesara (2013). A computer-aided diagnosis system for malignant melanomas. *Neural Computing and Applications*, 23(7–8), 2059–2071.

[3] V.V. Estrela, and A.E. Herrmann (2016). Content-based image retrieval (CBIR) in remote clinical diagnosis and healthcare. In Encyclopedia of E-Health and Telemedicine (pp. 495–520). IGI Global.

[4] N. Razmjooy, M. Ramezani, and N. Ghadimi (2017). Imperialist competitive algorithm-based optimization of neuro-fuzzy system parameters for automatic red-eye removal. *International Journal of Fuzzy Systems*, 19(4), 1144–1156.

[5] P. Moallem, and N. Razmjooy (2012). Optimal threshold computing in automatic image thresholding using adaptive particle swarm optimization. *Journal of Applied Research and Technology*, 10(5), 703–712.

[6] V.V. Estrela, and N.P. Galatsanos (1998, October). Spatially-adaptive regularized pel-recursive motion estimation based on cross-validation. In Proceedings 1998 International Conference on Image Processing. ICIP98 (Cat. No. 98CB36269) (Vol. 2, pp. 200–203). IEEE.

[7] B. Somayeh Mousavi, and F. Soleymani. (2014). Semantic image classification by genetic algorithm using optimised fuzzy system based on Zernike moments. *Signal, Image and Video Processing*, 8(5), 831–842.

[8] A. Martins Coelho, and V.V. Estrela (2016). EM-Based Mixture Models Applied to Video Event Detection. arXiv preprint arXiv:1610.02923.

[9] N. Razmjooy, B.S. Mousavi, M. Khalilpour, and H. Hosseini (2014). Automatic selection and fusion of color spaces for image thresholding. *Signal, Image and Video Processing*, 8(4), 603–614.

[10] F. Cappello, S. Ramasamy, and R. Sabatini, A low-cost and high performance navigation system for small RPAS applications. *Aerospace Science and Technology*, 58, 529–545, 2016.

[11] P. S. Bhagavatula, C. Claudianos, M. R. Ibbotson, and M. V. Srinivasan. Optic flow cues guide flight in birds, *Current Biology*, vol. 21, pp. 1794–1799, 2011.

[12] M. V. Srinivasan, "Honeybees as a model for the study of visually guided flight, navigation, and biologically inspired robotics," *Physiological Reviews*, 91, 413–460, 2011.

[13] H. Chao, Y. Gu, and M. Napolitano, "A survey of optical flow techniques for robotics navigation applications," *Journal of Intelligent and Robotic Systems: Theory and Applications*, 73, 361–372, 2014.

[14] M. V. Srinivasan, "An image-interpolation technique for the computation of optic flow and egomotion," *Biological Cybernetics*, vol. 71, pp. 401–415, 1994.

[15] R. Pericet-Camara, G. Bahi-Vila, J. Lecoeur, and D. Floreano, "Miniature artificial compound eyes for optic-flow-based robotic navigation," in 2014 13th Workshop on Information Optics (WIO), 2014, pp. 1–3.

[16] S. Hrabar and G. S. Sukhatme, "A comparison of two camera configurations for optic-flow based navigation of a UAV through urban canyons," in 2004 IEEE/RSJ International Conference on Intelligent Robots and Systems (IROS), 2004, pp. 2673–2680.

[17] S. Hrabar, G. S. Sukhatme, P. Corke, K. Usher, and J. Roberts, "Combined optic flow and stereo-based navigation of urban canyons for a UAV," in 2005 IEEE/RSJ International Conference on Intelligent Robots and Systems, IROS, 2005, pp. 302309.

[18] S. Hrabar and G. Sukhatme, "Vision-based navigation through urban canyons," *Journal of Field Robotics*, 26, 431–452, 2009.

[19] F. Ruffier and N. Franceschini, "Optic flow regulation in unsteady environments: a tethered MAV achieves terrain following and targeted landing over a moving platform," *Journal of Intelligent & Robotic Systems*, 79, 275–293, 2015.

[20] B. Herissé, T. Hamel, R. Mahony, and F.-X. Russotto, "Landing a VTOL unmanned aerial vehicle on a moving platform using optical flow," *IEEE Transactions on robotics*, 28, 77–89, 2012.

[21] H. Chao, Y. Gu, J. Gross, G. Guo, M. L. Fravolini, and M. R. Napolitano, "A comparative study of optical flow and traditional sensors in uav navigation," in American Control Conference (ACC), 2013, 2013, pp. 3858–3863.
[22] L. Mejías, S. Saripalli, P. Campoy, and G. S. Sukhatme, "Visual servoing of an autonomous helicopter urban areas using feature tracking," *Journal of Field Robotics*, 23, 185–199, 2006.
[23] J. Conroy, G. Gremillion, B. Ranganathan, and J. S. Humbert, "Implementation of wide-field integration of optic flow for autonomous quadrotor navigation," *Autonomous robots*, 27, 189–198, 2009.
[24] H.W. Ho, C. De Wagter, B.D.W. Remes, and G.C.H.E. de Croon, Optical-flow based self-supervised learning of obstacle appearance applied to MAV landing, *Robotics and Autonomous Systems*, 100, 78–94, 2018, ISSN 0921-8890, https://doi.org/10.1016/j.robot.2017.10.004.
[25] W. Ding, J. Wang, S. Han, *et al.*, "Adding optical flow into the GPS/INS integration for UAV navigation," in Proceedings of International Global Navigation Satellite Systems Society Symposium, 2009, pp. 1–13.
[26] H. Chao, Y. Gu, J. Gross, M. Rhudy, and M. Napolitano, "Flight-test evaluation of navigation information in wide-field optical flow," *Journal of Aerospace Information Systems*, 2016;13:419–432.
[27] P. Corke, Robotics, Vision and Control: Fundamental Algorithms In MATLAB® Second, Completely Revised vol. 118, Springer, 2017.
[28] D. Floreano, J.-C. Zufferey, A. Klaptocz, J. Germann, and M. Kovac, "Aerial locomotion in cluttered environments," in Robotics Research, Springer, 2017, pp. 21–39.
[29] A. Beyeler, J. C. Zufferey, and D. Floreano, "Vision-based control of near-obstacle flight," *Autonomous Robots*, 27, 201–219, 2009.
[30] M. A. Garratt and J. S. Chahl, "Vision-based terrain following for an unmanned rotorcraft," *Journal of Field Robotics*, vol. 25, pp. 284–301, 2008.
[31] J. S. Humbert, "Bio-inspired visuomotor convergence in navigation and flight control systems," *California Institute of Technology*, 2005.
[32] D. Scaramuzza and F. Fraundorfer, "Visual odometry [tutorial]," *IEEE Robotics & Automation Magazine*, 18, 80–92, 2011.
[33] J. R. M. Veth, "Fusion of low-cost imagining and inertial sensors for navigation," in International Technical Meeting of the Satellite Division of The Institute of Navigation, 2006, pp. 1093–1103.
[34] X. Song, L. D. Seneviratne, and K. Althoefer, "A Kalman filter-integrated optical flow method for velocity sensing of mobile robots," *IEEE/ASME Transactions on Mechatronics*, 16, 551–563, 2011.
[35] F. Fraundorfer and D. Scaramuzza, "Visual odometry: Part II: Matching, robustness, optimization, and applications," *IEEE Robotics & Automation Magazine*, 19, 78–90, 2012.
[36] T. J. Steiner, R. D. Truax, and K. Frey, "A vision-aided inertial navigation system for agile high-speed flight in unmapped environments: Distribution

statement A: Approved for public release, distribution unlimited," in Aerospace Conference, IEEE, 2017, pp. 1–10.

[37] M. Hwangbo and T. Kanade, "Visual-inertial UAV attitude estimation using urban scene regularities," in 2011 IEEE International Conference on Robotics and Automation (ICRA), 2011, pp. 2451–2458.

[38] C. Troiani, A. Martinelli, C. Laugier, and D. Scaramuzza, "Low computationalcomplexity algorithms for vision-aided inertial navigation of micro aerial vehicles," *Robotics and Autonomous Systems*, 69, 80–97, 2015.

[39] J. M. Roberts, P. I. Corke, and G. Buskey, "Low-cost flight control system for a small autonomous helicopter," in IEEE International Conference on Robotics and Automation, 2003 (ICRA'03), 2003, pp. 546–551.

[40] R. J. Moore, S. Thurrowgood, D. Bland, D. Soccol, and M. V. Srinivasan, "A stereo vision system for uav guidance," in IEEE/RSJ International Conference on Intelligent Robots and Systems, 2009. IROS 2009.2009, pp. 3386–3391.

[41] M. Warren, P. Corke, and B. Upcroft, "Long-range stereo visual odometry for extended altitude flight of unmanned aerial vehicles," *The International Journal of Robotics Research*, 35, 381–403, 2016.

[42] Y. Song, S. Nuske, and S. Scherer, "A multi-sensor fusion MAV state estimation from long-range stereo, IMU, GPS and barometric sensors," *Sensors*, 17, 11, 2016.

[43] A. A. Proctor and E. N. Johnson, "Vision-only approach and landing," in AIAA Guidance, Navigation, and Control Conference and Exhibit, 2005.

[44] A. Wu, E. N. Johnson, and A. A. Proctor, "Vision-aided inertial navigation for flight control," *JACIC*, 348–360, 2005.

[45] N. Trawny, A. I. Mourikis, S. I. Roumeliotis, A. E. Johnson, and J. F. Montgomery, "Vision-aided inertial navigation for pin-point landing using observations of mapped landmarks," *Journal of Field Robotics*, 24, 357–378, 2007.

[46] J. Shang and Z. Shi, "Vision-based runway recognition for uav autonomous landing," *International Journal of Computer Science and Network Security*, 112–117, 2007.

[47] P. D. Groves, *Principles of GNSS, Inertial, and Multisensor Integrated Navigation Systems*. Artech House, 2013.

[48] G. Conte and P. Doherty, "Vision-based unmanned aerial vehicle navigation using geo-referenced information," *EURASIP Journal on Advances in Signal Processing*, 2009, 10, 2009.

[49] F. Lindsten, J. Callmer, H. Ohlsson, D. Törnqvist, T. B. Schön, and F. Gustafsson, "Geo-referencing for UAV navigation using environmental classification," in 2010 IEEE International Conference on Robotics and Automation (ICRA), 2010, pp. 1420–1425.

[50] Á. Gómez-Gutiérrez, J.J. Sanjosé-Blasco, J. Lozano-Parra, F. Berenguer-Sempere, and J.D. Matías-Bejarano (2015). Does HDR pre-processing improve the accuracy of 3D models obtained by means of two conventional SfM-MVS software packages? The Case of the Corral del Veleta Rock Glacier. *Remote Sensing*, 7, 10269–10294.

[51] R. Gomez-Ojeda, Z. Zhang, J. Gonzalez-Jimenez, and D. Scaramuzza (2018). Learning-based image enhancement for visual odometry in challenging HDR environments, in 2018 IEEE International Conference on Robotics and Automation (ICRA), pp. 805–811.

[52] S. Poddar, R. Kottath, and V. Karar (2018). Evolution of Visual Odometry Techniques. CoRR, abs/1804.11142.

[53] D.J. Hemanth, and V.V. Estrela (2017). *Deep learning for image processing applications, advances in parallel computing series*, vol. 31, IOS Press, ISBN 978-1-61499-821-1 (print), ISBN 978-1-61499-822-8 (online).

[54] M.A. de Jesus, V.V. Estrela, O. Saotome, and D. Stutz (2018) Super-resolution via particle swarm optimization variants. In: Hemanth J., Balas V. (eds) *Biologically Rationalized Computing Techniques For Image Processing Applications. Lecture Notes in Computational Vision and Biomechanics*, vol. 25. Springer, Cham.

[55] C. Wang, Z. Sun, X. Zhang, and X. Zhang (2017). Autonomous navigation of UAV in large-scale unknown complex environment with deep reinforcement learning. 2017 IEEE Global Conf. on Signal and Information Processing (GlobalSIP), 858–862.

[56] J.A. Paredes, F.J. Álvarez, T. Aguilera, and J.M. Villadangos (2017). 3D indoor positioning of UAVs with spread spectrum ultrasound and time-of-flight cameras. Sensors.

[57] F. Santoso, M.A. Garratt, and S.G. Anavatti (2017). Visual–inertial navigation systems for aerial robotics: Sensor fusion and technology. *IEEE Transactions on Automation Science and Engineering*, 14, 260–275.

[58] K. Ribeiro-Gomes, D. Hernández-López, J.F. Ortega, R. Ballesteros, T. Poblete, and M.A. Moreno (2017). Uncooled thermal camera calibration and optimization of the photogrammetry process for UAV applications in agriculture. Sensors, 2017;17(10):2173.

[59] T. Adão, J. Hruska, L. Pádua, *et al.* (2017). Hyperspectral imaging: A review on UAV-based sensors, data processing and applications for agriculture and forestry. *Remote Sensing*, 9, 1110.

[60] H. Li, W. Ding, X. Cao, and C. Liu (2017). Image registration and fusion of visible and infrared integrated camera for medium-altitude unmanned aerial vehicle remote sensing. *Remote Sensing*, 9, 441.

[61] S. Siewert, V. Angoth, R. Krishnamurthy, *et al.* (2016). Software defined multi-spectral imaging for arctic sensor networks.

[62] M. Jaud, N.L. Dantec, J. Ammann, *et al.* (2018). Direct georeferencing of a pushbroom, lightweight hyperspectral system for mini-UAV applications. *Remote Sensing*, 10, 204.

[63] R.J. Aroma, and K. Raimond (2019) Intelligent Land Cover Detection in Multi-sensor Satellite Images. In: Ane B., Cakravastia A., and Diawati L. (eds), *Proceedings of the 18th Online World Conference on Soft Computing in Industrial Applications (WSC18). WSC 2014. Advances in Intelligent Systems and Computing*, vol. 864. Springer, Cham.

[64] C. Li, S.X. Yang, Y. Yang, *et al.* Hyperspectral remote sensing image classification based on maximum overlap pooling convolutional neural network. *Sensors* 2018, 18, 3587.

[65] H. Zhao, Z. Wang, G. Jia, X. Li, and Y. Zhang (2018). Field imaging system for hyperspectral data, 3D structural data and panchromatic image data measurement based on acousto-optic tunable filter. *Optics Express*, 26, 13, 17717–17730.

[66] A. Lioulemes, G. Galatas, V. Metsis, G.L. Mariottini, and F. Makedon (2014). Safety challenges in using AR. Drone to collaborate with humans in indoor environments. PETRA.

[67] P. Schmuck (2017). "Multi-UAV collaborative monocular SLAM," in 2017 IEEE International Conference on Robotics and Automation (ICRA), pp. 3863–3870.

[68] X. Li, and T. Chen (2017). "Collaborative visual navigation for UAVs in blurry environment," in 2017 IEEE International Conference on Signal Processing, Communications and Computing (ICSPCC), pp. 1–6.

[69] S. Zhao, Z. Li, R. Cui, Y. Kang, F. Sun, and R. Song (2017). Brain–machine interfacing-based teleoperation of multiple coordinated mobile robots. *IEEE Transactions on Industrial Electronics*, 64, 5161–5170.

[70] S.M. Mahi (2018). Multi-Modal Multi sensor Interaction between Human and Heterogeneous Multi-Robot System. ICMI.

[71] C.T. Recchiuto, A. Sgorbissa, and R. Zaccaria (2016). Visual feedback with multiple cameras in a UAVs human-swarm interface. *Robotics and Autonomous Systems*, 80, 43–54.

[72] A. I. Mourikis and S. I. Roumeliotis, "A multi-state constraint Kalman filter for vision-aided inertial navigation," in Proc. 2007 IEEE International Conference on Robotics and automation, 2007, pp. 3565–3572.

[73] D. D. Diel, P. DeBitetto, and S. Teller, "Epipolar constraints for vision-aided inertial navigation," in 2005 Seventh IEEE Workshops on Application of Computer Vision. WACV/MOTIONS'05, 2005;1:221–228.

[74] J. W. Langelaan, "State estimation for autonomous flight in cluttered environments," *Journal of Guidance Control and Dynamics*, 30, 1414, 2007.

[75] M. B. Rhudy, Y. Gu, H. Chao, and J. N. Gross, "Unmanned aerial vehicle navigation using wide-field optical flow and inertial sensors," *Journal of Robotics*, 2015, 1, 2015.

[76] F. Kendoul, I. Fantoni, and K. Nonami, "Optic flow-based vision system for autonomous 3D localization and control of small aerial vehicles," *Robotics and Autonomous Systems*, 57, 591–602, 2009.

[77] M. E. Antone and S. Teller, "Automatic recovery of relative camera rotations for urban scenes," in 2000 IEEE Conference on Computer Vision and Pattern Recognition, 2000, 282–289.

[78] C. Demonceaux, P. Vasseur, and C. Pégard, "UAV attitude computation by omnidirectional vision in urban environment," in 2007 IEEE International Conference on Robotics and Automation, 2007, pp. 2017–2022.

Chapter 5
Modelling and simulation of UAV systems

Narendran Muraleedharan[1] *and Daniel S. Cohen*[2]

The modelling of a system's dynamics and performing simulation are commonly used techniques and almost critical components in the modern development of manned and unmanned systems alike. This chapter addresses the need for modelling and simulation of unmanned air vehicle systems and reviews historical developments in the field, current techniques used and future developments.

5.1 Need for modelling and simulation

The advancements in technology associated with robotic systems, computational power and unmanned aircraft have driven a strong need for dynamic modelling and simulation. The ability to perform fast real-time computations has led to dynamical simulation to be used in designing and rapidly prototyping controllers, evaluate the core design of robots, simulate virtual sensors and develop predictive model controllers [1,2].

Additionally, due to the growth of the unmanned systems industry, and the movement towards using unmanned systems for more complex and expensive tasks, it has become vital to test these systems in simulation prior to deployment [3]. Testing in simulation has been shown to present very high returns on investment and provide major cost savings in the aerospace industry [4].

Following are three key applications of modelling and simulation in the unmanned air vehicle (UAV) industry.

5.1.1 Control systems design

Dynamic modelling of systems is widely used in control systems design for multiple applications. Model predictive control (MPC) is a promising control strategy to map control inputs over non-linear systems with multi-dimensional state spaces. Accurate knowledge of the system dynamical model and equations of motion are necessary for MPC [5]. Various MPC strategies, including partial feedback linearisation and computed torque control, are used for control of robotic systems and UAVs.

Simulation is also an important aspect in control systems design due to its ability to provide a cost-effective and time-saving alternative to rapidly prototype

[1]Aptus Engineering, Inc., Scottsdale, AZ 85260, USA
[2]Embry-Riddle Aeronautical University, Prescott, AZ 86301, USA

and test control systems [1–3]. The use of simulation systems allows for developers and engineers to tune control system gains, parameters and even experiment with different control strategies in very short periods.

5.1.2 Operator training

Engineers and project development personnel are not the only ones who benefit from modelling and simulations. Similar to the use of flight simulation in manned flight systems and pilot training, simulation of UAV operations provide vital training to operators, remote pilots and payload specialists. Training in the simulation provides significant cost savings as the physical UAV need not be deployed. Additionally, emergency scenarios can be simulated without the risk of damaging equipment or property [6].

Using sensor models, virtual reality or augmented data can be integrated with UAV simulators to train payload operators as well as UAV pilots.

Simulators for UAV operator training at universities

Leading aviation-focussed universities around the world have adopted unmanned aerial system programmes and majors in the growing industry. These programmes range from design and developing of UAVs to training pilots and operators. Boeing(Insitu), the developers of the RQ-21A, Scan Eagle and Integrator UAVs, has launched university training programmes for their ICOMC2 ground control station software, providing software simulation models for students [7]. Dedicated UAV payload simulation software companies, such as Simlat, have also worked with universities from around the world including Drexel University, Embry-Riddle Aeronautical University and Macquarie University. All three universities have renowned UAS programmes to provide their simulation software for operator training and research labs [8–10].

5.1.3 Sub-system development and testing

Simulation of UAV systems is also used for development and testing of payloads and sub-systems. The simulator provides a platform with similar responses and interfaces to that of the physical system allowing for its use to emulate flight. Recently, camera systems have been tested, and computer vision-based applications have been developed through hardware-in-the-loop (HIL) simulation [11].

Simulation systems are also being used for testing radio data links, ground stations and other interfaces [6].

5.2 History and adoption

The following section speaks to the introduction of modelling and simulation in aviation, the development of computational simulation techniques and platforms, the entry of UAVs into service and how simulation moulded the design and testing of modern drones.

5.2.1 Early aviation

Creating models and simulations of flight platforms have been at the core of aviation since its inception. The first real models analysed were birds in flight. These live models shaped the early manned and unmanned glider designs by Sir George Cayley. He was one of the Europe's leading scientists in aeronautical research during the early 1900s, and many of his engineering endeavours would mirror the flight of birds [12], including his scale model gliders with vertical and horizontal stabilisers [13]. In a typical scientific manner, engineers in those days regularly reported new findings of what did and did not work and formed aviation-based societies to share ideas with others.

The need for better ways of testing and simulation was all to appear during the dawn of aviation as many of the designers would perish in their creations. This drove the Wright brothers to build their wind tunnel, allowing them to safely record never before seen data on macro-relationships between wing shapes and lift. This wind tunnel gave them the data to create the controllable rudder, twisting wing and elevators [13].

Following in the footsteps of the Wright brothers National Advisory Committee for Aeronautics (NACA) engineers continued to build larger and more capable wind tunnels to test more advanced aircraft design [14]. During this time, NACA also focussed on the dreaded tailspin by creating dynamic free-flight models to compare with radio-controlled scale models [15].

Many early modelling methods of hand calculations and testing aircraft would catch up to the engineers and pilots of the day [16]. A breakthrough was needed to advance aviation into the modern era. This would be by using an analogue computer to simulate real-world conditions and verified by the outputted voltage or the movement of physical mechanisms [17].

5.2.2 First computerised simulations

In the 1940s, analogue computers could solve time-dependent engineering problems like differential equations. Inherently, they were adept at handling non-linear flight characteristics [18]. This would lead to Boeing's development of the Ground-to-Air Pilotless Aircraft. At the time Boeing created two analogue computers, one was used to simulate the missile flight characteristic, and the other was used directly to assist engineers in designing subsystem and equipment. The early adopters were the aerodynamics, mechanical, power plant, structure and electrical departments [19].

As the need for more advanced aircraft grew, simulation processes were found that allowed experimental aircraft programmes to grow with new requirements. The prototypical and most famous example is the X-15 Research Vehicle. The X-15 modelling, simulation and control laid the sound foundation of the Mercury and Apollo programme [20].

The modelling was initially crude, only looking at research vehicles in three-dimensional space. The early models were quickly replaced by five-degree-of-freedom simulations. This model was running simultaneously with limited six-degree-of-freedom simulations to provide clear control attributes. However, this again would prove to be inadequate and the ultimate six-degree-of-freedom simulation. A key feature of the final iteration was piloted and unmanned tested [20].

Modelling methods would adapt and grow as newer technology entered the market and digital computing proliferated. These methods would be built around linear, time-invariant dynamics models [21].

5.2.3 Entry of UAVs into service

Although the term UAV was coined in the late 1950s and notably defined by the ability to return successfully after a sortie, the subject of unmanned aircraft date back to World War I [22].

These early vehicles were purpose-built for delivering explosive ordinances in wartime. The Kettering Bug was the first and prime example developed by Orville Wright and Charles F. Kettering [23]. Kettering, an Electrical Engineer and founder of Dayton Engineering Laboratories Company (DELCO), spearheaded the aircraft development, which would utilise a mechanical computer, barometer and gyroscope [24]. The Kettering Bug could not be controlled mid-flight, and therefore the operator set an early version of an autopilot [23].

After the First World War, the fascination and necessity grew for the unmanned system. However, at this time, they would typically be used as artillery target drones. The OQ-2 made by Radio-plane was mass-produced and widely successful [25]. Although lucrative UAV still suffered from lingering issues that stop them from being used in more application, they could not return safely. Charles Stark Draper would change this with his inertial navigation system. Draper would improve the gyroscope and accelerometers of the day also to measure minute changes in the attitude of an object and make suggested control inputs [26]. His new autopilot would be proved during a transcontinental flight with a U.S. Air Force B-29. This innovation coupled with a global positioning system, would lead to the first flight of the predator UAV and a new era of aviation [27].

5.2.4 Commercial and consumer drones

In the modern-day, UAVs are extensively used for commercial applications including agriculture, mapping, forest service, recreational, cinematography and law enforcement. Modelling and simulation are mostly used in developing autopilots and flight control systems for such drones. HIL simulation has become a very popular tool for testing and simulation of commercial drones [28].

5.3 Modelling of UAV dynamics

Various techniques are used to develop dynamic models for UAVs. Many have been derived from methods employed for modelling rigid bodies, manned aircraft and robotic systems. This section delves into terminology, representation, reference frames, generalised equations of motion and specific models that are commonly used to model both fixed-wing and rotary-wing UAV dynamics.

5.3.1 Model representation methods

Definition 1: A dynamic model of a system is a set of mathematical equations that represent the system behaviour over time.

Dynamic models can be represented in many different forms of which differential equations or systems of linear equations are commonly used. These equations capture the variability and process of system states over time and changing inputs.

5.3.1.1 Differential equations

For successful modelling of physical systems, quantitative mathematical models must be obtained. Therefore, it is imperative to characterise the dynamic parameters under scrutiny so simulations can be created.

As aircraft are dynamic systems, they can typicality be viewed as differential equations. In general, aircraft dynamics take the form time-invariant differential equation used to start the modelling process can be found in the following equation:

$$\frac{d^n y}{dt^n} + a_{n-1}\frac{d^{n-1}y}{dt^{n-1}} + \cdots + a_1\frac{dy}{dt} + a_0 y = b_0 u \tag{5.1}$$

where the left-hand terms describe the physical rates of systems being modelled, like $d^{n-1}y/dt^{n-1}$ multiplied by a constant a_{n-1}, and the right-hand terms characterise an input to the system.

Producing accurate differential equations provides the first of many critical steps system modelling so designers may suggest potential improvements.

5.3.1.2 State-space representation

The state-space representation of a dynamic system can play a pivotal role in modelling by performing simplifications on more complex dynamics equations. State-space representation is very commonly used to design linear control systems, including the commonly used theoretically optimal linear quadratic regulator controller.

The prototypical case can be found in the following equations (where $t \geq t_0$):

$$\dot{\overline{x}} = A\overline{x} + B\overline{u} \tag{5.2}$$

$$\overline{y} = C\overline{x} + D\overline{u} \tag{5.3}$$

where $\overline{x}$ is the state, $\overline{y}$ is the output vector, $\overline{u}$ is the input or control vector, A is the system matrix, B is the input matrix, C is the output matrix and D is the feed-forward matrix.

State-space modelling aims to define “phase variables” where the subsequent state variables can be defined as and by the derivative of the previous state variable. Using the state-space equations, the second-order differential equations can be built into coupled first-order differential equations.

5.3.2 Common reference frames

It is important to understand the various frames of reference used to derive the equations of motion of a system. A common technique used in most robotic systems involves an inertial frame, along with a coordinate frame for each rigid body. In most robotics scenarios, the robot's base frame is considered as an inertial frame, and the rotation of the Earth is ignored. However, in high-speed aerospace and mostly orbital systems, inertial frames are different from rotating Earth frames due to the effect of Coriolis and centripetal forces from the Earth's rotation.

5.3.2.1 Inertial frame of reference

In an inertial frame of reference, a body with zero net force acting upon it has zero acceleration, that is, the body is either at rest or moving at a constant speed in a straight line [29].

Definition 2: An inertial frame of reference is a reference frame that describes time and space homogeneously, isotropically and in a time-independent manner [30].

Derivations of system equations of motion are done with reference to an inertial frame since Newton's second law is expressed with respect to an inertial frame [31]. Inertial reference frames can be placed at arbitrary location – the origin of the heliocentric inertial frame is at the centre of the Sun and is used commonly in orbital mechanics, while a geocentric inertial frame is at the centre of the Earth [32].

5.3.2.2 Earth-centre frames of reference

An inertial frame can be placed at the centre of the Earth without any rotation – known as the geocentric inertial or Earth-centred inertial (ECI) frame; however, since the Earth rotates about its centre axis, it is also useful to fix the frame orientation on the Earth – known as the Earth-centred earth-fixed (ECEF) frame.

In orbital mechanics, the ECI frame is a frame aligned with fixed stars. However, the errors when using an ECEF frame as the inertial frame of reference are negligible in the lower regions of the Earth's atmosphere. Therefore, an ECEF frame is assumed to be sufficiently "inertial" to validate Newton's second law [31].

Figure 5.1 depicts the two Earth-centred frames – the ECI frame, labelled with superscript i, and the ECEF frame, labelled with superscript e.

5.3.2.3 Navigation frame of reference

A local frame of reference, also known as the geodetic or navigation (n) frame, is used to represent the UAV's attitude, heading and velocity to a reference point on the Earth's surface projected at the UAV position [34]. A commonly used navigation frame is the North-East-Down (NED) frame, where:

- the origin is placed at the centre of the UAV;
- the x-axis points to true North;
- the y-axis points to East; and
- the z-axis points downwards to the centre of the Earth.

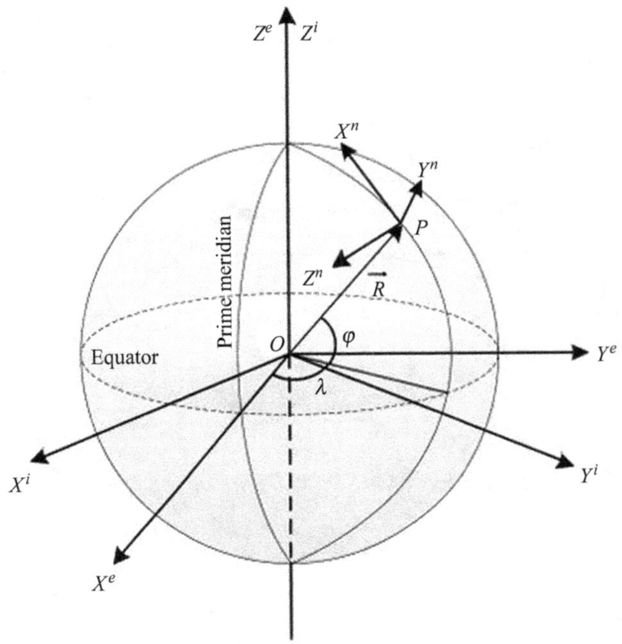

Figure 5.1 ECI (i), ECEF (e) and Navigation (n) reference frames [33]

Figure 5.1 depicts navigation (n) reference frame at longitude λ and latitude ϕ. The transformation from the ECEF frame to the n frame is $R_z(\lambda)R_x(\varphi)$. Rotation matrices are discussed in more detail in Section 5.3.3.2.

5.3.2.4 Body frames of reference

Establishing body fix frames in physical systems allows modellers to measure the positions and orientations relative to other frames and an inertial axis. Flight characteristic parameters on the aircraft body such as the centre of gravity, the centre of pressure and aerodynamic centre are generally measured with respect to the body frame. Commonly, the right-wing-down reference is used on fixed-wing UAVs, with the x-axis of the frame pointed forward, the y-axis pointed to the right of the aircraft and the z-axis pointed down. This allows for easy transformation to the NED navigation frame.

Figure 5.2 depicts a UAV body frame of reference.

5.3.3 Representation of state variables

As dynamic models have grown in complexity, the need for more succinct notation becomes a necessity. Generalised coordinates allow dynamic modellers to identify dynamic parameters uniquely suited for use with Lagrangian and Newton–Euler mechanics. As with many models, one goal is to create detailed rules so others may follow the work.

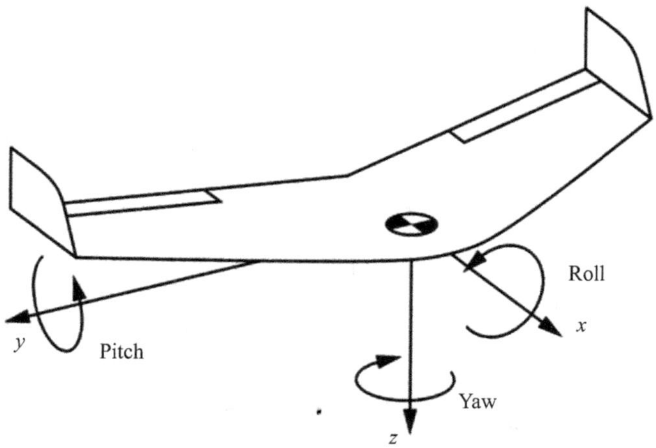

Figure 5.2 Representation of the body frame of reference on a UAV [35]

The number of generalised coordinates for a physical system is equal to the number of degrees-of-freedom of the system. For example, a quadcopter would have 11 unique degrees-of-freedom, so that results in 11 generalised coordinates. The Greek letter γ is traditionally assigned, as the variable for the vector of generalised coordinates. In the case of the quadcopter example, the gamma could be written as $\overline{\gamma} = [x, y, z, \phi, \theta, \psi, \theta_1, \theta_2, \theta_3, \theta_4]$, where the UAV's position is in terms of x, y and z. The rotations ϕ, θ and ψ are, respectively, about roll, pitch and yaw axes. The four remaining degrees-of-freedom are the positions of fixed pitch propellers.

The physical positions and respective rates in the state space comprise the details the important dynamic parameters used to discover the underlying mathematics to the system.

$$\overline{x} = [\overline{\gamma}, \dot{\overline{\gamma}}]^T. \tag{5.4}$$

Using state-space representation, the modellers can use the positions and velocities to create a better-defined model. Many times, dynamics and controls will speak about the relative nature of the aircraft's frames, typically centred on the centre of gravity of the aircraft. Looking back at the quadcopter example again, models must be able to describe the state space in usable terms that will provide tangible relationships. Two proliferated techniques unitise Euler angles and Quaternions, which look to describe the orientations of body-fixed frames.

5.3.3.1 Euler angles

Euler angles describe the position of one frame relative to another. As each frame contains three orthonormal axes, there are three possible Euler angles. However, there can be redundant answers to describe the process from rotating from one frame to the next. Thus, modellers seek a set of rotations that indicate the final orientation. This is denoted as the minimal representation of the orientation as seen in the following equation [36]:

$$\overline{\Phi} = \left[\phi, \theta, \psi\right]^{T}. \tag{5.5}$$

Traditionally, ϕ, θ and ψ are not limited to being the sole variables of rotational about a single axis. However, a compact way of indicating the first, second and third rotation.

Moreover, to describe the orientation of a frame attached to fixed body modellers, it is necessary to calculate the sequential rotations by way of rotation matrices. Such as *ZYZ* or *ZYX*, where rotations in the z-, y- and x-axis can be seen in (5.6)–(5.8), respectively [36]:

$$R_z(\phi) = \begin{bmatrix} \cos(\phi) & -\sin(\phi) & 0 \\ \sin(\phi) & \cos(\phi) & 0 \\ 0 & 0 & 1 \end{bmatrix} \tag{5.6}$$

$$R_y(\theta) = \begin{bmatrix} \cos(\theta) & 0 & \sin(\theta) \\ 0 & 1 & 0 \\ -\sin(\theta) & 0 & 1\cos(\theta) \end{bmatrix} \tag{5.7}$$

$$R_x(\psi) = \begin{bmatrix} 1 & 0 & 0 \\ 0 & \cos(\psi) & -\sin(\psi) \\ 0 & \sin(\psi) & \cos(\psi) \end{bmatrix} \tag{5.8}$$

In many dynamic modelling fields, computing composite rotation matrices by successive multiplication of ZYZ matrices is sufficient. However, aeronautical modellers, many times prefer to use ψ, θ and ϕ, respectively, Z, Y and X rotations. When these are multiplied accordingly in order, they produce a final orientation of an aircraft relative to an internal axis.

5.3.3.2 Rotation matrices

To represent the relative position and orientation of rigid bodies, its relation to other coordinate frames must be assigned at points of interest to describe physical relationships that will further the study and will lead to a better result and simulation. To characterise coordinate frames relative to one another, closed-form analytics solutions are employed.

In the case of ZYX rotations, it can be written in more detail in (5.9) [36]. For a compact notation, the following shorthand may be used: cos (λ) and sin (λ) will be shorted to c_λ and s_λ, where λ is an arbitrary rotation angle [36]:

$$\begin{aligned} R(\Phi) &= R_z(\phi)R_y(\theta)R_x(\psi) \\ &= \begin{bmatrix} c_\phi c_\theta & c_\phi s_\theta s_\psi - s_\phi c_\psi & c_\phi s_\theta s_\psi + s_\phi s_\psi \\ s_\phi c_\theta & c_\phi s_\theta s_\psi + c_\phi c_\psi & s_\phi s_\theta c_\psi - c_\phi s_\psi \\ -s_\theta & c_\theta s_\psi & c_\theta c_\psi \end{bmatrix} \end{aligned} \tag{5.9}$$

where Φ denotes the rotations relative to the fixed frame connected to the centre of mass of the aircraft [36]. Equation (5.9) shows the relatively simple nature of rotation matrices. However, Euler angle-based rotation matrices can misinterpret the orientation of the aircraft if a gimbal lock occurs.

5.3.3.3 Quaternions

Although Euler angles are widely used along with rotation matrices, there are many times they are not optimal due to inherent problems with Euler angles. The most prominent issues are related to the gimbal lock conditional. In a gimbal lock condition, at least two of the three axes of rotation are aligned. When inertial measurement units (IMUs) and mechanical systems are unable to handle the alignment logically, the gimbal lock takes effect [37]. Once in effect, attitude reference systems no longer report usable data.

The use of a four-parameter model helps to overcome any flaws. A model is based on the unit quaternion can describe the orientation of a body gimbal without alignment issues. Although a single Euler angle can initially define them, θ and unit vector of rotation $r = [r_x, r_y, r_z]^T$ are receptive to an origin frame. Equation (5.11) shows the relationship between θ, r and the unit quaternions:

$$Q = \{\eta, \varepsilon\} \tag{5.10}$$

where Q is defined by the following equations [36]:

$$\eta = \cos\left(\frac{\theta}{2}\right) \tag{5.11}$$

and

$$\varepsilon = \cos\left(\frac{\theta}{2} r\right) = [\varepsilon_x, \varepsilon_y, \varepsilon_z]^T \tag{5.12}$$

Equation (5.11) calculates the scaler portion quaternion, and (5.12) calculates the vector portion. It should be noted (5.10)–(5.12) are considered by the following equation [36]:

$$\eta^2 + \varepsilon_x^2 + \varepsilon_y^2 + \varepsilon_z^2 = 1 \tag{5.13}$$

This unit relationship provides the namesake of the unit quaternion and shows that unit quaternions are invariant with respect to the rotation matrix and its transpose.

Many times modellers like to view the rotation matrix of a given quaternion. This can be seen in the following equation [36]:

$$R(\eta, \varepsilon) = \begin{bmatrix} 2(\eta^2 + \varepsilon_x^2) - 1 & 2(\varepsilon_x\varepsilon_y - \eta\varepsilon_z) & 2(\varepsilon_x\varepsilon_z + \eta\varepsilon_y) \\ 2(\varepsilon_z\varepsilon_y - \eta\varepsilon_z) & 2(\eta^2 + \varepsilon_y^2) - 1 & 2(\varepsilon_y\varepsilon_z - \eta\varepsilon_x) \\ 2(\varepsilon_x\varepsilon_z - \eta\varepsilon_y) & 2(\varepsilon_y\varepsilon_z - \eta\varepsilon_x) & 2(\eta^2 + \varepsilon_x^2) - 1 \end{bmatrix} \tag{5.14}$$

Furthermore, closed-form solutions to the inverse problem can be found. Wherein, using (5.14), one can find η and ε. This can be done by utilising the following equations [36]:

$$R = \begin{bmatrix} r_{11} & r_{12} & r_{13} \\ r_{21} & r_{22} & r_{23} \\ r_{31} & r_{32} & r_{33} \end{bmatrix} \tag{5.15}$$

$$\eta = \frac{1}{2}\sqrt{r_{11} + r_{22} + r_{33} + 1} \tag{5.16}$$

$$\varepsilon = \frac{1}{2}\begin{bmatrix} \operatorname{sgn}(r_{32} - r_{23})\sqrt{r_{11} - r_{22} - r_{33} + 1} \\ \operatorname{sgn}(r_{13} - r_{31})\sqrt{r_{22} - r_{33} - r_{11} + 1} \\ \operatorname{sgn}(r_{32} - r_{23})\sqrt{r_{33} - r_{11} - r_{22} + 1} \end{bmatrix} \tag{5.17}$$

It should be noted that $\operatorname{sgn}(x) = 1$ when $x \geq 0$ and $\operatorname{sgn}(x) = -1$ when $x < 0$. Also, it is implied $\eta \geq 0$, which is equivalent to $\theta \in [-\pi, \pi]$. Therefore, all rotations will not cause singularity [36].

Like in the case of Euler angle-based rotation matrices, quaternion rotations can be used in sequence to model the orientation between frames. However, this must be done by way of the quaternion product operator "*". The expanded view of quaternion-based matrix multiplication can be seen in the following equation [36]:

$$Q_1 * Q_2 = \{\eta_1\eta_2 - \varepsilon_1^T\varepsilon_2, \eta_1\varepsilon_2 + \eta_2\varepsilon_1 + \varepsilon_1 \times \varepsilon_2\} \tag{5.18}$$

Although at first quaternion can seem daunting, they are immune to the flaws of Euler angle-based rotations, which is a clear motivator for dynamicists.

Example 1 (quaternion identity) : In the following example, we will cover the identity element for a quaternion. Like in Euler angle-based rotation matrices, the identity property can be related via the invert of the rotation matrix is equal to the original matrix. Therefore, when $Q_1^{-1} = Q_2$, the quaternion is $\{1,0\}$.

Representation of states, reference frames and orientation are critical to being able to model the dynamics of a system. With these, rigid body properties and equations can be set up to derive the system's equations of motion.

5.3.4 Deriving the system equations of motion

The system's equations of motion are a set of mathematical equations – generally a set of differential equations, which can also be represented as a system of linear equations. The equations of motion describe the system's response to various inputs and states over time. Various methods are used to derive a system's equations of motion based on the conservation of linear and angular momentum. Popular methods, including the Euler–Lagrange and Newton–Euler methods are also based on the core concept of Newton's second law.

Certain assumptions are made in order to simplify the modelling process [31].

Assumption 1: The UAV is assumed to be a rigid body, i.e., the distance between any two arbitrary points on the UAV is fixed and time-invariant.

Assumption 2: The UAV mass is assumed to be constant. Even for combustion and turbine engines, this assumption holds for short periods of time.

Assumption 3: The mass distribution through the UAV is assumed to be time-invariant.

5.3.4.1 Conservation of momentum

The conservation of linear and angular momentum equations start with Newton's second law: $\overline{f} = m\overline{a}$, where $\overline{f}$ is the vector of forces on a body of mass m with respect to an inertial reference frame and $\overline{a}$ is the vector of accelerations on the body with respect to the same reference.

From Newton's second law, following through with assumption 2, the relationship below is reached:

$$m\frac{d\overline{V}_P}{dt} = m\overline{g} + \overline{F}_A + \overline{F}_T \tag{5.19}$$

where $\overline{V}_P$ is the velocity of the body, $\overline{g}$ is the acceleration due to gravity, $\overline{F}_A$ is the aerodynamic force on the body and $\overline{F}_T$ is the thrust force on the body [31].

Conservation of linear momentum

Equation (5.19) is with respect to the ECI frame and is more useful with respect to the UAV body frame. This transformation can be made in order to derive the conservation of linear momentum equations in the body frame:

$$m(\dot{U} + QW - RV) = mg_x + F_{A,x} + F_{T,x} \tag{5.20}$$

$$m(\dot{V} + UR - PW) = mg_y + F_{A,y} + F_{T,y} \tag{5.21}$$

$$m(\dot{W} + PV - QU) = mg_z + F_{A,z} + F_{T,z} \tag{5.22}$$

where U, V and W are the velocity components in the x, y and z directions of the body frame, respectively, and P, Q and R are rotation rates about the x-, y- and z-axes of the body frame respectively [31].

Conservation of angular momentum

The conservation of angular momentum equations is derived from (5.19) as well. The following equation is reached:

$$\dot{P}J_{xx} - \dot{R}J_{xz} - PQJ_{xz} + RQ(J_{zz} - J_{yy}) = L_A + L_T \tag{5.23}$$

$$\dot{Q}J_{yy} - PR(J_{xx} - J_{zz}) + (P^2 - R^2)J_{xz} = M_A + M_T \tag{5.24}$$

$$\dot{R}J_{zz} - \dot{R}J_{xz} + PQ(J_{yy} - J_{xx}) + QRJ_{xz} = N_A + N_T \tag{5.25}$$

where J is the inertia tensor of the aircraft; L, M and N are the moments about the x, y and z directions of the body frame, respectively; sub-script A indicates moments due to aerodynamic effects and sub-script T indicates moments due to thrust [31].

5.3.4.2 Euler–Lagrange method

For more complex systems with multiple bodies (where assumption 1 is not valid), other approaches need to be taken. The Euler–Lagrange method for dynamic modelling is an approach that also results in closed-form equations of motion.

The Euler–Lagrange method is capable of modelling the equations of motion of a system that has multiple bodies with complex dynamical coupling. The Euler–Lagrange formulation itself is:

$$\overline{f} = \left(\frac{d}{dt}\left(\frac{\partial L}{\partial \dot{\overline{\gamma}}}\right) - \frac{\partial L}{\partial \overline{\gamma}}\right)^T \tag{5.26}$$

where L is the Lagrangian of the system, defined by the difference between the total kinetic energy K and potential energy U and $\overline{f}$ is a vector of generalised forces along with the generalised coordinates $\overline{\gamma}$ [37].

The kinetic energy of a system can be computed with reference to an inertial reference frame using the equation:

$$K = \sum_i \left(\frac{1}{2} m_i \dot{\overline{r}}_i^T \dot{\overline{r}}_i + m_i \dot{\overline{r}}_i^T {}^I\dot{T}_i \overline{r}_{i,cm} + \frac{1}{2}\text{trace}\left({}^I\dot{T}_i \hat{J}_i {}^I\dot{T}_i^T\right)\right) \tag{5.27}$$

where $\overline{r}_i$ is a vector from the origin of the inertial frame of reference to the frame of body i, m_i is the mass of body i, ${}^I\dot{T}_i$ is the transformation matrix from the body frame i to the inertial reference frame, $\overline{r}_{i,cm}$ is the vector from the origin of the body frame i to the centre of mass of body i measured with respect to the body frame and $\hat{J}_i$ is the pseudo-inertia tensor of body i, which is defined as follows:

$$\hat{J}_i = \frac{1}{2}\text{trace}(J_i)I - J_i \tag{5.28}$$

where J_i is the true inertia tensor, and I is a three-dimensional identity matrix [37].

The potential energy of the system can be computed with the equation:

$$\sum_i \left([0, 0, g_z]\left(\overline{r}_i + {}^I T_i \overline{r}_{i,cm}\right)\right) m_i \tag{5.29}$$

where g_z is the gravitational acceleration.

The Euler–Lagrange formulation in (5.26) can be re-arranged to the form:

$$\overline{f} = H(\overline{\gamma})\ddot{\overline{\gamma}} + \overline{c}(\overline{\gamma}, \dot{\overline{\gamma}}) + \overline{g}(\overline{\gamma}) \tag{5.30}$$

where $H(\overline{\gamma})$ is the system mass matrix written as a function of the generalised coordinates, $\overline{d}(\overline{\gamma}, \dot{\overline{\gamma}})$ is the vector of Coriolis and centripetal forces, and $\overline{g}(\overline{\gamma})$ is the vector of gravitational forces on the system [37]. Equation (5.30) can be rearranged to solve for the generalised accelerations $\ddot{\gamma}$ as follows:

$$\ddot{\gamma} = H^{-1}(\overline{\gamma})(\overline{f} - \overline{c}(\overline{\gamma}, \dot{\gamma}) - \overline{g}(\overline{\gamma}) \tag{5.31}$$

The generalised accelerations can then be integrated to simulate the model of the system.

Either the momentum conservation equations (5.20)–(5.25) or the equations derived using the Euler–Lagrange formulation (5.31) can be used for most UAV systems. If the UAV has many complex dynamically coupled bodies in serial or if a numerical integration method is required, the Newton–Euler approach can be taken.

5.3.4.3 Newton–Euler recursive method

The Newton–Euler method for dynamic modelling is based on the general framework of the Newton equation for the translational motion:

$$\overline{f}_i - \overline{f}_{i+1} + m_i\overline{g} = m_i\ddot{\overline{r}}_{i,cm} \tag{5.32}$$

and the Euler equation for rotational motion:

$$\overline{\tau}_i + \overline{\tau}_{i+1} + \overline{f}_i \times_{i-1}\overline{r}_{i,cm} - \overline{f}_{i+1} \times \overline{r}_{i,cm} = J_i\dot{\overline{\omega}}_i \times J_i\overline{\omega}_i \tag{5.33}$$

where $\overline{\tau}_i$ is the vector of generalised torques on body i, ${}_{i-1}\overline{r}_{i,cm}$ is the vector from the origin of the previous frame to the centre of mass of body i measured with respect to frame i and $\overline{\omega}_i$ is the vector of angular velocities from the previous frame to the current frame i [38].

The Newton–Euler method is performed with a forward recursion starting from the inertial frame and going up the chain to compute linear and angular accelerations of the end of each joint, and then with a backward recursion starting from the end effector going back to the inertial frame to compute the vector of forces and torques on each joint using the Newton and Euler equations shown in (5.32) and (5.33). These equations are used to solve the direct dynamics of the system. A call of the Newton–Euler dynamics function $NE0(\overline{\gamma}, \dot{\gamma}, \ddot{\gamma}, \overline{g})$ performs the steps of the Newton–Euler method and returns the vector of forces and torques mapped in the generalised coordinates γ [38].

The same format of the equations of motion-defined in (5.31) can be generated using the Newton–Euler recursive method. The system mass matrix $H(\overline{\gamma})$ is computed using the following equation:

$$H(\overline{\gamma}) = [\mathrm{NE0}(\overline{\gamma}, \overline{0}, \overline{e}_1, \overline{0}_3), \mathrm{NE0}(\overline{\gamma}, \overline{0}, \overline{e}_2, \overline{0}_3), \ldots, \mathrm{NE0}(\overline{\gamma}, \overline{0}, \overline{e}_n, \overline{0}_3)] \tag{5.34}$$

with each column of the system mass matrix being formed by running through the Newton–Euler method with the inputs for the generalised acceleration as a unit vector in the direction of the i-th dimension for column i. The vector of Coriolis, centripetal, gravitational and frictional forces is constructed by running through the Newton–Euler recursive equations with the zero vector passed as the generalised accelerations [38]:

$$\overline{c}(\overline{\gamma}, \dot{\gamma}) + \overline{g}(\overline{\gamma}) = \mathrm{NE0}(\overline{\gamma}, \dot{\gamma}, \overline{0}, \overline{g}) \tag{5.35}$$

The Newton–Euler recursive method can be used to integrate large complex systems numerically where closed-form equations of motion are not viable due to memory or computational resource availability.

5.3.5 Flight physics models

Any of the above-discussed methods can be used to derive the system's generalised equations of motion. However, specific forces and moments such as the forces and moments in each axis due to aerodynamic effects and thrust need to be estimated. Analytical or empirical physical models can be used to develop sets of equations for the specific UAV that model its forces and moments.

5.3.5.1 Fixed-wing flight

The aerodynamic forces and moments of fixed-wing flight can be modelled using many different methods, including aerodynamic equations, wind tunnel analysis, computational fluid dynamics and system identification. Napolitano's Aircraft Dynamics: From Modelling to Simulation describes the various aerodynamic forces and moments and equations that can be used to model them in detail [31].

Wind tunnel analysis

The design and build of the first modern wind tunnel are credited to Frank H. Williams, a member of the Aeronautical Society of Great Britain. The Wright brothers have used wind tunnels to design their aircraft and NASA has been using wind tunnels since the 1920s for testing and analysis of aircraft [39].

Wind tunnels provide high fidelity and low-risk approach to approximate aerodynamic effects. Wind tunnels allow for actual and mostly scaled models to be tested in order to model realistic flight. However, wind tunnels are very expensive and obtaining useful data is very time consuming [31].

Computational fluid dynamics

With the recent advance of computational systems and the ability to perform a large number of simultaneous calculations in parallel, such systems can be used to simulate fluid flow. Computational fluid dynamics (CFD) analyses use numerical methods to solve and analyse problems involving fluid flow. CFD problems are based on the Navier–Stokes equations, which define single-phase flow. Two-dimensional solutions and codes have been used for aerodynamic analysis since the 1980s, and two-dimensional codes have been recently introduced, which provide a very powerful and low-cost tool to approximating flight models [31].

System identification

A common approach to estimating model parameters when a set of model equations are available is system identification, also known as parameter identification. System identification is also used in robotics to estimate static and dynamic parameters of multi-link robotic manipulators [36].

System identification for modelling of fixed-wing flight is a very recent approach to modelling the aerodynamic behaviour of aircraft. The approach entails developing mathematical models based on estimates and general models and then using data collected from flight to fit the models. NASA has used empirical model matching techniques since the 1950s, and researchers at the NASA Dryden centre started using system identification techniques to approximate flight models. Maximum likelihood estimation was the first system identification method used which have also aided in modelling the dynamics and developing simulations for the space shuttle [31].

These modelling and approximation methods can be used for and are to date being used for UAVs similar to manned aircraft, which these methods have been developed for.

5.3.5.2 Multi-rotors and VTOL

Multi-rotors and Vertical Take-Off and Landing (VTOL) UAVs are fledgelings in the industry. Therefore, many new approaches have not been explicitly developed to model their aerodynamics. The assumption of low-speed flight, however, allows us to develop an approximate flight dynamics model. Due to the physical nature of multi-rotors and the multiple rigid bodies, a robotics approach can be taken to modelling the dynamics.

A robotics approach

On setting up the coordinate frames for multi-rotors and VTOL aircraft, the system kinetic and potential energy equations can be derived using (5.27) and (5.28). The Euler–Lagrange method can then be used to derive the equations of motion as done for classical robotic systems.

Modelling of forces and moments

Forces and moments of multi-rotors and VTOL aircraft can be modelled using simple propeller equations, free-body diagrams and simplified aerodynamic drag equations. For instance, propeller thrust and torque can be mapped to motor revolutions per minute (RPM) via a simple quadratic relation.

5.4 Flight dynamics simulation

A dynamical model of a system can be used for model-predictive control simulation; however, for almost all other applications, simulation is necessary. The model is essentially a series of equations describing dynamic behaviour and responses of the system. The equations can be rearranged to be used to compute accelerations and integrated for velocities and positions.

5.4.1 Integration of the equations of motion

In the previous sections, dynamic models have been used to find the equations of motions of a physical system. These equations are typically shown as second-order differential equations such as Newton's second law $\overline{f} = m\overline{a}$. These are the rates of the motion, which rely on the current state and external factors such as time. Since modellers are concerted with not only the acceleration but the velocity and position as a function as the time it is necessary to find ether closed-form analytical solutions or computational methods [40].

In either case, once the solution to equations of motions is solved for, they can be used by modellers for simulation purposes. Many use these simulations to experiment by testing new inputs and variables to determine how the model responses. Moreover, simulations are lower-cost programmes to finance then designing and building full systems or deploying code on an experiential flying vehicle.

The following subsections will outline techniques used in the industry to generate solutions to dynamic models.

5.4.1.1 Euler method

Prior to modern computational methods, analytical integration used a pen and paper to find solutions to the differential equations. The explicit Euler method was solved by hand before early computational methods. To be solved by modern methods, the equations must be discretised and then evaluated at different time steps. Classically, this method was used to solve the initial value problem and the associated differential equations [40].

For a differential equation such as (5.36), a solution can be formed by rearranging the definition of derivative found in (5.37):

$$\frac{dy}{dt} = \dot{y} = f(t, y(t)), y(t_0) = y_0 \tag{5.36}$$

$$\dot{y}(t) = \lim_{\delta \to 0} \frac{y(t+\delta) - y(t)}{\delta} \tag{5.37}$$

To simplify (5.37), δ can be defined as a small positive time step. As such, this approximation of 1.38n can be used.

$$\dot{y} \approx \frac{y(t+\delta) - y(t)}{\delta} \tag{5.38}$$

Therefore, after rearranging (5.38) into (5.39), and substituting $t_{n+1} = t_n + \delta$ and $y_n = y_n(t_n)$, the solution y_{n+1} can be found in (5.40) [41] as follows:

$$y(t+\delta) - y(t) \approx \dot{y}\delta \tag{5.39}$$

$$y_{n+1} \approx \dot{y}_{n+1}\delta + y_n \tag{5.40}$$

By evaluating (5.40) over a series of time steps, the integral of the original differential equation can be found. Using this method, a second time starting with a second-order differential equation in a dynamic model yields the position of the system.

5.4.1.2 Runga–Kutta methods

A more modern technique for handling numerical integration comes from the motivation to gain independence from using the Taylor methods on specific initial value problems. These techniques do not rely on differentiating the function being studied and make it possible to be used as a general-purpose ordinary differential equation solver.

The simplest Runga–Kutta (RK) method it aptly named the RK-2, which encompass two first lower order terms of the Taylor series explanation about t_0.

Starting with the initial value problem, see (5.36) as an estimate of the derivative of (5.36) can be found. Assuming the initial time t_0 in which initial conditions are known, we obtain the following equation:

$$k_1 = f(t_0, y^*(t_0)) \tag{5.41}$$

where $y^*(t)$ is the approximate of the derivative at time $t = t_0 + \delta$ (endpoint method) or $t = t_0 + \frac{\delta}{2}$ (midpoint method). The next step is to use k_1 to estimate k_2, the following equations for the RK-2 algorithm will outline the use of the midpoint method [42]:

$$y_1\left(t_0 + \frac{\delta}{2}\right) = y^*(t_0) = k_1 \frac{\delta}{2} \tag{5.42}$$

Then, estimating the derivative produces k_2 as follows:

$$k_2 = f\left(t_0 \frac{\delta}{2}, y_1\left(t_0 \frac{\delta}{2}\right)\right). \tag{5.43}$$

Using k_1 and k_2 the estimate for the full-time step can be found.

$$y(t_0 + \delta) = y^*(t_0) + k_2\delta \tag{5.44}$$

To fully realise the RK-2 algorithm, k_1, y_1, k_2 and y^* must be recursively calculated throughout the simulation [41].

The more prolific RK method is the RK-4, which encompass higher order terms from the Taylor series explanation about t_0 by finding k_3 and k_4.

Once again, the method starts with the initial value problem from in (5.36) as an estimate of the derivative of (5.36), which produces the terms k_1, k_2, k_3 and k_4. Assuming the initial time t_0 in which initial conditions are known, the over the time period of the, then:

$$k_1 = f(t_0, y^*(t_0)) \tag{5.45}$$

$$k_2 = f\left(t_0 + \frac{\delta}{2}, y^*(t_0) + k_1 \frac{\delta}{2}\right) \tag{5.46}$$

$$k_3 = f\left(t_0 + \frac{\delta}{2}, y^*(t_0) + k_2 \frac{\delta}{2}\right) \tag{5.47}$$

$$k_4 = f(t_0 + \delta, y^*(t_0) + k_3\delta) \tag{5.48}$$

Lastly, the equation to estimate $y(t_0 + \delta)$ can be found by using the following equations [41]:

$$y(t_0 + \delta) = y(t_0) + \frac{k_1 + 2k_2 + 2k_3 + k_4}{6}\delta. \tag{5.49}$$

Using the RK-4 integration sequentially on second-order differential equations derived from dynamic models allows for simulations of physical bodies implemented in more heuristic terms for modellers and engineers.

5.5 Conclusion

This chapter reviewed the need for modelling and simulation in the unmanned aircraft industry, its development, how it is used today and the fundamental concepts that allow its use. Furthermore, technological advancements in computing power bring about many more use cases and improvements in the simulation. Dynamic modelling and simulation has become a critical component of any UAV design and development project and continues to provide many cost and time benefits.

References

[1] Ivaldi S, Padois V, and Nori F. Tools for dynamics simulation of robots: A survey based on user feedback. Paris, France: Sorbonne Universite; 2014. Report No.: arXiv:1402.7050vl. Supported by the EU Project CODYCO (FP7-ICT-2011-9, No. 600716).

[2] Wu X, Figueroa H, and Monti A. Testing of digital controllers using real-time hardware in the loop simulation. In: 2004 IEEE 35th Annual Power Electronics Specialists Conference (IEEE Cat. No.04CH37551). vol. 5; 2004. p. 3622–3627.

[3] Muraleedharan N, Isenberg DR, and Gentilini I. Recreating planar free-floating environment via model-free force-feedback control. In: 2016 IEEE Aerospace Conference; 2016. p. 1–12.

[4] Carter JR. A business case for modeling and Simulation. U.S. Army Aviation and Missile Research, Development, and Engineering Center; 2011. Report No.: RD-AS-01-02.

[5] Pappu MR. Online Model Predictive Control of a Robotic System by Combining Simulation and Optimization [Master of Science in Technology]; 2015.

[6] Wu J, Wang W, Zhang J, *et al.* Research of a kind of new UAV training simulator based on equipment simulation. In: Proceedings of 2011 International Conference on Electronic Mechanical Engineering and Information Technology. vol. 9; 2011. p. 4812–4815.

[7] Insitu. Insitu – ICOMC2 – University Training Program; 2018. Available from: https://www.insitu.com/information.

[8] Simlat. Simlat has delivered an advanced UAS Training and Simulation Classroom to Embry-Riddle Aeronautical University campus at Prescott, AZ. 2017.

[9] Simlat. Drexel University to use Simlat C-STAR Simulation System in ASSURE human factors research. Drexel University at Philadelphia, PA. 2017.

[10] Simlat. Simlat delivered advanced simulation-based research lab to Macquarie University in Sydney, Australia. 2017.

[11] Ma O, Flores-Abad A, and Boge T. Use of industrial robots for hardware-in-the-loop simulation of satellite rendezvous and docking. Acta Astronautica. 2012;81(1):335–347.

[12] Ackroyd JAD. Sir George Cayley: The invention of the aeroplane near Scarborough at the time of Trafalgar. Journal of Aeronautical History. 2011;1(06):130–136.
[13] Bilstein RE. Flight in America. The Johns Hopkins University Press; 1994.
[14] Air SN, Museum S. The NACA/NASA Full Scale Wind Tunnel; 2015. Available from: https://www.airandspace.si.edu/stories/editorial/nacanasa-full-scale-wind-tunnel.com.
[15] Chambers J. The Role of Dynamically Scaled Free-Flight Models in Support of NASA's Aerospace Programs. Library of Congress; 2009.
[16] Day DA. Computers in Aviation; 2002. Available from: https://www.centennialofflight.net/essay/Evolution_of_Technology/Computers/Tech37.htm.
[17] Bochannek A. Why Analog Computers? 2013. Available from: https://www.computerhistory.org/atchm/why-analog-computer/.com.
[18] Wallmann R. Analog Computer; 2007. Available from: https://www.smecc.org/analog_computers.htm.
[19] Small JS. The Analogue Alternative: The Electronic Analogue Computer In Britain and USA, 1930–1975. Routledge; 2001.
[20] Copper NR. X-15 Analog Flight Simulation Program; 1961. Available from: https://www.omputer.org/csdl/proceedings/afips/1961/5058/00/50580623.pdf.
[21] Stengel RF. Flight Dynamics. Princeton University Press; 2004.
[22] Stamp J. World War 1:100 Years later; 2013. Available from: https://www.smithsonianmag.com/arts-culture/uumanned-drones-have-been-around-since-world-war-i16055939/.
[23] Miller R. The First Drones, Used in World War 1; 2003. Available from: https://io9.gizmodo.com/the-first-drones-used-in-world-war-i-453365075.
[24] Jeffries Z. Charles Franklin Kettering 1876–1958. National Academy of Sciences; 1960.
[25] Turcker FJR. The Pioneers Aviation and Aeromodelling-Interdependent Evolution and Histories; 2005. Available from: www.cite.monash.edu.au/hargrave/righter6_part1.html.
[26] Diaz M. The Father of Inertial Guidance; 2005. Available from: http://www.nmspacemuseum.org/halloffame/index.php?type=search.
[27] Harris-Para B. 100 Years of Sustained Power Flight History; 2003. Available from: https://www.faa.gov/education/educators/curriculum/k12/media/K-12_One_Hundred_Years_of_Sustained_Flight.pdf.
[28] Jung D, and Tsiotras P. Modeling and Hardware-in-the-Loop Simulation for a Small Unmanned Aerial Vehicle. In: AIAA Infotech Aerospace 2007 Conference and Exhibit; 2007.
[29] Fields D, editor, and Galilean Relativity. University of New Mexico; 2015.
[30] Landau LD, and Lifshitz EM. Mechanics. Pergamon Press; 1960.
[31] Napolitano MR. Aircraft Dynamics: From Modeling to Simulation. John Wiley and Sons, Inc.; 2012. Physics 262-01.
[32] Colasurdo G. Astrodynamics. Politecnico di Torino, Space Exploration and Development Systems; 2006.

[33] Chu T, Guo N, Backen S, *et al.* Monocular Camera/IMU/GNSS Integration for Ground Vehicle Navigation in Challenging GNSS Environments. 2012 12;12:3162–85.

[34] Noureldin A, Karamat TB, and Georgy J. Fundamentals of Inertial Navigation, Satellite-based Positioning and their Integration. Springer; 2013.

[35] Beyeler A, Jean-Christophe, and Floreano Z. Vision-based control of near-obstacle flight. Autonomous Robots. 2009;27:201.

[36] Sciavicco L, and Siciliano B. Modelling and Control of Robot Manipulator. Springer; 2000.

[37] Muraleedharan N, Cohen DS, and Isenberg DR. Omnidirectional locomotion control of a pendulum driven spherical robot. In: SoutheastCon 2016; 2016. p. 1–6.

[38] Muraleedharan N. Development of an Emulated Free-Floating Environment for on-Earth Testing of Space Robots [Master of Science in Mechanical Engineering].

[39] Baals DD, and Corliss WR. Wind Tunnels of NASA. Independently Published. 2018.

[40] Myers AC. Introduction to Numerical ODE Integrations Algorithms; 1999. Available from: pages.physics.cornell.edu/myers/teaching/Computational Methods/ComputerExercises/Pendulim/NumericalODE.pdf.

[41] Sauer T. Numerical Analysis. Person; 2012.

[42] Bishop RCDR. Modern Control Systems. Person; 2011.

Chapter 6

Multisensor data fusion for vision-based UAV navigation and guidance

Suraj Bijjahalli[1] *and Roberto Sabatini*[1]

Significant advances in sensor technology, along with economies of scale due to large production volumes, have supported the miniaturisation of navigation sensors, allowing widespread low-cost integration on unmanned aircraft systems (UAS). In small-size UAS applications, standalone sensors are not a viable option since the reduction in navigation sensor form-factor, weight and cost typically results in lowered accuracy and precision. The fusion of multiple sensor measurements in UAS navigation systems can support greater accuracy, integrity and update rates than is achievable employing individual sensors. This chapter introduces the fundamentals of state-estimation methods employed on UAS and presents different sensor integration architectures, along with an assessment of their advantages and trade-offs. Attention is devoted primarily to recursive optimal estimation algorithms such as the Kalman filter and its variants owing to its prolific employment in various classes of UAS. The need to support robust navigation performance in the global navigation satellite system denied environments, and the proliferation of visual sensors has led to the development of numerous methods for integrating visual sensor measurements (primarily) with inertial sensors. Therefore, the reader is introduced to the most popular system architectures for visual-inertial sensor integration in order to provide an understanding of the current state-of-the-art and to support the identification of future research pathways.

6.1 Introduction

A large number of commercial use-cases for small size, weight, power and cost (SWaP-C) unmanned aircraft systems (UAS) have emerged over the last five years. A number of these applications, including package delivery and surveillance typically involve a flight over densely populated areas, thereby incurring the risk of injury and/ or liability in the event of degraded navigation system performance. Most UAS rely on the global navigation satellite system (GNSS) to support drift-free absolute

[1]RMIT University, Bundoora, Australia

positioning and can be examined using the cyber-physical system paradigm [1]. However, GNSS-based positioning is subject to several errors and failure modes that compromise its accuracy, integrity and availability. In particular, operations in dense urban environments are vulnerable to outright loss of service due to low satellite visibility and misleading information due to signal multipath, consequently posing a higher risk of unstable guidance. Over the past decade, visual sensors have emerged as desirable candidates for inclusion in UAS sensor suites. Highly miniaturised colour CMOS cameras are quickly becoming low-cost, high-performance commodities with low weight and form-factor, allowing their deployment on small aerial platforms. A large body of knowledge has developed on the implementation of visual navigation sensors on unmanned aircraft either as standalone systems or more frequently as part of integrated multi-sensor systems. The typical strategy is to augment conventional sensor suites that comprise GNSS, inertial sensors, magnetometer and barometric altimeter in order to improve state observability and robustness to individual sensor failures. This chapter commences by presenting the fundamentals of data-fusion algorithms most commonly employed in unmanned aircraft such as the extended Kalman filter (EKF), and to a lesser extent, the unscented Kalman filter (UKF). Different sensor integration architectures are presented, primarily focusing on the fusion of visual sensors. A review of the literature in this domain was conducted in order to capture seminal work and significant outcomes in order to highlight the current state of the art and to future research.

6.2 Data-fusion algorithms

6.2.1 Extended Kalman filter

The Kalman filter (KF) is an algorithm that outputs an optimal estimate of a system using a series of measurements from multiple sensors corrupted by noise, essentially achieving accuracy greater than would be possible using a standalone sensor. The method utilises Bayesian inference recursively to maximise the joint probability of the states and measurements. Since the inception of the method in 1960, it has been widely applied to the navigation of aerial platforms, typically implemented in a discrete-time form. The algorithm runs in two steps, a prediction step that predicts the navigation state vector and its associated uncertainty by propagation through a model of the underlying system dynamics, and a measurement step, where data from the sensors is incorporated to correct the prediction and output an optimal estimate. The optimality is achieved by weighting the measurements and prediction according to the associated uncertainties. Despite its obvious benefits, the original KF is subject to two assumptions that limit its practical usage. The first is that the states, process noise and measurement noise must be Gaussian. The second is that the system model and the sensor measurement models must be linear, rendering the basic algorithm inapplicable to most navigation problems. The EKF addresses this limitation by linearising the models describing the system dynamics and sensor measurements. This procedure necessitates the computation of a matrix of partial derivatives (the Jacobian) for the state-transition and measurement models. The EKF equations are now described in state-space notation by a set of first-order non-linear differential equations:

$$\dot{x} = f(x, w) \tag{6.1}$$

where $x \in \mathbb{R}^n$ denotes the system state vector, $f(x)$ is the non-linear function of the states and $w \in \mathbb{R}^n$ represents a zero-mean random process. The matrix of the process noise, $Q \in \mathbb{R}^{n \times n}$ is given by:

$$Q = E\left(ww^T\right) \tag{6.2}$$

The measurement equation is considered to be a non-linear function of the measured states and is given by:

$$z = h(x, v) \tag{6.3}$$

where $v \in \mathbb{R}^m$ is a zero-mean random process described by the matrix of measurement noise $R \in \mathbb{R}^{m \times m}$:

$$R = E\left(vv^T\right) \tag{6.4}$$

It is possible to rewrite the non-linear equation of measurements for systems with discrete-time measurements as:

$$z_k = h(x_k, v_k) \tag{6.5}$$

As the measurement equations are non-linear, it is necessary to linearise them by the first-order approach to obtain the dynamic matrix of the system, F, and the measurement matrix, H. These matrices are related to the non-linear equations and are expressed as:

$$F = \left.\frac{\partial f(x)}{\partial x}\right|_{x=\hat{x}} \tag{6.6}$$

$$H = \left.\frac{\partial h(x)}{\partial x}\right|_{x=\hat{x}} \tag{6.7}$$

where $\hat{x}$ represents the mean value. The fundamental matrix is approximated by the Taylor series expansion as follows:

$$\Phi_k = I + FT_s + \frac{F^2 T_s^2}{2!} + \frac{F^3 T_s^3}{3!} + \cdots \tag{6.8}$$

where T_s is the sampling time, and I is the identity matrix. The Taylor series is often truncated to first order:

$$\Phi_k \approx I + FT_s \tag{6.9}$$

The overview of the steps in an iteration of the EKF is illustrated in Figure 6.1.

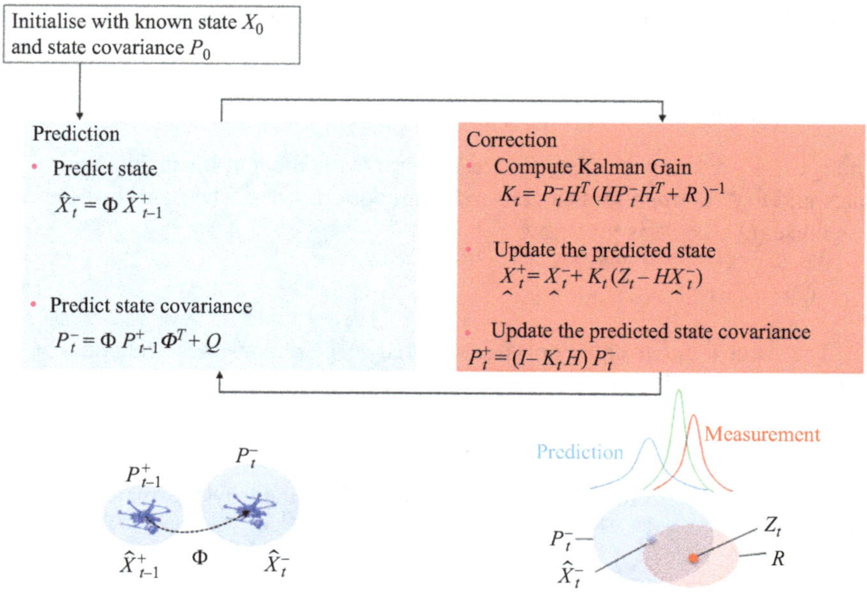

Figure 6.1 EKF algorithm overview

The prediction of the navigation state x and its covariance P at time step k from the previous time step through the state-transition matrix $\mathbf{\Phi}$ is given by:

$$\hat{x}_k^- = \Phi_k \hat{x}_{k-1}^+ \tag{6.10}$$

$$P_k^- = \Phi_k P_{k-1}^+ \Phi_k^T + Q \tag{6.11}$$

where the $(-)$ and $(+)$ superscripts denote the process before and after incorporating the measurement, respectively. The ($\hat{}$) symbol denotes an estimate of the state. Q is the covariance of the process noise. The computation of the Kalman gain K_k is given by:

$$K_k = P_k^- H^T \left(HP_k^- H^T + R\right)^{-1} \tag{6.12}$$

which is a function of the state covariance, the measurement noise covariance R and the measurement matrix H. The measurement is incorporated in the correction step to obtain the optimal estimate of the state $\hat{x}_k^+$ and its covariance P_k^+:

$$\hat{x}_k^+ = \hat{x}_k^- + K_k\left(z_k - h_k\left(\hat{x}_k^-\right)\right) \tag{6.13}$$

$$P_k^+ = (I - K_k H)P_k^- \tag{6.14}$$

$\hat{x}_k^+$ and P_k^+ are then used recursively in the next iteration of the filter. Despite its widespread use in aircraft navigation, the EKF is subject to limitations, prompting the development of alternative implementations of the KF. One of these methods is the UKF.

6.2.2 Unscented Kalman filter

Although the EKF is one of the most widely used filtering methods for non-linear system estimation, this filter has two key drawbacks that make it challenging for a practical implementation [2–5].

First, highly non-linear state-transition models cause poor performance and instability. Second, the accuracy of the non-linearly propagated mean and the covariance is limited to the first order since the filter uses a first-order truncated Taylor series linearisation scheme.

The UKF provides derivative-free higher-order approximations with Gaussian distributions rather than arbitrary non-linear functions. In comparison to the UKF, the EKF is difficult to implement and tune and is mostly suboptimal for non-linear systems. The unscented transform (UT) is central to the UKF, which evaluates the statistics of a non-linear transformed random variable. A number of "Sigma Points" are generated to characterise the state vector and its uncertainty and are propagated via the state-transition model. The sigma points essentially are used to approximate the moments of the distributions representing the states. The unscented transform (UT) process is a mathematical function that is used to estimate a given non-linear transformation using probability distribution that is characterised only in terms of a finite set of statistics. The most common use of the UT is in the non-linear projection of mean and covariance estimates in the context of non-linear extensions of the KF. The UT, especially as part of the UKF, has largely replaced the EKF in many non-linear filtering and control applications, including underwater [6], ground and air navigation [7] and spacecraft [5] applications.

Generally, the UKF performs with a higher order of computational expense than the EKF; however, it also demonstrates better performance. This can be attributed to increased accuracy of the moment approximations in the UKF attributed to the UT process. The UKF achieves accuracy comparable to the second-order EKF, which requires the derivation of Jacobian and Hessian matrices. A top-level representation of the UKF algorithm is shown in Figure 6.2.

- Determine sigma points $\chi_{k-1}^1, \ldots, \chi_{k-1}^s$ and weights $w^1, \ldots, w^s$ to match a mean $\hat{x}_{k-1|k-1}$ and covariance matrix $P_{k-1|k-1}$.
- Compute the transformed sigma points $\chi_k^i = f(\chi_{k-1}^i)$, $i = 1, \ldots, s$.
- Compute the predicted state statistics:

$$\hat{x}_{k|k-1} = \sum_{i=1}^{s} w^i \chi_k^i$$

$$P_{k|k-1} = Q_k + \sum_{i=1}^{s} w^i \left(\chi_k^i - \hat{x}_{k|k-1}\right)\left(\chi_k^i - \hat{x}_{k|k-1}\right)^T$$

- Determine the sigma points $\chi_k^1, \ldots, \chi_k^s$ and weights $w^1, \ldots, w^s$ to match mean $\hat{x}_{k|k-1}$ and covariance matrix $P_{k|k-1}$.
- Compute the transformed sigma points $y_k^i = h(\chi_k^i), i = 1, \ldots, s$

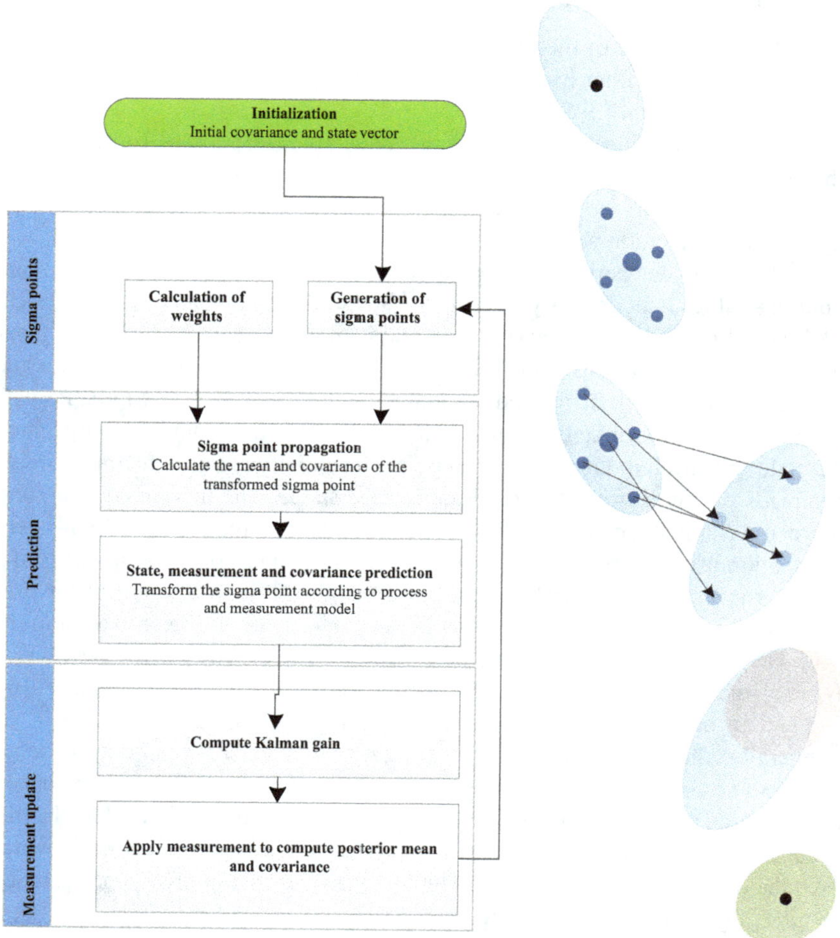

Figure 6.2 UKF algorithm

- Compute the predicted measurement statistics:

$$\hat{y}_{k|k-1} = \sum_{i=1}^{s} w^i y_k^i$$

$$S_k = R_k + \sum_{i=1}^{s} w^i \left(y_k^i - \hat{y}_{k|k-1}\right)\left(y_k^i - \hat{y}_{k|k-1}\right)^T$$

$$\Psi_k = \sum_{i=1}^{s} w^i \left(\chi_k^i - \hat{x}_{k|k-1}\right)\left(y_k^i - \hat{y}_{k|k-1}\right)^T$$

- Compute the posterior mean and covariance matrix:

$$\hat{x}_{k|k} = \hat{x}_{k|k-1} + \Psi_k S_k^{-1}\left(y_k - \hat{y}_{k|k-1}\right)$$
$$P_{k|k} = P_{k|k-1} - \Psi_k S_k^{-1}\Psi_k^T$$

6.2.3 *Integration architectures*

Integration architectures differ in the type of measurements extracted from the constituent sensors and the method by which corrections are formed and applied to the solution. Although there is no standard definition of different architecture categories, broad distinctions can be made depending on how measurements are fused and corrections fed back to sensors/sub-systems. Accordingly, an architecture can be categorised as cascaded, centralised, federated or hybrid. Each category has different advantages and limitations, and designs tend to vary widely between different platforms.

Cascaded architectures, as illustrated in Figure 6.3, are simple to implement. Each sensor passes its data to a dedicated local navigation processor, which outputs a local navigation solution and its associated covariance to the integration or "master" filter, which can be a snapshot least-squares algorithm, or more commonly a KF variant. The integration filter outputs an integrated navigation solution and can also be designed to feedback corrections to each of the sensor-specific navigation processors.

This architecture is suited to meet a requirement of integrating multiple sensors/sub-systems from different suppliers. Under this framework, each sensor/sub-system is essentially treated as a black box without tapping into raw sensor data. The most commonly adopted implementation is to employ KF variants for both the local navigation processors and the integration filter. Owing to the distribution of functions, the architecture is not as computationally intensive as centralised architectures. However, the primary disadvantage is due to the fact that KFs are

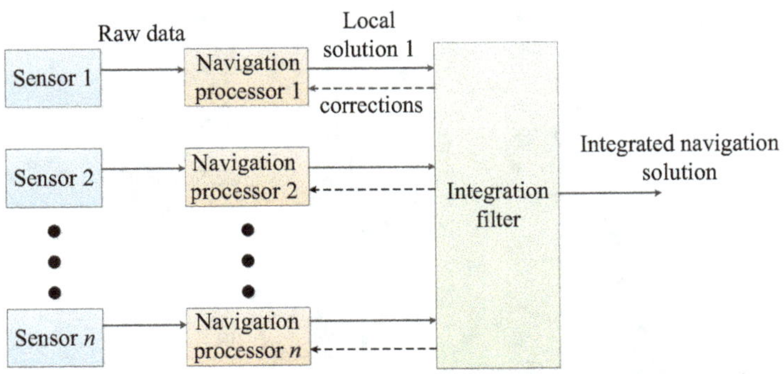

Figure 6.3 Cascaded architecture

placed in series. A fundamental assumption of Kalman filtering is that the measurement noise is white-noise (uncorrelated). However, the use of local Kalman filters introduces time-correlated noise in the inputs to the integration Kalman filter, which can cause instability and in extreme cases, filter divergence. The common practice to address this is through one of the following ways: increase measurement update intervals to match the correlation time of the inputs or assume a higher measurement noise covariance to down-weight the Kalman gain. In practice, this addresses stability issues but renders the process sub-optimal.

Centralised architectures, as shown in Figure 6.4, are characterised by an integration filter that accepts raw sensor data from all available navigation sensors. The filter can also be designed to feedback sensor-specific corrections. The errors and noise of each sensor are modeled in a single central filter. This method allows for all error correlations to be accounted for and all measurements to be optimally weighted. The elimination of cascaded Kalman filters allows higher gains to be used before the onset of instability. Given high-fidelity models of navigation sensors, centralised integration provides the optimal navigation solution in terms of accuracy and robustness. However, this comes at an inherently higher computational cost.

A federated architecture is shown in Figure 6.5, which comprises a reference sub-system independently fed to a bank of KFs, each of which is dedicated to a specific aiding-sensor. The integration filter outputs the navigation solution and corrections to the reference sensor/sub-system, along with reset values. Depending on the type of reset information provided, a federated architecture can either be no-reset, fusion reset or zero-reset. When using fusion reset filters, the integration filter feeds back the navigation solution and its associated covariance to each local filter.

Zero-reset filters set the local filter outputs to zero, and the covariances to their initialisation values at the end of each cycle. In no-reset filters, there is no feedback from the integration filter, and each local filter retains its unique portion of the total system information.

Although centralised architectures present an optimal solution, most small UAS manufacturers typically implement federated/centralised architectures owing to the requirement to integrate multiple "black-box" sensors/sub-systems from

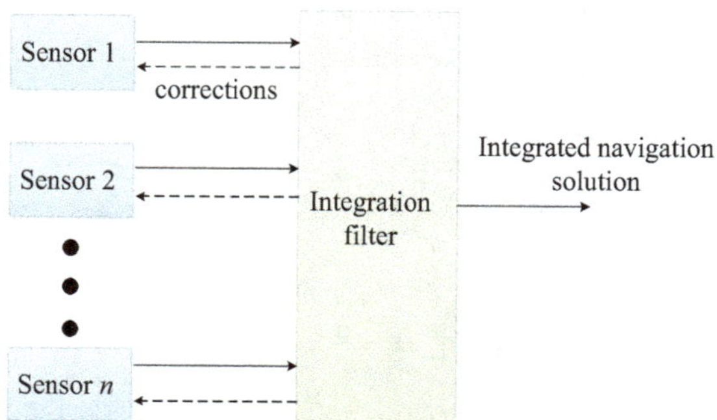

Figure 6.4 Centralized architecture

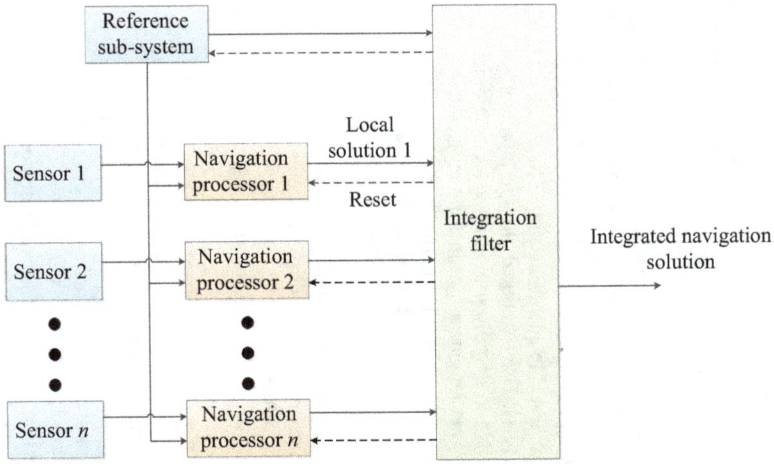

Figure 6.5 Federated architecture

different suppliers, and because computational overheads are typically lower. A common practice is to implement architectures that are a hybrid, i.e. architectures wherein subsets of sensors/sub-systems are alternately cascaded and centralised.

6.3 Fusion of visual sensors

In order to leverage visual sensors, most UAS typically employ visual data to augment conventional sensor suites in one of the following ways.

Visual-inertial aiding: Visual and inertial sensors are used to aid each other, e.g. inertial sensors are typically used to reduce the search space for feature matching between frames. A common strategy is to utilise optical flow (OF) measurements to augment velocity estimation

Visual-inertial odometry: This can be thought of as a subset of visual-inertial aiding where 6DoF visual odometry (VO) is fused in either a loosely- or a tightly coupled sensor, typically with inertial measurements.

Table 6.1 highlights key research in each category of vision-based systems and identifies similar work in the domain. An exhaustive list of each implemented technique is impractical owing to the vast number of combinations and variations in techniques. Instead, the focus is on identifying and evaluating the most commonly employed methods in UAS vision-based navigation.

As in the case of integrated GNSS-INS systems, visual-inertial fusion architectures can usually be categorised as either loosely or tightly coupled depending on the data fused (Figure 6.6).

In loosely coupled systems, the inertial and visual systems output standalone solutions but assist each other. Motion estimates from the INS are used to predict visual feature motion between frames in order to make reduce the size of the search space and make feature correspondences robust. Velocity estimation from the

Table 6.1 Summary of commonly employed visual sensor fusion algorithms

Method	Literature	Description	Similar work
Visual-inertial aiding	Navigation and guidance system architectures for small unmanned aircraft applications [8]		[9]
	Unmanned aerial vehicle navigation using wide-field optical flow and inertial sensors [10]	Wide-field optical flow-aided inertial navigation via an unscented information filter. State-estimation error statistics were recorded from flight trials to assess performance	
	An optic flow-based vision system for autonomous 3D localization and control of small aerial vehicles [11]	SAD block matching to compute optical flow. Integration of optical flow with inertial measurements for UAV localization and SFM	
	Adding optical flow into the GPS/INS integration for UAV navigation [12]	Image interpolation algorithm (I2A). Use of optical flow to augment state estimation from a GPS/INS system by providing velocity aiding during GPS dropouts, and ground height in terrain following tasks	
	Fusion of low-cost imaging and inertial sensors for navigation [13]	Tight coupling of visual and inertial sensors. The inertial sensor output is used to predict and reduce search space for feature tracking. statistical characterization of positioning error was performed by Monte-Carlo simulations	[14,15]
	A multi-state constraint Kalman filter for vision-aided inertial navigation [16]	Development of an EKF measurement model that incorporates geometrical constraints that exist between multiple camera observations of a single static feature	[17]
	Visual-inertial UAV attitude estimation using urban scene regularities [18]	Exploits regularities in the urban scene in the form of line-segments to correct gyroscope rates	[19,20]
Visual-inertial odometry	The monocular vision for long-term micro-aerial vehicle state estimation: a compendium [21]	Keyframe-based parallel tracking and mapping outputs 6DoF VO pose which is loosely coupled with high-frequency inertial measurements	
	Low computational-complexity algorithms for vision-aided inertial navigation of micro aerial vehicles [22]	The algorithms rely on onboard IMU measurements to calculate the relative rotation between two consecutive camera frames and the reprojection error to detect the inliers and discard wrong estimations in a two-point RANSAC algorithm for efficient VO	[13]
	Long-range stereo VO for extended altitude flight of unmanned aerial vehicles [23]	Modified bundle adjustment method to account for deformations in the stereo transform and thereby ensure accurate triangulation of structure and robust pose estimation where traditional rigid-stereo VO methods fail	[24]

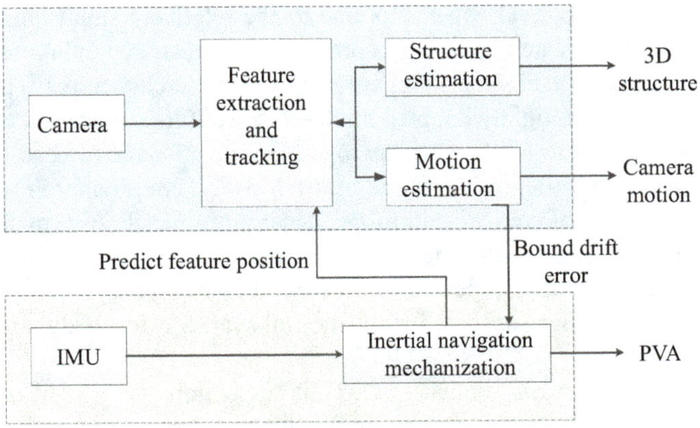

Figure 6.6 Loosely coupled visual-inertial fusion

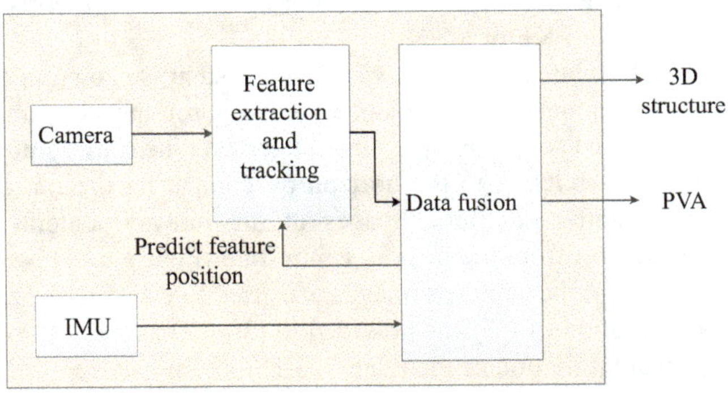

Figure 6.7 Tightly coupled visual-inertial fusion

visual sub-system is used to bound integration errors in the INS. A rigorously formulated loosely coupled approach is presented in [21].

In tightly coupled integration, raw data from the sensors is input to a single centralised data-fusion module which outputs aircraft state estimates and feeds back inter-frame feature displacements to the image-processing module (Figure 6.7).

Image processing is a branch of computer knowledge that is concerned with processing a digital signal representing images taken with a digital camera or scanned by a scanner [25–31].

Most visual-inertial odometry implementations for low Swap-C platforms are typically based on loosely coupled monocular systems wherein the visual sub-system is essentially a 6DoF pose estimator that outputs the relative pose and attitude (between two subsequent frames). In recent years, small platforms have almost exclusively utilised monocular cameras owing to the obvious gains in weight-saving and form factor as opposed to stereo-vision cameras. The approach is usually sub-optimal as compared to a tightly coupled system but is considerably simpler to implement and place less demanding computational requirements.

The saving in computational expense is due to the relatively small number of elements in the estimated state vector. This comparison is illustrated in Figure 6.8, which compares the typically estimated states between the two architectures. The inclusion of visual features in the tightly coupled architecture renders the architecture optimal but ultimately difficult to implement due to the computational overhead.

In [8], a hybrid (centralised/federated) loosely coupled architecture was developed which employed a vision-based navigation sub-system to augment attitude estimation of a fixed-wing UAV during the approach and landing phase. The top-level architectural pipeline is illustrated in Figure 6.9. The architecture comprises the following sensors: visual-inertial-GNSS-aircraft dynamics model (VIGA).

An aircraft dynamics model (ADM) is essentially a virtual sensor (i.e. knowledge-based module) that augments the state vector by computing the UAV trajectory and attitude motion. Both three degrees-of-freedom (3-DoF) and six degrees-of-freedom (6-DoF) ADM can be employed in the virtual sensor design. Two integration schemes were developed differing primarily in terms of the integration filters, either an EKF or a UKF.

The conceptual architecture of the visual sub-system is shown in Figure 6.10.

The sub-system comprises basic image-processing routines to extract dominant edges from the observed scene, namely, the horizon and the runway markers. Pitch and roll are estimated relative to the horizon by comparing the images captured online with *priori*-stored keyframe images of the runway. Lateral position is computed as a deviation from the runway centre-line (Figure 6.11).

OF extraction from the captured images was used to estimate the aircraft body rates. A formal definition of OF was provided in [30–32] as the 2-D velocity field describing the apparent motion or the relative 3-D motion between the camera and the environment. The OF generated from movement is the result of translational and rotational motions. Tracking the changes or shifts in the images using these two components can give information about the motion of the camera and the shape of the environment it is viewing.

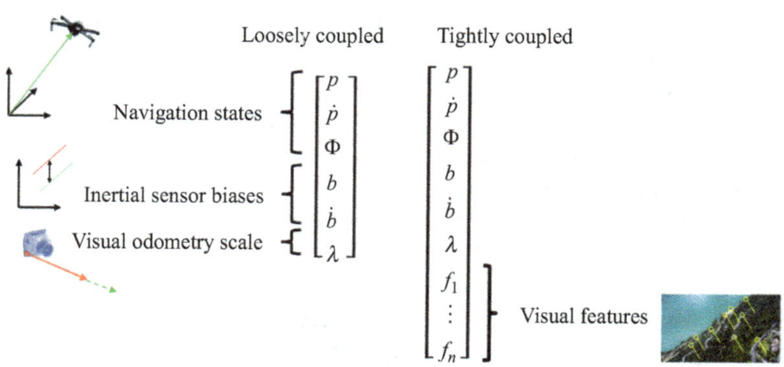

Figure 6.8 Comparison between state-vector dimensions in loosely and tightly coupled systems

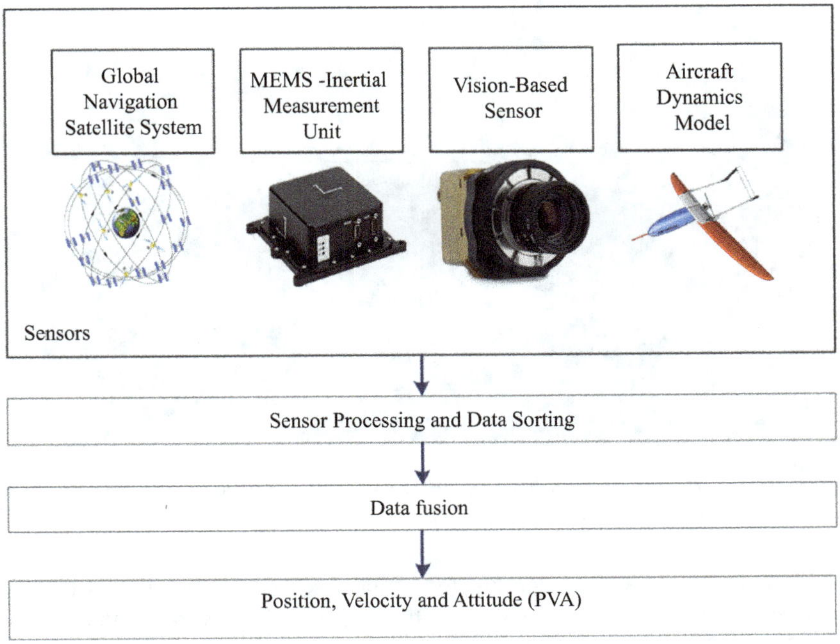

Figure 6.9 Multi-sensor navigation system architecture for a small UAV (adopted from [8])

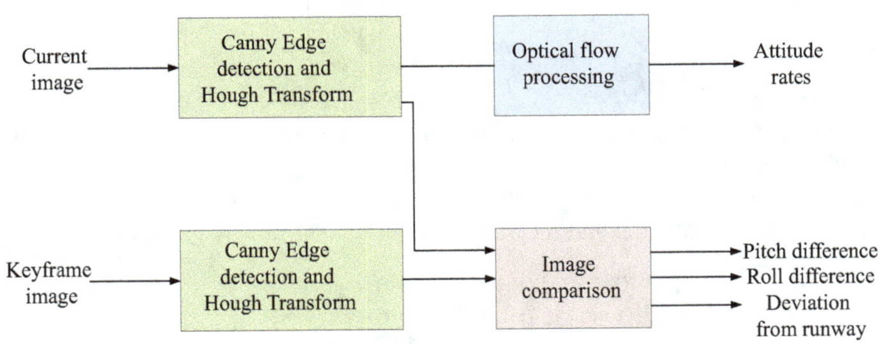

Figure 6.10 Functional architecture for vision-based attitude and lateral positioning [8]

The well-known OF equation is given by:

$$\begin{bmatrix} \dot{u} \\ \dot{v} \end{bmatrix} = \frac{f}{z}\begin{bmatrix} 1 & 0 & \dfrac{-u}{f} \\ 0 & 1 & \dfrac{-v}{f} \end{bmatrix}\begin{bmatrix} \dot{r}_x^w \\ \dot{r}_y^w \\ \dot{r}_z^w \end{bmatrix} + \begin{bmatrix} \dfrac{uv}{f} & -\left(f+\dfrac{u^2}{f}\right) & v \\ \left(f+\dfrac{v^2}{f}\right) & -\dfrac{uv}{f} & -u \end{bmatrix}\begin{bmatrix} \omega_x \\ \omega_y \\ \omega_z \end{bmatrix} \tag{6.15}$$

Figure 6.11 Attitude and relative position estimation based on edge extraction and image comparison [8]

where u and v are the pixel coordinates of an observed 3D point; f is the focal length of the camera and ω and r are the vectors of body rates and position, respectively. If the observed point is on the horizon, z is large, and the first term on the right-hand side of (6.15) tends to zero leading to a simplified form:

$$\begin{bmatrix} \dot{u} \\ \dot{v} \end{bmatrix} = \begin{bmatrix} \frac{uv}{f} & -\left(f+\frac{u^2}{f}\right) & v \\ \left(f+\frac{v^2}{f}\right) & -\frac{uv}{f} & -u \end{bmatrix} \begin{bmatrix} \omega_x \\ \omega_y \\ \omega_z \end{bmatrix} \tag{6.16}$$

The sub-system was used to provide measurement updates to augment an EKF-based estimator.

The measurement model in the EKF incorporating direct measurements of VBS-based roll and pitch and measurements of optical flow is given by:

$$\begin{bmatrix} \phi \\ \theta \\ \dot{u}_1 \\ \dot{v}_1 \\ \dot{u}_2 \\ \dot{v}_2 \\ \cdots \\ \dot{u}_i \\ \dot{v}_i \end{bmatrix} = \begin{bmatrix} 1 & 0 & 0 & 0 & 0 \\ 0 & 1 & 0 & 0 & 0 \\ 0 & 0 & \frac{u_1 v_1}{f} & -\left(f+\frac{u_1^2}{f}\right) & v_1 \\ 0 & 0 & \left(f+\frac{v_1^2}{f}\right) & -\frac{u_1 v_1}{f} & -u_1 \\ 0 & 0 & \frac{u_2 v_2}{f} & -\left(f+\frac{u_2^2}{f}\right) & v_2 \\ 0 & 0 & \left(f+\frac{v_2^2}{f}\right) & -\frac{u_2 v_2}{f} & -u_2 \\ \cdots & \cdots & \cdots & \cdots & \cdots \\ 0 & 0 & \frac{u_i v_i}{f} & -\left(f+\frac{u_i^2}{f}\right) & v_i \\ 0 & 0 & \left(f+\frac{v_i^2}{f}\right) & -\frac{u_i v_i}{f} & -u_i \end{bmatrix} \begin{bmatrix} \phi \\ \theta \\ \omega_x \\ \omega_y \\ \omega_z \end{bmatrix} \tag{6.17}$$

The inherent sub-optimality of loosely coupled systems necessitates a robust sensor fault-detection and measurement rejection capability to avoid filter contamination. The visual sub-system in [8] was implemented in a flexible, modular architecture, as illustrated in Figure 6.12 [9].

Raw sensor data is pre-processed through a Boolean decision logic layer that prioritises and gates sensors depending on their performances in real-time. The architecture leverages knowledge of the vehicle dynamics by implementing an ADM as a virtual sensor to augment the filter and provide additional robustness to individual sensor failures (Figure 6.13).

Future onboard and off-board architectures for computing exploring computational intelligence and cloud computing will allow alternative ways to handle the existing constraints without the need to develop special hardware [1,25–29].

A rigorously developed tightly coupled framework was presented in [11] that uses OF fused with inertial sensor data to perform joint motion and environmental structure estimation. The navigation system architecture is presented in Figure 6.14.

The architecture employs a cascaded architecture with EKF's being employed for the following steps:

1. Optical flow computation: Inertial data and SFM module outputs (velocity and depth) are used to aid the frame-to-frame feature matching that is performed to compute optical flow. More specifically, the solution from the previous time step is used along with inertial measurements to predict the displacement of a given patch (block of pixels) in the subsequent images and to adapt the patch shape. This has the effect of reducing the search area, resulting in lowered computational expense compared to conventional block-matching algorithms (Figure 6.15).

   This method of inertial measurements to minimise the search space and efficiently detect correspondences between features in subsequent frames is a commonly employed inertial-aiding technique [13].
2. Use of inertial measurements and computed OF to estimate the vehicle rotation and translational motion: the observed OF field is a combination of the effects of rotation and translation. The second EKF in the cascade was used to compute the translational component of the OF and the vehicle rotation. The measurements in this instance are the OF vectors from the previous step and the angular rate measurements from the IMU.
3. The translational OF computed from the previous step serves as input to the third and final EKF in the cascade to jointly estimate structure (parameterised as a depth map) and vehicle motion.

It must be pointed out that careful tuning is required to implement this architecture without causing filter instability owing to the cascading of multiple EKFs. From the perspective of implementation, OF measurements can be easily integrated in a loosely coupled sense owing to advancements in manufacturing technology and miniaturisation. The opportunity to integrate low-cost, lightweight visual sensors has led to efforts being made to implement computationally efficient VO routines in low SWaP-C platforms.

VO incrementally updates pose by estimating the frame-to-frame motion on the basis of visual features that are detected and tracked across consecutive images

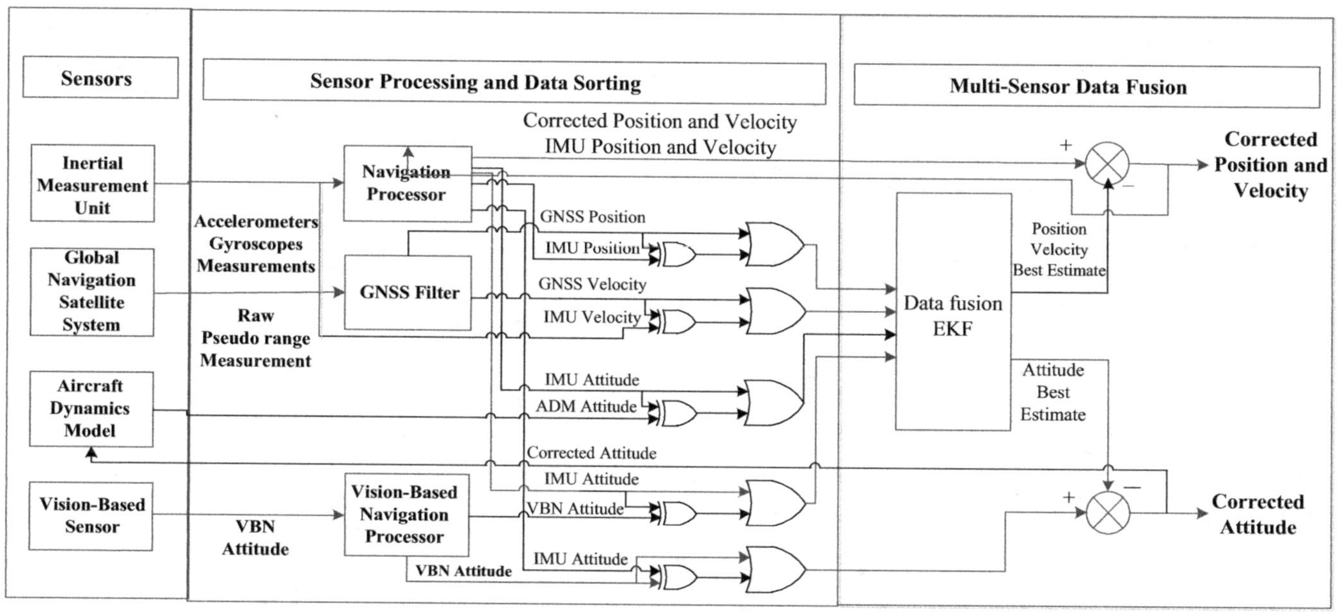

Figure 6.12 Visual-inertial-GNSS-aircraft dynamics (VIGA) architecture [9]

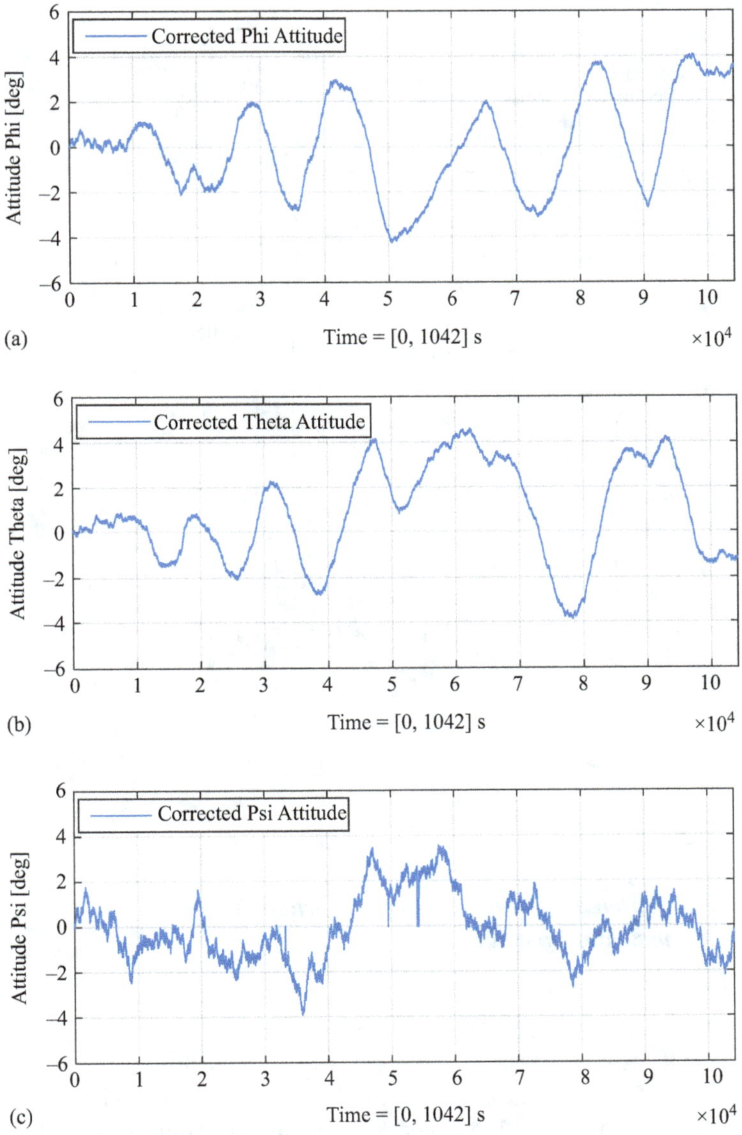

Figure 6.13 Error time series of VIGA attitude estimation during the approach and landing phase: (a) roll error; (b) pitch error; and (c) yaw error

according to their similarity. The primary error source in VO is incorrect data association, mainly because changes in viewpoint, occlusions, image noise, illumination changes and image noise are not accounted for by feature-matching techniques. In order to achieve robust motion estimation, outlier removal is a key step in any VO pipeline and is typically very expensive from a computational point

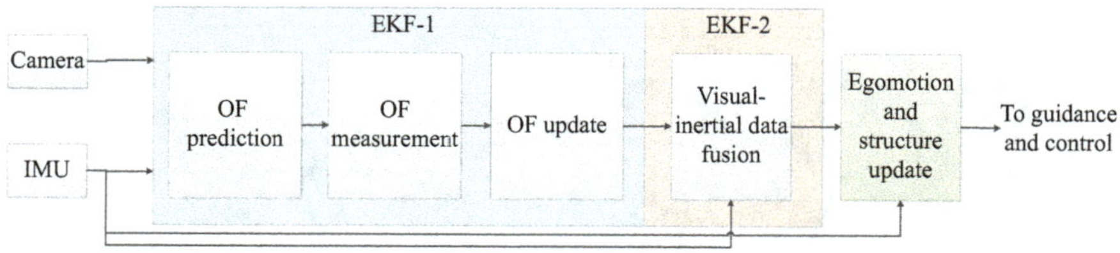

Figure 6.14 Optical-flow-based state estimation architecture [11]

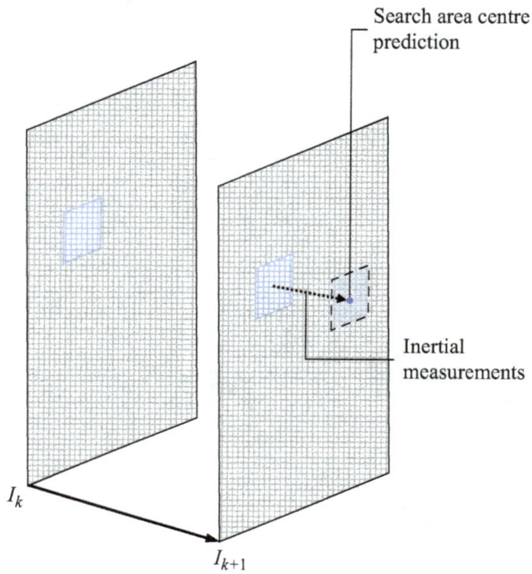

Figure 6.15 Search area prediction using inertial measurements to reduce block matching computation

of view. The standard method for model estimation from a set of data affected by outliers is a RANdom SAmple Consensus (RANSAC). It consists of randomly selecting a set of data points, computing the corresponding model hypothesis and verifying this hypothesis on all the other data points. The solution is the hypothesis with the highest consensus. The number of iterations (N) necessary to guarantee a robust outlier removal is:

$$N = \frac{\log(1 - p)}{\log[1 - (1 - \varepsilon)^s]} \tag{6.18}$$

where s is the number of data points from which the model can be computed, ε is the percentage of outliers in the dataset and p is the probability of success requested. Figure 6.16 shows the number of iterations (N) with respect to the number of points necessary to estimate the model (s). The values are computed for

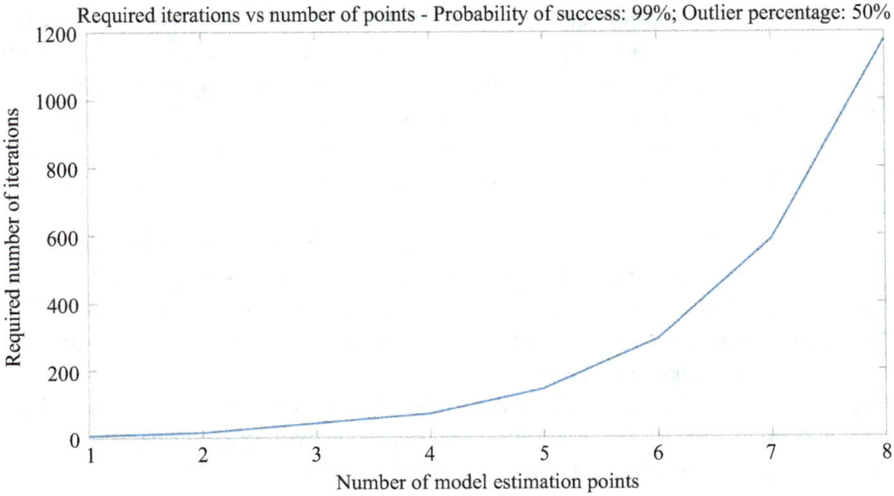

Figure 6.16 Required number of RANSAC iterations against a number of required correspondences to achieve a given probability of success and outlier percentage

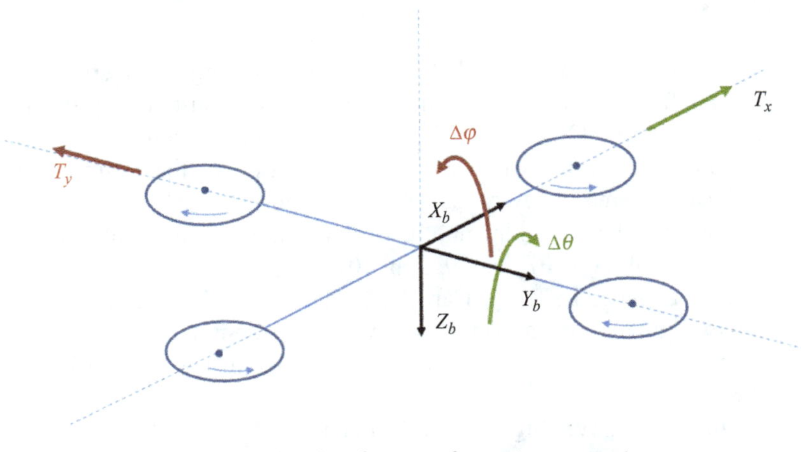

Figure 6.17 Quadcopter dynamic constraints

$p = 0.99$ and $\varepsilon = 0.5$. It is apparent that N is exponential in the number of data points s, making it important for low SWaP-C platforms to reduce the number of correspondences to parameterize the motion model.

In [22], the IMU-estimated attitude is also employed to reduce the model complexity so that fewer feature matches are required to compute a motion hypothesis for the RANSAC. The work exploits the fact that multirotor translational and angular velocity are strongly coupled to attitude (Figure 6.17).

A positive translation in the *x*-axis of the body frame requires a negative pitching motion, i.e. a negative rotation about the *y*-axis, and a negative translation in the *y*-axis requires a negative roll, i.e. a negative rotation about the *x*-axis. The motion constraints were used to discard wrong estimations in a RANSAC-based outlier detection approach. Translational and rotational increments provided by the IMU measurements were used to check the consistency of the motion hypothesis.

This chapter provided an introduction to the integration of visual sensors in multi-sensor systems, along with a review of significant research efforts in the domain. The research area has matured owing to the need to operate UAS in GNSS-denied environments. Computational constraints on low SWaP-C platforms and the requirement of system designers to integrate sensors/sub-systems from different suppliers have prompted development efforts to focus on loosely coupled EKF-based systems mainly. However, this trend may shift with greater efficiency gains in single-board computers in the market. More widespread adoption of the UKF is likely, owing to its greater accuracy for highly dynamic flight, and its simplicity of implementation as compared to the EKF. System robustness remains a primary point of ongoing research in order to protect against drop-outs and/or filter contamination due to glare, occlusion and pattern repetition.

References

[1] V. V. Estrela, O. Saotome, H. J. Loschi, *et al.* Emergency response cyber-physical framework for landslide avoidance with sustainable electronics. Technologies, 6, 42, 2018. doi:10.3390/technologies6020042.

[2] S. J. Julier and J. K. Uhlmann, "New extension of the Kalman filter to nonlinear systems," in AeroSense'97, 1997, pp. 182–193.

[3] S. J. Julier and J. K. Uhlmann, "Unscented filtering and nonlinear estimation," Proc. of the IEEE, vol. 92, pp. 401–422, 2004.

[4] J. G. C. Lozano, L. R. G. Carrillo, A. Dzul, and R. Lozano, "Spherical simplex sigma-point Kalman filters: A comparison in the inertial navigation of a terrestrial vehicle," in Proc. of 2008 American Control Conference, 2008, pp. 3536–3541.

[5] R. Van Der Merwe, "Sigma-point Kalman filters for probabilistic inference in dynamic state-space models," Oregon Health & Science University, 2004.

[6] S. Challa, "Fundamentals of object tracking." Cambridge University Press, 2011.

[7] S.-M. Oh, "Nonlinear estimation for vision-based air-to-air tracking," Georgia Institute of Technology, 2007.

[8] R. Sabatini, C. Bartel, A. Kaharkar, T. Shaid, and S. Ramasamy, "Navigation and guidance system architectures for small unmanned aircraft applications," Int'l Journal of Mechanical and Mechatronics Engineering, vol. 8, pp. 733–752, 2014.

[9] F. Cappello, S. Bijjahalli, S. Ramasamy, and R. Sabatini, "Aircraft dynamics model augmentation for RPAS navigation and guidance," Journal of Intelligent & Robotic Systems, vol. 91, pp. 709–723, 2018.

[10] M. B. Rhudy, Y. Gu, H. Chao, and J. N. Gross, "Unmanned aerial vehicle navigation using wide-field optical flow and inertial sensors," Journal of Robotics, vol. 2015, p. 1, 2015.

[11] F. Kendoul, I. Fantoni, and K. Nonami, "Optic flow-based vision system for autonomous 3D localization and control of small aerial vehicles," Robotics and Autonomous Systems, vol. 57, pp. 591–602, 2009.

[12] W. Ding, J. Wang, S. Han, *et al.*, "Adding optical flow into the GPS/INS integration for UAV navigation," In Proc. of Int'l Global Navigation Satellite System Society Symposium, 2009, pp. 1–13.

[13] J. R. M. Veth, "Fusion of low-cost imagining and inertial sensors for navigation," in International Technical Meeting of the Satellite Division of The Institute of Navigation, 2006, pp. pp. 1093–1103.

[14] X. Song, L. D. Seneviratne, and K. Althoefer, "A Kalman filter-integrated optical flow method for velocity sensing of mobile robots," IEEE/ASME Transactions on Mechatronics, vol. 16, pp. 551–563, 2011.

[15] J. W. Langelaan, "State estimation for autonomous flight in cluttered environments," J. of Guidance Control and Dynamics, vol. 30, p. 1414, 2007.

[16] A. I. Mourikis and S. I. Roumeliotis, "A multi-state constraint Kalman filter for vision-aided inertial navigation," In Proc. 2007 IEEE Int'l Conference on Robotics and Automation, 2007, pp. 3565–3572.

[17] D. D. Diel, P. DeBitetto, and S. Teller, "Epipolar constraints for vision-aided inertial navigation," in Seventh IEEE Workshops on Application of Computer Vision, 2005. WACV/MOTIONS'05, Volume 1, 2005, pp. 221–228.

[18] M. Hwangbo and T. Kanade, "Visual-inertial UAV attitude estimation using urban scene regularities," in Proceedings of 2011 IEEE International Conference on Robotics and Automation (ICRA), 2011, pp. 2451–2458.

[19] M. E. Antone and S. Teller, "Automatic recovery of relative camera rotations for urban scenes," In Proc. IEEE Conference on Computer Vision and Pattern Recognition 2000, 2000, pp. 282–289.

[20] C. Demonceaux, P. Vasseur, and C. Pégard, "UAV attitude computation by omnidirectional vision in urban environment," In Proc. 2007 IEEE International Conference on Robotics and Automation, 2007, pp. 2017–2022.

[21] S. Weiss, M. W. Achtelik, S. Lynen, *et al.*, "Monocular vision for long-term micro aerial vehicle state estimation: a compendium," Journal of Field Robotics, vol. 30, pp. 803–831, 2013.

[22] C. Troiani, A. Martinelli, C. Laugier, and D. Scaramuzza, "Low computational-complexity algorithms for vision-aided inertial navigation of micro aerial vehicles," Robotics and Automation Systems, vol. 69, pp. 80–97, 2015/07/01/2015.

[23] M. Warren, P. Corke, and B. Upcroft, "Long-range stereo visual odometry for extended altitude flight of unmanned aerial vehicles," International Journal of Robotics Research, vol. 35, pp. 381–403, 2016.

[24] Y. Song, S. Nuske, and S. Scherer, "A multi-sensor fusion MAV state estimation from long-range stereo, IMU, GPS and barometric sensors," Sensors, vol. 17, p. 11, 2016.

[25] N. Razmjooy, B. S. Mousavi, M. Khalilpour, and H. Hosseini. Automatic selection and fusion of color spaces for image thresholding. Signal, Image and Video Processing, vol. 8(4), pp. 603–614, 2014.

[26] D. J. Hemanth, and V. V. Estrela. Deep learning for image processing applications, Adv in Parallel Computing Series, vol. 31, IOS Press, 2017. ISBN 978-1-61499-821-1 (print), ISBN 978-1-61499-822-8 (online).

[27] V. V. Estrela, H. A. Magalhaes, and O. Saotome. "Total variation applications in computer vision." In Handbook of Research on Emerging Perspectives in Intelligent Pattern Recognition, Analysis, and Image Processing, pp. 41–64. IGI Global, 2016.

[28] B. Mousavi, F. Somayeh, N. Razmjooy, and F. Soleymani. Semantic image classification by genetic algorithm using optimised fuzzy system based on Zernike moments. Signal Image and Video Processing, vol. 8, no. 5, pp. 831–842, 2014.

[29] N. Razmjooy, V. V. Estrela, and H. J. Loschi. "A survey of potatoes image segmentation based on machine vision." In Applications of Image Processing and Soft Computing Systems in Agriculture, pp. 1–38. IGI Global, 2019.

[30] M. A. de Jesus, and V. V. Estrela. Optical flow estimation using total least squares variants. Oriental Journal of Computer Science and Technology, 2017;10(3):563–579. http://dx.doi.org/10.13005/ojcst/10.03.03

[31] V. V. Estrela, and A. M. Coelho. State-of-the-art motion estimation in the context of 3D TV. In R. Farrugia, and C. Debono (Eds.), Multimedia Networking and Coding (pp. 148–173). Hershey, PA: IGI Global, 2013. http://dx.doi.org/10.4018/978-1-4666-2660-7.ch006

[32] H. R. Marins, and V. V. Estrela. On the use of motion vectors for 2D and 3D error concealment in H.264/AVC Video. In Feature Detectors and Motion Detection in Video Processing (pp. 164–186). Hershey, PA: IGI Global, 2017. http://dx.doi.org/10.4018/978-1-5225-1025-3.ch008

Chapter 7

Vision-based UAV pose estimation

Paulo Silva Filho[1], Elcio Hideiti Shiguemori[2], Osamu Saotome[1], and Jairo Panetta[1]

As the use of unmanned aerial vehicles increased, studies regarding their autonomous flight became an academic field of great interest for researchers. Until recently, most studies based their developments using an inertial measurement unit (IMU) and a global navigation satellite system (GNSS) as the main sensors to calculate and estimate the UAVs pose. These sensors, on the other hand, have several limitations, which can affect the navigation, therefore, the fully autonomous aspect of the system. Images captured during flight, computer vision algorithms, and photogrammetry concepts have become a core source of data to estimate the UAVs pose in real-time, therefore, composing new alternative or redundant navigation systems. Several algorithms have been proposed in the scientific community, each one working better in specific situations and using different kinds of imaging sensors (active and passive sensors). This chapter describes the main visual-based pose estimation algorithms and discusses where they best apply and when each fails. Fresh results depict the development of new strategies that will overcome the remaining challenges of this research field.

7.1 Introduction

Unmanned aerial vehicles (UAVs) became a highly popular technology in recent years. Their applicability has expanded to several civil applications such as monitoring [1], agriculture [2], search and rescue missions [3], remote sensing [4], photogrammetry [5], construction inspections [6], entertainment [7], and many others. It became then a significant and strategic technology due to all those applications, drawing the attention of the scientific community to develop studies on how to make these platforms even more stable, intelligent, and reliable and mostly autonomous.

In terms of stability and reliability, the navigation system (NS) of these vehicles plays an important part, and it is essential in order to develop a fully autonomous UAV. A good NS must provide to the control system of the UAV precise

[1]DCTA-ITA-IEEA, Sao Jose dos Campos, SP, Brazil
[2]IEAv – Institute of Advanced Studies, S. J. Campos, SP, Brazil

state information and interpret this information without human intervention [8]. It has to estimate the UAVs state (position, velocity, attitude, airspeed, angle of attack, sideslip angle, and rotation – pitch, roll, and yaw – rates) and give this information to a control and guidance system, which will allow it to manoeuvre in a way consistent with its planned route. Obtaining this state information is not a simple task. One can deduce then that the basis for the autonomous navigation is the state estimation of the aircraft.

Sensors embedded in the aircraft are used to estimate its state for the control system. Nowadays, the main sensors used are gyrometers and accelerometers, which compose the inertial measurement unit (IMU) or inertial navigation system (INS) and the global navigation satellite system (GNSS). It is possible to operate a UAV autonomously in most situations combining these sensors. There are, though, several situations in which those sensors fail or are unable to determine an accurate state. Because of the situations in which the IMU + GNSS does not work well, new and different navigation systems (NSs) are needed. The most promising techniques rely on computer vision (CV).

This chapter presents and discusses the central CV NSs being developed. It will explain the main drawbacks of the IMU + GNSS navigation systems and the most challenging aspects of any vision-based system. After that, it will discuss the different vision models.

7.2 INS–GNSS drawbacks

INS and GNSS revolutionised the development of autopilots for many types of vehicles and artefacts that have to follow a precise trajectory, with little or no human interaction. This is the basis for any autonomous system. It is essential first to understand how the INS and GNSS work in order to be aware of when these sensors may fail and how to overcome those drawbacks.

7.2.1 Inertial navigation systems

The INS is a self-contained navigation technique that uses the measurements provided by sensors contained in an IMU to estimate the position and velocity of an object relative to a known starting point, orientation, as well as velocity [9]. An IMU combines accelerometers (motion sensors), gyroscopes (rotation sensors), and, sometimes, magnetometers to estimate the linear acceleration, the angular rate, and the reference heading of the UAV, respectively. These are some of the state variables the NS must provide. The INS then uses those state variables to estimate the linear velocity, attitude, and position from the IMU readings, using an integration method in addition to the initial reference value for each (initial position, initial attitude, and initial velocity). The great advantage of an INS is that it does not require any other external signal to estimate the position, orientation, or velocity once it has been initialised, so it is immune to any deception or jamming.

The main issue with the INS is that over time, the estimation accuracy degrades, affecting the safety of the flight and also the accomplishment of the desired route [10].

Figure 7.1 INS flight route compared to a planned route

This loss in position accuracy is known as the INS drift. Each INS has a different degrading rate. Typically, the smaller the IMU in the INS, the more significant the decrease in accuracy [10]. The IMUs used on UAVs usually lose precision in a matter of seconds, because they have to be small and light. Figure 7.1 compares the flight of an aircraft with only the IMU with a planned route.

7.2.2 Global navigation satellites systems

The INS is still being used as a source of state information for autopilot systems in most human-crewed vehicles. The advent of more unmanned vehicles and applications that demanded even more precise state information demanded other navigation systems or new INS technologies that would reduce or correct its accumulated errors. If there were a system that could estimate precisely the position and velocity of a vehicle in some instant, this information could be used as the new initial position and velocity for the INS, correcting its accumulated error, therefore maintaining a more precise INS estimation through the pursued trajectory.

GNSS is the technology that came to tackle this aspect. Based on the Doppler effect, it was observed that satellites could be tracked from the ground, by measuring the frequency of the radio signals they emitted [11]. With several satellite positions, it is also possible to obtain the position of the radio receiver on the ground (Figure 7.2).

The modern GNSSs are composed of a network of satellites on the lower Earth orbit, with periods between 10 and 15 h. Nowadays, there are several different GNSS from different regions and countries. Some examples are the global position system (GPS) from EUA, which is the most popular; Galileo from the European Union; the global navigation satellite system (GLONASS) from Russia; and Beidou from China [10]. Their operation is quite similar, differing only in the orbit periods, the number of

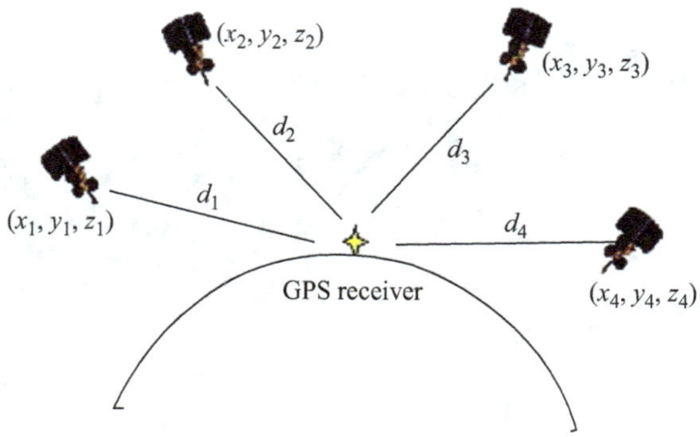

Figure 7.2 GPS triangulation scheme. Source: [12]

satellites in the constellation, and the number of orbits. The GPS, for example, is composed of 24 satellites divided into 6 orbits, with 4 in each orbit [13]. These different orbits cause the geometry between satellites to change constantly.

The GNSS estimated position is the solution of a triangulation problem as in (7.1), obtaining the position (x, y, z) and the GNSS clock offset time T_b. The offset time conforms to the atomic clock embedded in the satellite, and it is used to maintain synchronisation between all of the satellites in the network. (x_i, y_i, z_i) with $i = 1, 2, 3, 4$ are the satellites exact positions, and c is the speed of light.

From the system, one can infer that it needs at least four satellites in a good geometry to estimate the position. Each satellite emits a radio signal, with a specific frequency that is received by a sensor in the UAV. Nowadays, the satellite broadcasts a radio signal that contains orbital data (from which the position of the satellite can be calculated) and the precise time the signal was transmitted [11]. The receiver (an antenna) must be on a direct view of all four satellites. From this signal, the distances d_i between the sensor and each satellite are calculated, and, from the system equations, the UAV position is estimated:

$$\begin{cases} \sqrt{(x-x_1)^2+(y-y_1)^2+(z-z_1)^2}+cT_b=d_1 \\ \sqrt{(x-x_2)^2+(y-y_2)^2+(z-z_2)^2}+cT_b=d_2 \\ \sqrt{(x-x_3)^2+(y-y_3)^2+(z-z_3)^2}+cT_b=d_3 \\ \sqrt{(x-x_4)^2+(y-y_4)^2+(z-z_4)^2}+cT_b=d_4 \end{cases} \tag{7.1}$$

The dependency on external signals for position estimation is the main vulnerability of the GNSS. Accuracy depends on the number of satellites in a clear, direct view, and on the geometry formed by them, poor geometry or a small amount of satellites means an imprecise position. For example, in environments such as indoors or in dense urban areas, the GNSS signal will be weak and may also be blocked or reflected by the buildings, causing signal multipath [10]. The UAV will operate in a highly hazardous area, running the risk of getting lost or colliding with obstacles [14]. This is one of the main concerns for UAV operations in urban areas.

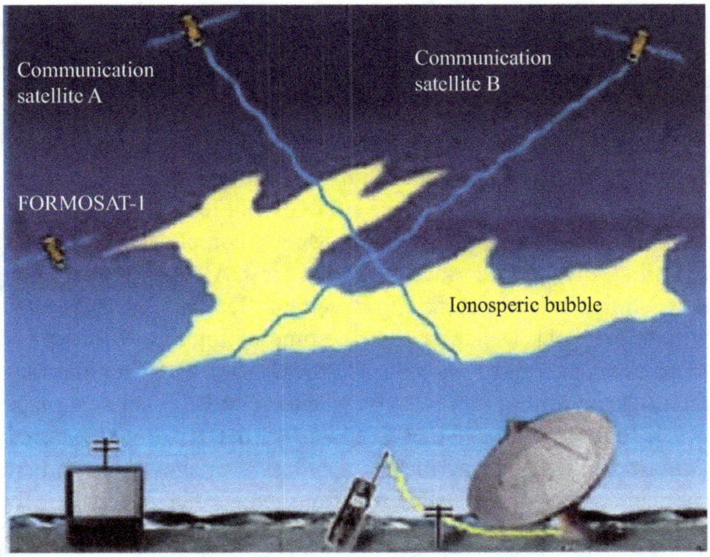

Figure 7.3 Equatorial plasma bubbles blocking radio signals. Source: [16]

Besides the geometry and multipath already mentioned, if for any reason, the sensor does not receive the satellite signal, it will not be able to estimate the position. Even in the clear satellite view, the GNSS signal can be blocked. A natural signal blocker is the equatorial plasma bubbles that are formed in the ionosphere [15]. They tend to delay radio waves and degrade the performance of GNSS readings. Figure 7.3 shows how the plasma bubbles block signals.

GNSS signals can be interfered with by malicious attacks also, which will degrade the signals readings, affecting the navigation like jamming and spoofing. Jamming is the process of blocking or interfering with a radio signal by decreasing the signal-to-noise ratio so that the signal is completely blocked. For the UAV GNSS signal receiver, it seems that there are no GNSS satellites in the area affected by the jamming device. Spoofing, on the other hand, is similar to jamming, but instead of blocking the signal, it deceives the receiver, by falsifying the data, so that the position estimated is completely wrong.

Attacks of such nature have already happened in recent years as the highjack of a Lockheed RQ-170 drone [17]. It is expected that the more popular UAVs become, the more hacking attacks as such can happen. An NS depending only on external signals is not suitable when safety, stability, and reliability are wanted.

7.3 Visual navigation: A viable alternative

Images have always been used as a source for information, but they were difficult to be processed by a computer until the development of more efficient and reliable CV algorithms. Since that, extracting motion and pose information from images became an important and viable strategy for new NSs. Images captured during flight can be compared with previous ground information to estimate the UAV pose [18], which can be used to eliminate the INS error, the same way GNSS readings do.

Images as an alternative for the GNSS, or also as a complementary NS, has several advantages: (i) the system does not depend on an external signal [19] and (ii) the technology involved. CV algorithms use mostly embedded passive sensors, which are cheaper than the maintenance of a GNSS [20].

On the other hand, several challenging aspects complicate the development of a visual NS. First, the particularities on which the UAV may fly are quite diverse and complex, so that it is impossible to develop a flexible and unique solution for all flight conditions [21]. Depending on the time of flight, the weather condition, and the scene, the sensor used to obtain the image for the navigation must change, and, therefore, the algorithm must also be adapted [19,21]. Table 7.1 maps some of the variables a flight might have that would affect the visual navigation system.

Each environmental set may impose a different image sensor to be used. For example, optical sensors in the visible spectrum may not be adequate if the aircraft route is over a lake or if the flight is at night. Other sensors with responses on different bands of the electromagnetic spectrum or active sensors would be more suitable for that. Different images would be produced, and, then, different approaches would be necessary for the navigation system.

Any solution proposed for the autonomous navigation using images must take into consideration how the image is acquired in each sensor it is going to use: the velocity of the sensor's shutter, the luminosity of the scene, the environment of the scene, the scale, rotation and projective factors, and many other physical aspects of the sensor. Besides that, aspects related to the algorithm itself, its complexity, the onboard computer that will process it, the payload of the aircraft, for example, also make restrictions on which method will be most effective [8].

Most visual navigation systems developed until now tries to determine the specific condition they will operate in order to choose which sensor they are going to use. They usually use passive sensors, such as RGB, thermal, multispectral, or hyperspectral cameras, mostly because of the more accessible sensors of these types with size, weight, and energy consumption compatible with the UAVs' limitations.

Passive image sensors are electronic devices that capture and convert the electromagnetic radiation reflected by the objects in the scene (or emitted by the objects, in the case of the thermal radiation) into electronic signals, which will form the image. Their pixels register the amount of electromagnetic radiation that falls on them, converting this amount into a corresponding number of electrons [22].

Table 7.1 Various flight conditions

Condition class	**Conditions**			
Period of flight	Morning	Afternoon	Night	
Weather	Cloudy/foggy	Sunny	Rainy	Snowy
Scene	Water bodies	Urban areas	Forest	Rural areas

The main difference between passive sensors is the electromagnetic band they capture.

The visual or RGB sensor works on radiation with wavelengths from 390 to 700 nm. The two main technologies for optic sensors are the charge-coupled device [23] and the complementary metal-oxide semiconductor [24,25]. Most works develop the visual NS using these sensors [20,26–32] since they are the most common sensor and capture exactly what one sees.

Multispectral and hyperspectral sensors are more sensitive to a broader range of wavelengths than the visible spectre. They are used more when details of the scene are needed, such as materials, types of plants, and other more specific information. The only difference between those sensors is that hyperspectral sensors capture contiguous spectral bands, covering a wide range of wavelengths in a fine resolution, and multispectral sensors, on the other hand, measure spaced spectral bands [33].

A thermal imaging camera is a sensor that forms an image using infra-red radiation, similar to a common visual sensor, but it is sensible to wavelengths as long as 14,000 nm [34]. An environment where these sensors would be used for the image NS would be in flights at night, for example, when optical (visual, multispectral, and hyperspectral) sensors would not respond properly. In [35], thermal information is used for the NS to estimate the position of the UAV. The thermal image captured during flight is compared with optical images previously stored in the UAV.

It is known, though, that not all flight conditions can be tackled with passive sensors. As mentioned before, at night, most of the passive sensors do not work correctly. That is because these sensors only capture the radiation that was reflected by the objects. If one does not have a source of radiation to illuminate the objects in the scene and then reflect that radiation to the sensor, the image will not be formed. The only exception for passive sensors is the thermal sensors because all objects are emitting heat, so they all are a source for this specific spectral band. However, for example, in a cloudy or rainy weather condition, even the thermal sensor would not work suitably, because the clouds would pose as a barrier for the image to form. Also, if the flight were over a lake or the sea, all the passive sensors would respond with a similar and plain image, which could not be used to estimate the aircraft's pose.

Active sensors are a possible solution in order to gather visual information inflight conditions when there is no source of radiation. They are devices with two complementary parts: a transmitter that sends out a specific signal, wavelength, or electrons to be reflected by a target; and also a data-gathering sensor that will register the signal reflected by the target. In order words, they have both the radiation source and the sensor for that specific radiation. The two most used sensors in nowadays investigation work in visual navigation are LIDAR and Synthetic aperture radar (SAR).

LiDAR [36] is an acronym for light detection and ranging. It refers to a remote sensing technology that emits intense, focussed beams of light and measures the time it takes for the reflections to be detected by the sensor. This information is used to compute ranges, or distances, to objects. The three-dimensional coordinates of the target objects and the intensity of the returned pulse are computed from the time difference between the laser pulse being emitted and returned, the angle at which the pulse was emitted and the absolute location of the sensor on or above the surface of the Earth [36]. Works have used LiDAR for navigation and remote sensing [37] when the UAV is flying over shallow waters [38,39] and forests [40].

SAR [41] is a form of radar which is used to create images (2D or 3D images) of a scene using the motion of the radar antenna over a targeted region to provide finer spatial resolution than is possible with conventional beam-scanning radars. To create a SAR image, successive pulses of radio waves are transmitted to illuminate a target scene, and the echo of each pulse is received and recorded. The pulses are transmitted, and the echoes received using a single beamforming antenna, with wavelengths of a metre down to several millimetres. As the SAR device onboard the aircraft moves, the antenna location relative to the target changes with time. Signal processing of the successively recorded radar echoes allows the combination of the recordings from these multiple antenna positions and allows the creation of higher resolution images than would otherwise be possible with a given physical antenna. The multiple antenna positions are what is called a synthetic antenna aperture. In [42], the algorithm proposed uses SAR images for visual navigation, for example.

7.4 Visual navigation strategies

Due to the variety of flight conditions already discussed, several visual NSs have been proposed. All systems, though, can be classified into three main core strategies, without loss of generality: visual odometry (VO), template matching, and landmark recognition. Each strategy demands certain conditions to be used, which can pose as an advantage or disadvantage, depending on what the flight is intended.

In this section, each strategy will be described in a general manner, with several works as examples. We will focus mostly on examples with optical sensors since they are used in most works. This simplification in the explanation does not lose generality also, because nowadays one of the leading research lines in visual navigation systems is adapting those same strategies for different types of sensors and images. Nevertheless, before exploring each strategy, first, we will discuss how pose information can be extracted from images, which is the core research on photogrammetry.

7.4.1 Photogrammetry: Extracting pose information from images

One of the areas in photogrammetry is focused on extracting exact metrics from the image and other complementary data sensors, determining relative positions between points in the image, distances, angles, areas, and sizes, among other proprieties. It also produces all sorts of metric maps, such as orthophotos, mosaics, and altimetry maps [43]. These products and information are obtained by associating the physical world with the image. Figure 7.4 shows a Cartesian system [44]. At the same time, the physical world is associated with a Cartesian space, using the latitude, longitude, and altitude, or similar geodetic reference systems.

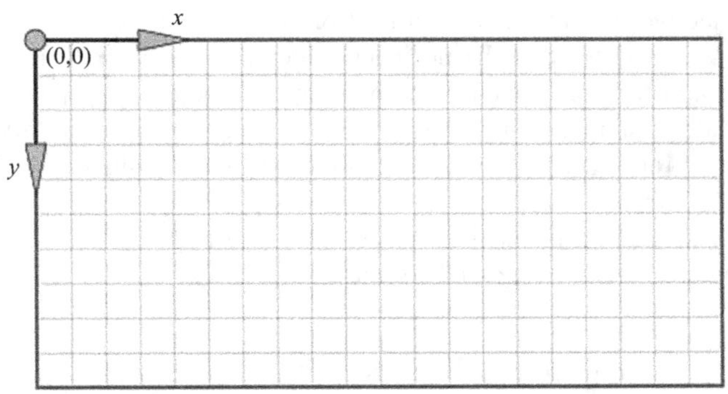

Figure 7.4 Cartesian system for a digital image

Georeferencing [45] associates the Cartesian image coordinates to the object space of the physical world. A suitable representation of the object space in the reference system, then, is needed, and a mathematical model must be generated to associate both reference systems.

There are different kinds of functions that can map one reference system to another. These functions are called geometric transformation. They can map a 3D system with a 2D one, or to another 3D one, in case a digital surface model (DSM) is taken into consideration [46], or they can associate two different 2D systems of two different images of the same scene. The most common geometric transformations are the affine general transformation and projective transformation.

The general affine transform is a first-degree polynomial transformation. The affine transformations assume that the relation between the two Cartesian systems is based on a translation (Δx, Δy), a rotation (β), and a scale factor on both axes (λ_x, λ_y). It assumes that both systems are contained in parallel planes [47]:

$$\begin{Bmatrix} x' \\ y' \end{Bmatrix} = \begin{Bmatrix} \cos\beta & -\sin\beta \\ \sin\beta & \cos\beta \end{Bmatrix} \begin{Bmatrix} x\lambda_x \\ y\lambda_y \end{Bmatrix} + \begin{Bmatrix} \Delta x \\ \Delta y \end{Bmatrix} \tag{7.2}$$

The equation can be rewritten in the following linear model as in (7.3), presented by [48]:

$$\begin{cases} x'(a_0, a_1, a_2, b_0, b_1, b_2) = a_0 + a_1 x + a_2 y \\ y'(a_0, a_1, a_2, b_0, b_1, b_2) = b_0 + b_1 x + b_2 y \end{cases} \tag{7.3}$$

From the linearised equation, it is possible to observe that it is necessary at least six equations to obtain the parameters for the affine transformation. Three pairs of corresponding points at least are necessary then. These points can also be called control points for the transformation. Each pair of points gives two equations for the linear equation system to be solved, and the answer to this system are the parameters of the affine transformation.

The projective transformation comes from the collinearity principle in photogrammetry, which states that the perspective centre of camera C, point P on the object space and the correspondent point p on the photography plane are on the same line, building a point-to-point relation between the image space and the object space [46]. The relation between the coordinates can be obtained from Figure 7.5, as shown in (7.11).

Consider M the matrix for the rotation of a rigid body, where ω, φ, and κ are the Euler angles (Figure 7.6), and s a scale factor,

$$M = \begin{bmatrix} m_{11} & m_{12} & m_{13} \\ m_{21} & m_{22} & m_{23} \\ m_{31} & m_{32} & m_{33} \end{bmatrix} \tag{7.4}$$

and

$$\begin{cases} m_{11} = \cos\omega \cos\kappa \\ m_{12} = \cos\omega \sin\kappa + \sin\omega \sin\varphi \cos\kappa \\ m_{13} = \sin\omega \sin\kappa - \cos\omega \sin\varphi \cos\kappa \\ m_{21} = -\cos\varphi \sin\kappa \\ m_{22} = \cos\omega \cos\kappa - \sin\omega \sin\varphi \sin\kappa \, . \\ m_{23} = \sin\omega \cos\kappa + \cos\omega \sin\varphi \sin\kappa \\ m_{31} = \sin\varphi \\ m_{32} = -\sin\omega \cos\varphi \\ m_{33} = \cos\omega \cos\varphi \end{cases} \tag{7.5}$$

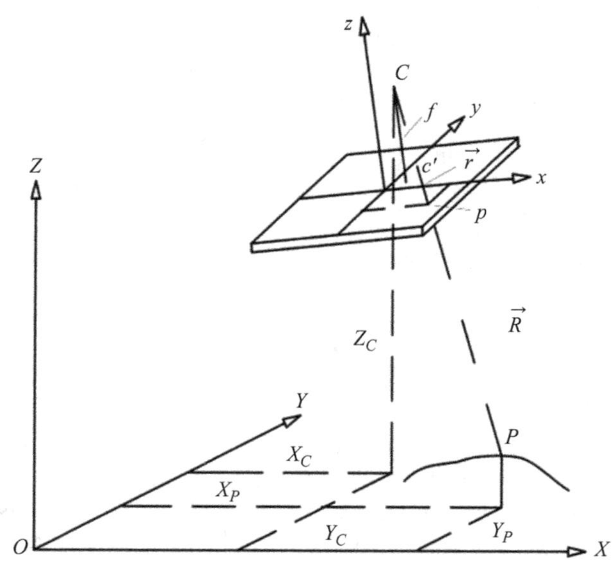

Figure 7.5 Collinearity equations scheme

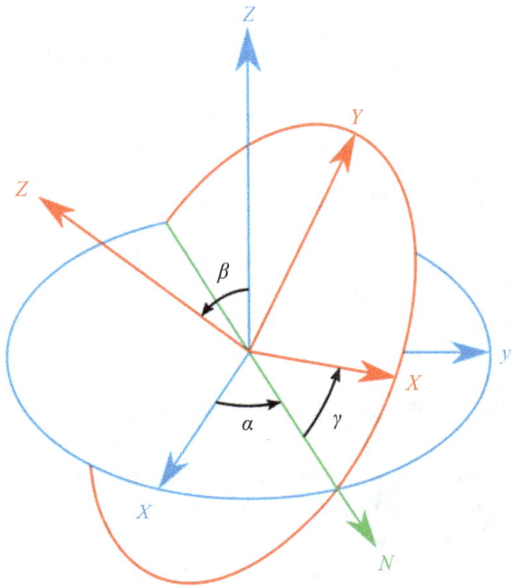

Figure 7.6 Euler angles scheme

From Figure 7.5, we also have that:

$$\vec{r} = \begin{Bmatrix} x_p - x_c \\ y_p - y_c \\ 0 - f \end{Bmatrix} \tag{7.6}$$

and

$$\vec{R} = \begin{Bmatrix} X_P - X_C \\ Y_P - Y_C \\ Z_P - Z_C \end{Bmatrix} \tag{7.7}$$

The collinearity condition extracts the following relation between $\vec{r}$ and $\vec{R}$, since P, p, and C must be on the same line:

$$\vec{R} = \frac{1}{s} M^T \vec{r}. \tag{7.8}$$

When we substitute the values in both equations:

$$\begin{cases} x_p - x_c = s[m_{11}(X_P - X_C) + m_{12}(Y_P - Y_C) + m_{13}(Z_P - Z_C)] \\ y_p - y_c = s[m_{21}(X_P - X_C) + m_{22}(Y_P - Y_C) + m_{23}(Z_P - Z_C)] \\ -f = s[m_{31}(X_P - X_C) + m_{32}(Y_P - Y_C) + m_{33}(Z_P - Z_C)] \end{cases} \tag{7.9}$$

$$\begin{cases} x_p - x_c = -f\dfrac{m_{11}(X_P - X_C) + m_{12}(Y_P - Y_C) + m_{13}(Z_P - Z_C)}{m_{31}(X_P - X_C) + m_{32}(Y_P - Y_C) + m_{33}(Z_P - Z_C)} \\ y_p - y_c = -f\dfrac{m_{21}(X_P - X_C) + m_{22}(Y_P - Y_C) + m_{23}(Z_P - Z_C)}{m_{31}(X_P - X_C) + m_{32}(Y_P - Y_C) + m_{33}(Z_P - Z_C)} \end{cases} \tag{7.10}$$

Rewriting on a matrix format:

$$\begin{cases} x' = -f\dfrac{M_1\bar{X}}{M_3\bar{X}} \\ y' = -f\dfrac{M_2\bar{X}}{M_3\bar{X}} \end{cases}, \tag{7.11}$$

where M_1, M_2, and M_3 are the rows of matrix M for the rotation of a rigid body, and $\bar{X} = (X_P - X - C, Y_P - Y_C, Z_P - Z_C)^T$.

The linear format of the equation, considering the planar projective equation ($Z_P - Z_C = 0$), the projective transformation is as in the following equation:

$$\begin{cases} x'(a_0, a_1, a_2, b_0, b_1, b_2, c_0, c_1) = xa_0 + ya_1 + a_2 - x'xc_0 - x'yc_1 \\ y'(a_0, a_1, a_2, b_0, b_1, b_2, c_0, c_1) = xb_0 + yb_1 + b_2 - y'xc_0 - y'yc_2 \end{cases} \tag{7.12}$$

To estimate the parameters of the projective transformation, it is necessary at least four pairs of points, which will result in a possible and determined linear system of equations. Observing the equations, we can see that the linear transformation is a particular and simplified model for the perspective transformation, for example.

It is possible to obtain the UAV pose using these equations and associations between images taken during the flight, or previously acquired images of the flight route with precise information. It is possible to obtain the UAV pose. Even though all strategies use photogrammetry aspects to compute the pose, each one of them has a different approach to how this information will be obtained.

7.4.2 Template matching

In this approach, the flight area is considered as known, and, for example, a georeferenced orthorectified image of that area is stored in the UAV visual NS. Any image captured by the aircraft's embedded sensor during the flight will be contained in that stored image. It is possible then to estimate the UAV position by finding where the captured image is in the previous georeferenced image. This short description is the core of the template matching strategy.

From that description, we can already infer the primary condition for the template matching: it needs previous information about the area of the flight, which can be a recent satellite image or a recent mosaic of images from the flight area. That image must be orthorectified and georeferenced, which means that it needs to consider the internal and external distortions of the image, and each pixel is correlated with a latitude, longitude, and altitude if a DSM is also available.

The challenge in template matching then is how to find the captured image in the stored image, since they can be quite different from each other. They are not necessarily from the same sensor, and it is not guaranteed that they will have the same orientation, scale, brightness level, and many other characteristics [20]. This poses a critical image-processing problem to be tackled. Different methods have been proposed for this problem, and most of them are described in image registration theory. Image registration has been an active research field for many years, and it is the process of overlaying two or more images of the same scene taken at different times, from different viewpoints and by different sensors and geometrically aligning these two images. In other words, we have to find a geometric transformation that the images may have the same viewpoint and similar characteristics.

One such method is the use of feature points, such as SIFT, SURF, LATCH, ORB, and AKAZE [49–53]. These techniques work in three main steps: the selection of points in each image, description of the point's neighbourhood, and matching of corresponding points in both images. Using the corresponding points, it is possible to state a projective transformation and then use it to estimate the UAV pose, as the stored image is a georeferenced image.

Considering F: $T(x, y)|\!\to G(\text{lat, long})$ the georeferencing relation from stored image T with the object space G, and K: $T(x, y)|\!\to Q(X, Y)$ the geometric transformation that maps the stored image T in the captured image Q, it is possible to build the georeferencing transformation H, from the query image Q, in which:

$$H : Q(X, Y) \to T(x, y)K^{-1} \to G(\text{lat}\, F, \text{long}) \tag{7.13}$$

That is, we can infer to each pixel of the captured image latitude and longitude position, thus, obtaining the UAV location. Associating the parameters of the projective transformation with the non-linear equation in (7.5) gives the pose.

Even though this is a possible and direct method, it is not yet computationally efficient. First, because the big size of a stored image of a flight area may demand a long time to obtain its feature points, and match them to the captured images points and, thus, exceeds the time required for the navigation system. Likewise, it is a method restricted to when we have images (stored image and captured image) from the same type of sensor and must be captured in a similar time of day, in order to eliminate luminosity differences.

Another method has been proposed to lower the execution time and compensate for all the restrictions. It needs to use complementary sensors such as a height sensor (a laser altimeter or sonar) and the INS, and the calibrated optical sensor's geometric parameters (focal distance, sensor size, and radial distortions) [54].

As the UAV fly, the images are taken by the embedded optical sensor every instant. In an instant, when the INS is in the limit of losing precision, the proposed template matching method starts the operation. It uses the INS attitude information, height information, and the sensors parameters to form a geometric transformation to compensate for the differences in scale, rotation, and disposition to rectify the captured image [13]. Now, the captured image has the same orientation and pixel size as the stored image. If the images are from the same sensor type, there is a high chance they can be found through a correlation method.

The correlation can measure the similarity of matching images in the spatial domain or a transform domain [8], which uses the intensity of the image pixels. Mutual information [55] and cross-correlation [56] are other metrics for matching images. Only the correlation is described above to simplify without losing generality.

Methods based on correlation can be implemented very efficiently and are suitable for real-time applications. They can also be applied in areas with no distinct landmarks. However, they are typically sensitive to differences between the sensed and the reference image [20] (Figure 7.7). They are performed as a mask, that does window operation as in (7.14) (squared difference correlation), (7.15) (classic correlation) or (7.16) (Pearson correlation), where T is the stored image, I is the transformed captured image, w and h are the width (number of columns) and height (number of rows) of the image I, respectively, as below:

$$R_1(x,y) = \frac{\sum_{x',y'}(T(x',y') - I(x+x',y+y'))^2}{\sqrt{(\sum_{x',y'}T(x',y')^2)(\sum_{x',y'}I(x+x',y+y')^2)}} \tag{7.14}$$

$$R_2(x,y) = \frac{\sum_{x',y'}(T(x',y')I(x',y'))}{\sqrt{\left(\sum_{x',y'}T(x',y')^2\right)\left(\sum_{x',y'}I(x+x',y+y')^2\right)}} \tag{7.15}$$

$$R_3(x,y) = \frac{\sum T'(x',y')I'(x',y')}{\sqrt{\left(\sum_{x',y'}T(x',y')^2\right)\left(\sum_{x',y'}I\left(x+x',y+y')^2\right)\right)}},$$
$$T'(x',y') = T(x',y') - \frac{1}{hw}\sum_{x'',y''}T(x'',y'') \tag{7.16}$$

(a) (b)

Figure 7.7 The use of the correlation metric to compare RGB images from the same sensor [57]

and

$$I'(x+x',y+y') = I(x+x',y+y') - \frac{1}{hw}\sum_{x'',y''} I(x+x'',y+y'')$$

The result of this window operation is a matrix of size ($W - w$, $H - h$), with W and H being the width and height of the stored image, respectively. Each pixel (x,y) of that matrix is the result of the correlation centred in that point. The maximum pixel value (or the minimum value, if the squared difference correlation is used) of this matrix is the location of the tested image (the transformed captured image) and therefore the location of the UAV. Even though the correlation is fastest of the two methods, obtaining and comparing feature points, testing the correlation in the whole stored image, though, can be quite time-consuming also. A strategy to reduce that time is using the INS position estimation, and reducing the searching space to a sub-image centred in this estimation, but also considering the INS accumulated error.

As mentioned before, the correlation can be used with the pixel value itself (spatial domain) or in a transform domain. When the spatial domain is used, the correlation tends to be quite sensitive to differences in luminosity and spectral responses from different sensors. It is recommended to use the correlation in the transform space, such as border detectors (Figure 7.8). The result of a border detector is a binary image where pixels of borders are 1, and the other pixels are 0. The Canny operator [58], the Sobel filter [59], or the artificial neural network (ANN) mask [60] are examples of how the borders can be extracted. The advantage of using border images is that they are less sensitive to luminosity changes and sensor changes [61]. For example, in [57], the template matching is performed for the UAV navigation, using an RGB satellite image as the stored image, but the images captured during the flight are thermal images.

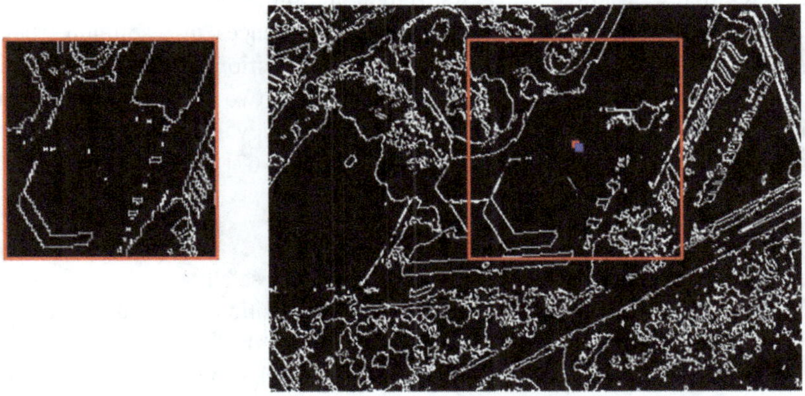

(a) (b)

Figure 7.8 Correlation and the comparison of images borders based on [57]

7.4.3 Landmark recognition

Landmark recognition is a natural approach for autolocalisation, since it is based on common practice for human pilots when the GNSS is not working properly. They usually look for distinct buildings or other human-made infrastructures that they already know from previous flights or a map, so they can determine the aircraft's position and then correct the INS [62] during flight. The aim of this strategy then is to employ a system to recognise landmarks captured by the onboard UAV vision in real-time when flying over specific locations [19].

Landmarks are structures that stand out on the field. It can be roads with intersections and crossings, rivers with crossing roads, runway and taxiway structures, shores of lakes, islands, large buildings, towers, bridges, and wood edges or isolated pieces of woodland, clearings in woods, and many others [61,63]. The location of the landmarks is well known, and they are selected during mission planning. Therefore, the same way as the template matching, the Landmark Recognition NS needs to know the area of the flight previously, but, instead of knowing the whole area, it needs to know only specific areas or objects. We consider knowing the landmarks as associating them (or pixels of a pattern image of the landmark) to position information: latitude, longitude, and altitude.

Once the landmarks are recognised in the image captured during the flight (it is known where the landmarks are in this image), their stored position information can be used and applied in (7.12) or in (7.3) to make the image georeferenced. As now all pixels in the captured image will have an associated position, the centre of the image can be considered as the centre of the camera attached to the UAV in a simplified manner. This position, then, can also be considered as the UAV position. More sophisticated photogrammetry equations can also be applied in order to obtain the centre of perspective of the camera and the attitude of the photo, which is even more precise.

So now, the challenge of the strategy is how to recognise those objects precisely. Although it is a non-trivial task, this has been the subject of study for many years in CV. There are several different algorithms proposed in the literature so far, and they all tend to work around the differences in resolution, rotating, translating, scale, luminosity, and others in the recognition process. As a matter of simplicity and didacticism, we will separate them in two situations as categories, both using concepts for data storage and image retrieval, such as content-based image retrieval and content-based video retrieval [64–67].

7.4.3.1 Knowing the exact landmark

When there are recent images from previous flights or satellite images of the route, landmarks can be selected and extracted with their position information from those small images and stored in the NS (aka pattern images).

The task then will be to find these patterns in the images that are captured during flight. It can be seen that this is a similar task to the template matching but in an inverted way: instead of looking for the captured image in the stored image, now we look for the stored images (pattern images) in the captured ones. This inversion makes use of the feature point algorithms described before possible, as the reduction of stored information makes them suitable for real-time applications depending on the number of landmarks to be tested. In order to reduce the amount of landmarks to be tested, it is recommended to use the INS position estimation considering its error to reduce the number of landmarks per flight region.

Once the landmarks are recognised, and their pixels are associated with the pixels from the captured image, it is possible to obtain the geometric transformation to georeference the captured image (7.13) and obtain the UAV position, considering it as the position of the centre of the image. An example appears in [26]. More sophisticated methods for pose estimation can be obtained the same way as described before, using (7.5).

The main disadvantages of this situation are related to both the limitations of the feature method used and the need for a recent image of the landmarks. If the landmark is removed or modified, the method will not be able to recognise it.

7.4.3.2 Identifying the landmarks' classes

If there are no images of the landmarks, but it is known that there are several types (or classes) of landmarks that the aircraft will pass by, instead of recognising the specific object, we can try to recognise any object of that type. For example, imagine that in the flight route, the aircraft must fly over a bridge, and this is the only bridge in that region. It is possible to develop an NS that will try to find any bridge in every image that is taken during the flight and then use this information for the position estimation. After that recognition, it will correct its navigation and then pass to recognise the next class of object in the route.

The methods in this category are related to the computer intelligence – aka soft computing (SC) – an area with methodologies such as ANNs, deep learning, and machine learning algorithms have been the core proposed solutions for this task [19,64,65,68–72].

Extra image processing, in addition to pattern recognition knowledge, can be integrated into the estimation models. The smarter the algorithm, the more challenging it is regarding computational requirements. SC reliefs the optimisation burden when used in place of hard computing strategies [64,70].

The photogrammetric information demands different image modalities with different resolutions that have to be registered [64,65,71,72]. Several particle swarm optimisation variants can help to work with different image resolutions along with the necessity [64].

As knowledge is constructed, future systems can fuse different multimodal data sources and solve problems using visual semantic information [65].

An example of the method appears in [68], where the histogram of oriented gradients associated with supporting vector machines with a Haar-like feature cascade and an LBP cascade are applied to recognise specific classes of objects, such as soccer fields and airports. Reference [69] present a method that creates classes of specific objects based on ontologies about the area and then uses this information for the recognition. In [19], on the other hand, the landmarks are recognised by the ANN and the Gabor transform.

The main problem with this situation is that the system is not able to separate landmarks of the same class. For example, if there are two different bridges in the route, the system will recognise both as bridges but will not be able to separate one from another, leading to probable errors in position estimation. This aspect has to be tackled during flight planning. The safest way is when each class of landmark appears only once, or if other techniques or other sensors information is employed. It is even possible to have the same class of landmarks on the same route, but they need to be far apart, and the planned route must have a strict sequence of landmarks to follow.

There are few works on aerial landmark recognition navigation systems in the literature yet. It is a promising field of study that suits the need for a reliable and precise method for the auto-localisation of the UAV, since it needs less information about the route than the template matching, and also because each estimation is independent of previous estimations.

7.4.4 Visual odometry

When the flight must be over an unknown area, neither of the strategies discussed before can be used. The flight starting position and the sequence of images captured during the flight will be the primary source of information for the pose estimation in these situations. The overlapping area between consecutive images can provide useful information for the navigation since it is possible to measure the motion between them. This strategy is known as VO.

The VO, though, estimates the UAV position regarding a scale factor [73], since it can only estimate the motion based on pixel values from the image [73]. This information needs to be in metres in order to be used to estimate the new position. It needs, then, other sensors in order to effectively obtain the UAV's movement, such as a laser altimeter, a barometer, or a velocimeter, and also the sensor calibrated parameters (focal distance, aperture, sensor size, among others).

The main aspect of the method is to look for corresponding points in both consecutive images [73] and use these points to estimate the transformation matrix (rotation and translation) (Figure 7.9). Consider P_1 the initial point of the flight, P_n as the destination point and the set of images $I_{1:n} = I_1, I_2, \ldots, I_n$ that were taken at each position $P_1, P_2, \ldots, P_n$. The main aspect of the odometry then is to estimate the set of transformations $T_{1:n} = T_1, T_2, \ldots, T_n$ that when applied sequentially to the initial position, can take the UAV from position P_1 to P_n, as in the following equation:

$$P_n = P_1(T_1 T_2 \ldots T_n) \tag{7.17}$$

Each transformation T_k is as in (7.18), where R_k is the rotation matrix, and t_k is the translation vector. They represent the sensor (thus, the UAV's) movement, and they can be obtained analysing the pair of sequence of images I_{k-1} and I_k that are taken in positions P_{k-1} and P_k, respectively. T_k becomes:

$$T_k = \begin{bmatrix} R_k & t_k \\ 0 & 1 \end{bmatrix} \tag{7.18}$$

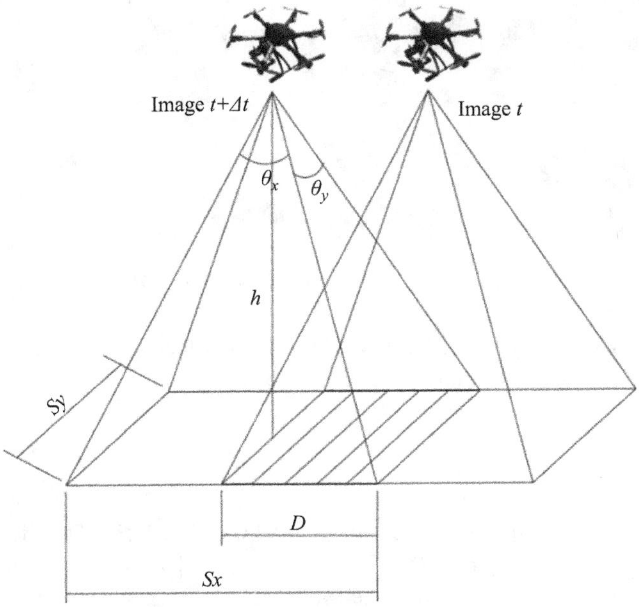

Figure 7.9 Scenario depicting how a UAV obtains images at t and t + Δt [74]

and to obtain both R_k and t_k, we need to determine the fundamental matrix F of each consecutive pair of images (I_{k-1}, I_k) [75]. The fundamental matrix is a 3×3 matrix that represents the epipolar geometry between two images of the same scene, which depends only on the parameters of the sensor and the positions each image was taken. The most important property of the fundamental matrix is that for every corresponding pair of points x_{k-1} and x_k in each image, according to:

$$x_k^T F_k x_k - 1 = 0 \tag{7.19}$$

Now, the problem becomes to find corresponding points between consecutive images. Again, feature point algorithms play an important part in solving it. They are in the core of most implementations of VO systems proposed so far. Any feature points algorithm used will result in a set of pairs of corresponding points that can be applied in (7.17) to form a linear system and obtain F.

Both the rotation matrix R_k and the translation vector t_k come from F, since there is a relationship between them, and the matrix containing the calibrated parameters of the sensor C. The relation from (7.20) contains C^{-T} as the inverse of the transpose of matrix C and $[t_k]_x$ is the skew-symmetric matrix representation of the cross-product of t_k as follows:

$$F_k = C^{-T} R_k [t_k]_x C^{-1} \tag{7.20}$$

Figure 7.10 Visual odometry displacement estimation [26]

F can also be expressed as in (7.21), where E is the essential matrix. Therefore, the definition of the essential matrix is as in (7.26) [75]:

$$F_k = C^{-T} E_k C^{-1} \tag{7.21}$$

$$E_k = R_k [t_k]_x \tag{7.22}$$

E can be decomposed into singular values (SVD) as in (7.23). This decomposition is vital because, as stated by [75], E and the matrix $[t_k]_x$ have the same singular values, since they are apart of only one rotation. From that statement, both $[t_k]_x$ and R_x are obtained by (7.24) and (7.25), where U and V^T are unitary matrices of the SVD of E, Σ is the diagonal matrix with the singular values in the diagonal, and W is the 90° rotation matrix in the z-axis. These entities are related as follows:

$$SVD(E) = U \Sigma V^T \tag{7.23}$$

$$R_k = U W^T V^T \tag{7.24}$$

and

$$[t_k]_x = U W \Sigma U^T \tag{7.25}$$

$$W = \begin{bmatrix} 0 & -1 & 0 \\ 1 & 0 & 0 \\ 0 & 0 & 0 \end{bmatrix} \tag{7.26}$$

The strategy can have a high rate of repeatability so that it can correct the INS frequently [8]. On the other hand, the VO tends to accumulate errors in the direction of movement. Therefore, the position is not well accurate in long flights [8]. This happens because, besides the first estimation, which uses the known first position, all other estimations use its previous estimation (and thus its implicit error) as input, accumulating errors similar to the INS. Some examples of VO systems are in [27,76–78] (Figure 7.10).

7.4.5 Combination of methods

Each of the strategies described before for an autonomous visual-NS has its advantages and disadvantages, which limits their applicability in several UAV missions. In order to overcome the disadvantages of the systems, most works nowadays tend to combine strategies in a single method. In this context, data fusion algorithms play an essential part in mixing the different estimations and their reliabilities to give a better pose for navigation. Systems with this aspect have proven to be more stable and reliable in most situations.

A first example is a work developed in [20,54]. Both use the template matching strategy combined with the VO. In [20], the position estimations and the INS information are fused using a Bayesian estimation framework, consisting of two parts: A standard Kalman filter and a point-mass filter [20]. Braga [54], on the other hand, uses a particle filter algorithm to estimate the new position.

Simultaneous localisation and mapping (SLAM) is another visual navigation that can be considered as a combination of strategies. It combines the VO with landmark recognition, and its goal is to localise a robot in the environment while mapping it at the same time [79]. SLAM nowadays is one of the most studied and developed methods because there is a high interest in using it in disastrous areas, for example, where GNSS signals are impossible, and the area has changed completely. In such a scenario, it is necessary to know how the environment is in order to plan how to act and rescue the survivors, as fast as possible.

In SLAM approaches, an internal representation of the world is built on-line mostly as a mosaic image, and a landmarks database is also formed, which is then used for localisation purposes. For indoor robotic applications, SLAM is already a standard [80]. More challenging is the use of such techniques in vast outdoor environments. It is one of the most promising techniques being developed for UAVs and aquatic UAVs [80].

There are two types of vision-based SLAM systems: the monocular, which uses only one sensor, and the binocular, which uses more than one sensor, and fuse their measurements for a stereo vision of the scene [20]. The binocular SLAM has less dependence on the hardware and is more flexible than the monocular SLAM. The binocular uses information from the sensors to form the stereovision pair and build the map of the scene [81]. The stereovision is used to obtain depth information of the environment. The monocular SLAM, on the other hand, uses the sequence of images also to obtain the depth information. It has, though, to be limited to the overlapping area between the two images.

In both SLAMs, the navigation is performed by the VO [81] at first. While the navigation is performed, the system also maps the region, automatically collecting data about landmarks encountered [82]. These data are stored in a database, and, as the reliability of the VO drops, the aircraft returns to previous areas already mapped to recognise the landmarks and eliminate the accumulated error of the visual [81]. This need to revisit already mapped places to identify some landmarks before continuing the route is the most significant disadvantage of SLAM, as it impacts on the autonomy of the vehicle for some missions [21].

7.5 Future developments on visual navigation systems

There are still several challenges to overcome in the visual navigation area though. An important difference to be pointed out is that indoor visual navigation problems are more studied and more developed than outdoor navigation. This happens because the GNSS flaws for outdoor aerial navigation are less frequent. The indoor navigation, on the other hand, is always under a severe vulnerability (the non-satellite visibility) and has a more controlled environment for the development. This research preference may change as UAVs become more popular, and the demands for safer and more reliable outdoor navigations increases.

Another possible future development relates to the sensor used for navigation. Most proposed methods until now have used mostly optical RGB sensors, which is quite limiting when we consider the vast possibility of flight scenes. New sensors and new adaptations of the strategies need to be studied and developed, in order to pursue a more reliable and broader visual NS. This also leads to studies on adaptive systems, which need first to understand their environment before deciding the best algorithm to produce the most reliable pose estimation.

At the same time, most methods yet developed have a high computational cost, which means that obtaining real-time conditions is always a challenge to be taken into consideration. One must not forget that the computation must be embedded, and there are certain limitations to the computer that will be on the aircraft, such as energy consumption and weight, which will affect the UAV's autonomy. Cloud computing and better communication links could be thought of as a solution at first [66,83–85]. On the other hand, they introduce vulnerabilities in the system due to data transmission, despite the use of cryptography and other security measures. Few works [54] nowadays are using hybrid computing to accelerate their methods. Field programmable gate array, multiple central processing units and graphical processing unit are combined, in order to obtain the most appropriate hardware architecture for the method in analysis.

Even though the navigation is discussed in this chapter as a task for UAVs individually, multi-UAV collaboration can also provide information for the swarm, and its location. This method is also not recommended since the communication between each UAV in the swarm would also introduce vulnerabilities in the system, as cloud computing would.

7.6 Conclusion

The high number of UAV applications has pushed the development of more reliable navigations systems in order to surpass the limitations and problems related to the GNSS + INS. Losing signals steadily, staying susceptible to jamming and spoofing and not being able to fly indoors became a high barrier to several important applications [83,85]. The scientific community then has been developing several visual NSs: landmark recognition, template matching, VO and the combinations of these strategies [84–87]. These strategies, therefore, have proven to be quite reliable so far, with results similar to the GNSS. There are several improvements yet to be made, which makes it a vast and rich research field to be explored.

References

[1] Mersheeva V, and Friedrich G. Multi-UAV monitoring with priorities and limited energy resources. In: ICAPS; 2015. p. 347–356.
[2] Costa FG, Ueyama J, Braun T, *et al.* The use of unmanned aerial vehicles and wireless sensor network in agricultural applications. In: 2012 IEEE Int'l Geoscience and Remote Sensing Symp. (IGARSS). IEEE; 2012. p. 5045–5048.
[3] Doherty P, and Rudol P. A UAV search and rescue scenario with human body detection and geolocalization. In: Australasian Joint Conf. on Artificial Intelligence. Springer; 2007. p. 1–13.
[4] Everaerts J, *et al.* The use of unmanned aerial vehicles (UAVs) for remote sensing and mapping. In: The Int'l Archives of the Photogrammetry, Remote Sensing and Spatial Inf. Sciences (IAPRSSIS). 2008;37. p. 1187–1192.
[5] Remondino F, Barazzetti L, Nex F, *et al.* UAV photogrammetry for mapping and 3D modeling–current status and future perspectives. In: IAPRSSIS. 2011;38(1). p. C22.
[6] Mader D, Blaskow R, Westfeld P, *et al.* Potential of UAV-based laser scanner and multispectral camera data in building inspection. In: IAPRSSIS. 2016. p. 41.
[7] Valavanis KP, and Vachtsevanos GJ. UAV applications: Introduction. In: Handbook of Unmanned Aerial Vehicles. Springer; 2015. p. 2639–2641.
[8] Silva CAO. Evaluation of Template Matching for UAV Location. UFMG; 2015.
[9] Woodman OJ. An introduction to inertial navigation. University of Cambridge, Computer Laboratory; 2007.
[10] LeMieux J. Alternative UAV Navigation Systems; 2012. Available from: http://electronicdesign.com/embedded/alternative-uav-navigation-systems [cited 12/05.2016].
[11] Parkinson BW, Enge P, Axelrad P, *et al.* Global positioning system: Theory and applications, Volume II. Am. Inst. of Aeronautics and Astronautics; 1996.
[12] Taiwan National Space Organization. NSPO Satellite Database.
[13] Goltz GAM. Artificial Neural Networks on Images for UAV Pose Estimation. Instituto Nacional de Pesquisas Espaciais; 2011.
[14] de Babo Martins F, Teixeira LF, and Nobrega R. Visual-inertial based autonomous navigation. In: Robot 2015: 2nd Ib. Rob. Conf. Springer; 2016. p. 561–572.
[15] Kintner PM, Kil H, Beach TL, *et al.* Fading timescales associated with GPS signals and potential consequences. Radio Sci. 2001;36(4):731–743.
[16] Allen PD. Modeling global positioning system effects in the Tlc/Nlc Model, 1994.
[17] Ruegamer A, and Kowalewski D. Jamming and spoofing of GNSS signals – An underestimated risk?! Proc. Wisdom Ages Chall. Mod. World. 2015; p. 17–21.
[18] Zhang C, Chen J, Song C, *et al.* An UAV navigation aided with computer vision. In: The 26th Chinese Control and Decision Conf. (2014 CCDC); 2014.

[19] Shiguemori EH, Martins MP, and Monteiro MVT. Landmarks recognition for autonomous aerial navigation by neural networks and Gabor transform. In: SPIE 2007 Electr. Imaging. SPIE; 2007. p. 64970R–64970R.

[20] Conte G, and Doherty P. Vision-based unmanned aerial vehicle navigation using geo-referenced information. In: EURASIP J. Adv. in Signal Proc. 2009;2009.

[21] Mathe K, and Busoniu L. Vision and control for UAVs: A survey of general methods and of inexpensive platforms for infrastructure inspection. Sensors. 2015;15(7):14887–14916.

[22] Gamal AE, and Eltoukhy H. CMOS image sensors. Circ. Dev. Mag. 2005; 21(3):6–20.

[23] Lees, AW, and Ryan WD. A simple model of a buried channel charge coupled device. Google Patents; 1974. US Patent 3,792,322.

[24] Cardoza S, Alexsander J, *et al.* Digital image sensors: CCD AND CMOS. In: VII CONNEPI-Congresso Norte Nordeste de Pesquisa e Inovacao; 2012.

[25] Sohn IY. CMOS active pixel sensor. Google Patents; 2002. US Patent 6,365,950. Available from: https://www.google.com/patents/US6365950.

[26] Silva Filho PFF. Automatic Landmark Recognition in Aerial Images for the Autonomous Navigation System of Unmanned Aerial Vehicles. Instituto Tecnologico de Aeronautica (ITA). Sao Jose dos Campos-SP; 2016.

[27] Roos DR. Machine learning applied to visual odometry for UAV pose estimation. UNIFESP. Sao Jose dos Campos; 2018.

[28] Rebecq H, Horstschaefer T, and Scaramuzza D. Real-time visual inertial odometry for event cameras using keyframe-based nonlinear optimization. In: British Machine Vis. Conf. (BMVC). vol. 3; 2017.

[29] Depaola R, Chimento C, Anderson ML, *et al.* UAV navigation with computer vision–flight testing a novel visual odometry technique. In: 2018 AIAA Guidance, Navigation, and Control Conference; 2018. p. 2102.

[30] Mansur S, Habib M, Pratama GNP, *et al.* Real time monocular visual odometry using Optical Flow: Study on navigation of quadrotors UAV. In: 2017 3rd Int'l Conf. Sc. Technology-Computer (ICST). IEEE; 2017. p. 122–126.

[31] Estrela VV, Magalhaes HA, and Saotome O. Total variation applications in computer vision. In Handbook of Research on Emerging Perspectives in Int. Pattern Rec., Analysis, and Image Proc. IGI Global; 2016, pp. 41–64.

[32] Estrela V., Rivera LA, Beggio PC, and Lopes RT. Regularized pel-recursive motion estimation using generalized cross-validation and spatial adaptation. In: Proc. SIBGRAPI; 2003. DOI: 10.1109/SIBGRA.2003.1241027

[33] Hagen NA, and Kudenov MW. Review of snapshot spectral imaging technologies. Optical Eng. 2013;52(9):090901.

[34] Gade R, and Moeslund TB. Thermal cameras and applications: A survey. Machine Vis. Appl. 2014;25(1):245–262.

[35] da Silva W, Shiguemori EH, Vijaykumar NL, *et al.* Estimation of UAV position with use of thermal infrared images. In: 2015 9th Int'l Conf. Sensing Technology (ICST). IEEE; 2015. p. 828–833.

[36] Carter J, Schmid K, Waters K, *et al.* Lidar 101: An Introduction to LIDAR Technology, Data, and Applications. National Oceanic and Atmospheric Administration (NOAA) Coastal Services Center; 2012.

[37] Yun S, Lee YJ, and Sung S. Range/optical flow-aided integrated navigation system in a strapdown sensor configuration. Int'l J. Contr., Aut. Sys. 2016; 14(1):229–241.

[38] Tuell G, Barbor K, and Wozencraft J. Overview of the coastal zone mapping and imaging lidar (CZMIL): A new multisensor airborne mapping system for the US Army Corps of Engineers. In: Int'l SPIE Def., Security, and Sensing. SPIE; 2010. p. 76950R–76950R.

[39] Braga JR, Velho HdC, and Shiguemori H. Estimation of UAV position using LiDAR images for autonomous navigation over the ocean. In: 2015 9th Int'l Conf. Sensing Technology (ICST). IEEE; 2015. p. 811–816.

[40] Wallace L, Lucieer A, Watson C, *et al.* Development of a UAV-LiDAR system with application to forest inventory. Remote Sensing. 2012;4(6): 1519–1543.

[41] Moreira A, Prats-Iraola P, Younis M, *et al.* A tutorial on synthetic aperture radar. Geosci. Remote Sensing Mag. 2013;1(1):6–43.

[42] Sineglazov V. Landmarks navigation system software. In: 2014 IEEE 3rd Int'l Conf. Meth. Syst. Nav and Motion Control (MSNMC). IEEE; 2014. p. 62–65.

[43] Slama CC, Theurer C, Henriksen SW, *et al.* Manual of Photogrammetry. 4th Ed. American Society of Photogrametry; 1980.

[44] Gonzalez RC, and Woods RE. Digital Image Processing. Prentice-Hall; 2005.

[45] Kumar A, Joshi A, Kumar A, *et al.* Template matching application in geo-referencing of remote sensing temporal image. Int'l J. Sig. Proc., Image Proc. and Patt. Recognition. 2014;7(2):201–210.

[46] Andrade JB. Photogrametry. SBEE. 1998; p. 98.

[47] Coelho L, and Brito JN. Fotogrametria digital. EDUERJ; 2007.

[48] Lima SAd, and Brito JLNeS. Digital image rectification strategies. In: Congresso Brasileiro de Cadastro Tecnico Multifinalitario; 2006.

[49] Cruz BF, de Assis JT, Estrela VV, and Khelassi A. A compact SIFT-based strategy for visual information retrieval in large image databases. Medical Technologies Journal. 2019;3(2):402–1, https://doi.org/10.26415/2572-004X-vol3iss2p402-412.

[50] Bay H, Tuytelaars T, and Gool LV. SURF: Speeded up robust features. In: Computer Vision – ECCV 2006. Springer; 2006. p. 404–417.

[51] Rublee E, Garage W, Park M, *et al.* ORB: An efficient alternative to SIFT or SURF. In: Int'l Conf. Computer Vision (ICCV). 2011; p. 2564–2571.

[52] Alcantarilla PF, Bartoli A, and Davison AJ. KAZE features. In: European Conf. on Computer Vision (ECCV). 2012. p. 214–227.

[53] Levi G, and Hassner T. LATCH: learned arrangements of three patch codes. In: 2016 IEEE Winter Conf. App. Comp. Vision (WACV). IEEE; 2016. p. 1–9.

[54] Braga JRG. UAV autonomous navigation by LiDAR images processing. Instituto Nacional de Pesquisas Espaciais (INPE). Sao Jose dos Campos; 2018. Available from: http://urlib.net/sid.inpe.br/mtc-m21c/2018/05.18.16.04.

[55] Maes F, Collignon A, Vandermeulen D, *et al.* Multimodality image registration by maximization of mutual information. IEEE Trans. Medical Imaging. 1997;16(2):187–198.

[56] Sarvaiya JN, Patnaik S, and Bombaywala S. Image registration by template matching using normalized cross-correlation. In: 2009 Int'l Conf. Advances in Comp., Control, and Telecommunication Techn. IEEE; 2009. p. 819–822.

[57] Silva Wd. Night UAV autonomous navigation by image processing of thermal infrared images. Instituto Nacional de Pesquisas Espaciais (INPE). Sao Jose dos Campos; 2017. Available from: http://urlib.net/sid.inpe.br/mtc-m21b/2016/08.08.20.34.

[58] Canny J. A computational approach to edge detection. IEEE Trans. Pattern Anal. Mach. Intell. 1986;8(6):679–698.

[59] Kanopoulos N, Vasanthavada N, and Baker RL. Design of an image edge detection filter using the Sobel operator. IEEE J. Solid-State Circ. 1988;23(2): 358–367.

[60] Suzuki K, Horiba I, and Sugie N. Neural edge enhancer for supervised edge enhancement from noisy images. IEEE Trans. Pattern Anal. Mach. Intell., 2003;25:1582–1596.

[61] Rivera LA, Estrela VV, and Carvalho PCP. Oriented bounding boxes using multiresolution contours for fast interference detection of arbitrary geometry objects. In: Proc. 12-th Int'l WSCG 2004; 2004.

[62] Epstein RA, and Vass LK. Neural systems for landmark-based wayfinding in humans. Philos. Trans. R. Soc. London B: Biol. Sc. 2014;369(1635):20120533.

[63] Michaelsen E, and Meidow J. Stochastic reasoning for structural pattern recognition: An example from image-based UAV navigation. Pattern Recog. 2014;47(8):2732–2744.

[64] de Jesus MA, Estrela VV, Saotome O, *et al.* Super-resolution via particle swarm optimization variants. In: Biologically Rationalized Computing Techniques For Image Processing Applications. Springer; 2018. p. 317–337. https://doi.org/10.1007/978-3-319-61316-1_14

[65] Estrela VV, and Herrmann AE. Content-based image retrieval (CBIR) in remote clinical diagnosis and healthcare. In: Encyclopedia of E-Health and Telemedicine. IGI Global; 2016. p. 495–520. https://doi.org/10.4018/978-1-4666-9978-6.ch039

[66] Estrela, VV, Monteiro ACB, Franca RP, Iano Y, Khelassi A., and Razmjooy N. Health 4.0: Applications, management, technologies and review. Med. Tech. J., 2019;2(4):262–276, https://doi.org/10.26415/2572-004X-vol2iss1p262-276.262.

[67] Estrela VV, and Coelho AM. State-of-the art motion estimation in the context of 3D TV. In: Multimedia Networking and Coding. IGI Global; 2013. p. 148–173. https://doi.org/10.4018/978-1-4666-2660-7.ch006.

[68] Cruz JEC. Object Recognition in Satellite Imagery with Descriptor-Classifier Approaches. INPE; 2014.

[69] Michaelsen E, Jager K, Roschkowski D, *et al.* Object-oriented landmark recognition for UAV-navigation. Patt. Rec. Im. An. 2011;21(2):152–155.

[70] Hemanth DJ, and Estrela VV. Deep Learning for Image Processing Applications. Vol. 31. IOS Press; 2017.

[71] Coelho AM, and Estrela VV. EM-based mixture models applied to video event detection. Principal Component Analysis—Engineering Applications. IntechOpen, 2012. https://doi.org/10.5772/38129

[72] Fernandes SR, Estrela VV, and Saotome O. On improving sub-pixel accuracy by means of B-spline. In: 2014 IEEE Int'l Conf. Imaging Syst. and Techn. (IST). IEEE; 2014. p. 68–72. https://doi.org/10.1109/IST.2014.6958448

[73] Fraundorfer F, and Scaramuzza D. Visual odometry: Part I: The first 30 years and fundamentals. IEEE Robotics Autom. Mag. 2011;18(4):80–92.

[74] Reboucas RA, Eller QDC, Habermann M, *et al.* Embedded system for visual odometry and localization of moving objects in images acquired by unmanned aerial vehicles. In: 2013 III Brazilian Symp. on Computing Systems Engineering (SBESC). IEEE; 2013. p. 35–40.

[75] Hartley R, and Zisserman A. Multiple view geometry in computer vision. Cambridge University Press; 2003.

[76] Quist EB, and Beard RW. Radar odometry on fixed-wing small unmanned aircraft. IEEE Trans. Aerosp. Electron. Syst. 2016;52(1):396–410.

[77] Post T. Precise localization of a UAV using visual odometry. 2015.

[78] More V, Kumar H, Kaingade S, *et al.* Visual odometry using optic flow for unmanned aerial vehicles. In: 2015 Int'l Conf. Cognitive Comp. and Information Processing (CCIP). IEEE; 2015. p. 1–6.

[79] Smith RC, and Cheeseman P. On the representation and estimation of spatial uncertainty. Int'l J. Robotics Res. 1986;5(4):56–68.

[80] Subramanya H. Monocular vision based simultaneous localization and mapping (SLAM) technique for UAV platforms in GPS-denied environments. Int'l J. Rob. Mechatron. 2016;2(1):37–43.

[81] Chudoba J, Kulich M, Saska M, *et al.* Exploration and mapping technique suited for visual-features based localization of MAVs. J. Int. Rob. Syst. 2016; 1–19.

[82] Wang L, Liu R, Liang C, *et al.* A new real-time visual SLAM Algorithm based on the improved FAST features. In: Human-Computer Interaction. Towards Intelligent and Implicit Interaction. Springer; 2013. p. 206–215.

[83] Estrela VV, Hemanth J, Saotome O, Grata EGH, and Izario DRF. Emergency response cyber-physical system for flood prevention with sustainable electronics. In: Iano Y. *et al.* (eds) Proc. BTSym 2017. Springer; 2019. https://doi.org/10.1007/978-3-319-93112-8_33

[84] Franca RP, Iano Y, Monteiro ACB, Arthur R, and Estrela VV. Betterment proposal to multipath fading channels potential to MIMO systems. In: Y. Iano *et al.* (eds) Proc. BTSym'18. Springer; 2019. https://doi.org/10.1007/978-3-030-16053-1_11

[85] Estrela VV, Khelassi A., Monteiro ACB, *et al.* Why software-defined radio (SDR) matters in healthcare? Med. Techn. J., 3(3), 2019, pp. 421–429, https://doi.org/10.26415/2572-004Xvol3iss3p421-429.

[86] Coelho AM, de Assis JT, and Estrela VV. Error concealment by means of clustered blockwise PCA. In: Proc. 2009 IEEE PCS. https://doi.org/10.1109/PCS.2009.5167442

[87] Razmjooy N, Ramezani M, Estrela VV, Loschi HJ, and do Nascimento DA. Stability analysis of the interval systems based on linear matrix inequalities. In: Y. Iano *et al.* (eds), Proc. BTSym'18. Springer; 2019. https://doi.org/10.1007/978-3030-16053-1_36

Chapter 8
Vision in micro-aerial vehicles

Navid Razmjooy[1], *Vania V. Estrela*[2] *and Roberto Sabatini*[3]

A micro-aerial vehicle (MAV) is a kind of minor version of unmanned aerial vehicles (UAVs), which has a size restriction. MAVs have a wide range of applications in commercial, research, government, and military purposes. Recently, the applications of bio-inspired MAVs have been increased following ecological, geological, meteorological, hydrological and human-induced disasters [1,2]. Indeed, animals utilise different strategies of movement in a varied environment, which allows them to adapt themselves to get food, ran away from a threat, etc., by using the least energy dissipation. Hence, designing and modelling UAVs inspired from animals not only can decrease the energy consumption of the robot for long-time applications but also it helps us to provide some kinds of robots, which can be utilised in different dangerous outdoors environment which even regular robots like UAVs cannot access them in *disaster research* [1,2].

UAVs are valuable tools due to some of their useful advantages such as safety, flexibility, relatively low cost of ownership, and ease of operation. Bio-inspired robots have been presented that can both fly and move on land, either by jumping or walking. One of the most critical parts of such vehicles (MAVs) is the vision (visual perception) system.

The vision system includes the ability of the robot to percept the environment by sensing light and image, which is inspired by the ability of the biological organism's visual perception. Visual perception is a natural ability of humans and animals to interpret and to communicate with their environment. This perception is a useful task of biological organisms, and it frequently appears in daily tasks. This perception can be found even in the smallest insects with their delicate bodies.

Even though a large number of researchers are continuously working on mimicking different skills of animals for robots, the natural vision has become one of the most successful and efficient than the other developed artificial systems. In

[1]Department of Electrical Engineering, Tafresh University, Tafresh, Iran
[2]Departament of Telecomunications, Federal Fluminense University, RJ, Brazil
[3]School of Engineering, RMIT University, Bundoora, Australia

this chapter, different technologies for designing the vision systems for bio-inspired MAVs will be reviewed.

8.1 Introduction

Micro-aerial vehicles (MAVs) possess the potential of accomplishing a wide range of limited duration missions. Their small size facilitates their transport to a launch site and allows them to remotely flown them by a single operator. This makes them ideal for reconnoitring dangerous situations (e.g., gas discharges, chemical spills, and so forth). Their size also makes them hard to be detected visually, making them useful critical surveillance missions. Therefore, MAVs provide a convenient and safe way to conduct missions that are presently unmanageable to do with any other existing device. When it is difficult for bigger unmanned air vehicles (UAVs) to operate, a MAV can be deployed both in confined and open-air spaces [3].

On the contrary, MAVs' limitations will be more visible when considering smaller sizes. Smaller wings decrease the dimensionless Reynolds number, which illustrates in fluid mechanics how turbulent or steady is a fluid flow passing a body or in a duct. The Reynolds number (*Re*) portrays the ratio between inertial forces and viscous forces and explains the amount of laminar or turbulent flow as follows:

$$\mathrm{Re} = \frac{\rho v_1}{\mu} = \frac{v_1}{\upsilon}, \tag{8.1}$$

where v is the fluid velocity in m/s, l is the characteristics length (or the cord width of an aerofoil) in m, ρ is the fluid density, μ is the dynamic fluid viscosity, and υ is the fluid kinematic viscosity in m^2/s. The systems that function at the same *Re* will have similar flow characteristics regardless of the fluid, speed as well as characteristic lengths vary.

They differ along with their means of producing propulsive thrust and lift fixed-wing MAV (FMAV) [4], rotary-wing MAV (RMAV), flapping-wing (also called biomimetic MAV or BMAV), and hybrid MAV as described in the following section [5,6] (Figure 8.1).

8.1.1 Fixed-wing MAVs

FMAVs look like and follow models for a plane with the fixed wings. FMAVs have higher flight speed and longer endurance as compared to the other categories. FMAVs face many distinctive challenges that complicate the design and development of a fully autonomous drone. The FMAVs operate in low Reynolds number (*Re*) regime, where many complex flow phenomena take place within the boundary layer.

The separation, transition, and reattachment of airflow can all occur within a short distance along the chord line of the wing [6]. Most FMAV designs produce thrust by using a propeller-driven electrical motor, although alternative systems

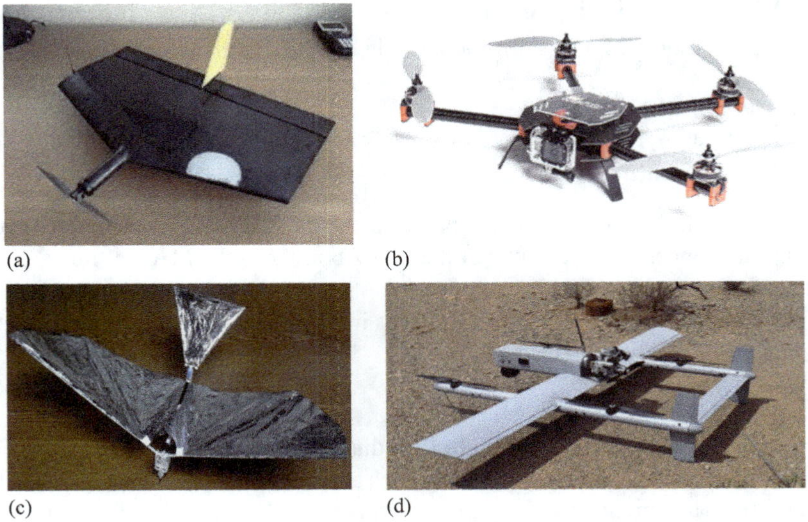

Figure 8.1 Different types of MAVs: (a) fixed-wing, (b) rotary-wing, (c) flapping-wing, and (d) hybrid MAVs

have also been used. Lift is generated by airflow over non-moving wings that have airfoil cross-sections. This generally limits the direction of flight to forwarding motion [7–9].

FMAVs often consist of rigid wing, fuselage, and tails, which use a motor and propeller as their propulsion system and can cover a wide range of possible operating environments, including a jungle, desert, urban, maritime, mountains, and arctic environments [10]. Because of their small dimensions compared to UAVs and low required power, FMAVs have low radar cross-section, are very difficult to detect and fly in environments, such as urban or forested areas, require short wings with low aspect ratios since UAVs with more extended wings are quite delicate, and likely to collide with obstacles [7,8].

High endurance and extensive range are critical in MAV applications such as data gathering or surveillance. Usually, FMAVs with more lift/drag values perform better than those with lower values. Furthermore, FMAVs have a more extended range and endurance and can fly at a higher altitude than the BMAVs and RMAVs, which usually perform indoor missions with slower flight speed. The wing planform is the wing silhouette when seen from above or beneath. Different types of platforms are possible: rectangular, elliptical, tapered wings with swept leading edges, Zimmerman, and inverse Zimmerman [10]. Some aircraft may have variable planform geometry during flight.

For mathematical modelling of FMAVs, some considerations can be performed: FMAVs can be modelled as a rigid body, curvature of the earth can be neglected, and FMAVs fly short distances [11]. By considering these assumptions, the model of an FMAV can be achieved by applying Newton's laws of motion.

8.1.1.1 Longitudinal dynamics

The dynamic model of MAV for the altitude can be described as follows [12]:

$$\begin{cases} \dot{\theta} = q \\ \dot{q} = M_q q + M_{\delta_e} \delta_e , \\ \dot{h} = V \sin(\theta) \end{cases} \tag{8.2}$$

where V is the magnitude of the aeroplane speed, θ is the pitch angle, Q is the pitch angular rate concerning the y-axis of the MAV body, h is the MAV altitude, and δe is the elevator deviation.

Mq and $M\delta e$ are the stability derivatives implicit in the pitch motion. Variables are shown in Figure 8.2(a).

8.1.1.2 Lateral dynamic

It generates the roll motion following by including a yaw motion (and vice versa). Then a natural coupling exists between the rotations about the axes of roll and yaw [14].

Different techniques can help to solve this issue. For instance, it is considered that there is a decoupling of yaw and roll movements [15]. Nonetheless, each movement can be controlled independently. The engine thrust affection is ignored [14]. The yaw angle is shown in Figure 8.2(b).

The yaw angle can be described as follows:

$$\begin{cases} \dot{\psi} = r \\ \dot{r} = N_r r + N_{\delta_r} \delta_r \end{cases} , \tag{8.3}$$

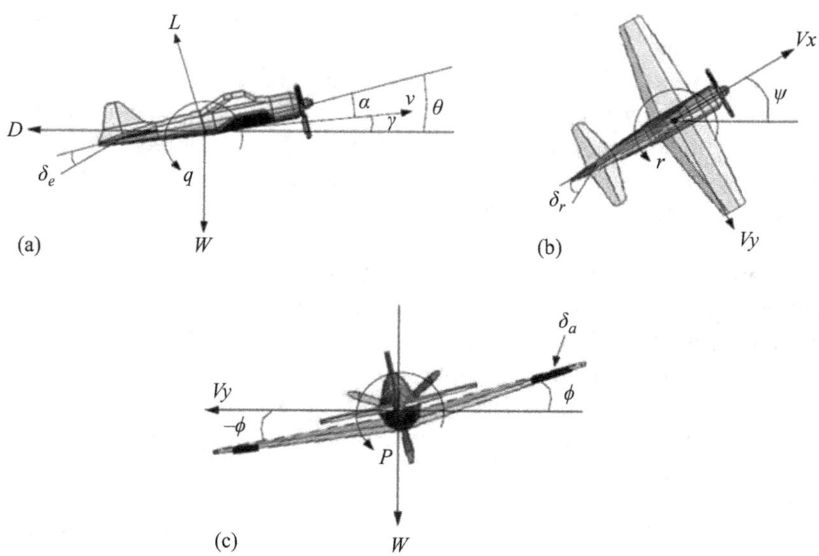

Figure 8.2 Dynamic motion modelling for (a) pure pitching motion, (b) pure yawing motion, and (c) pure rolling motion [13]

where δr is the rudder deflection, ψ is the angle of yaw, and r is the yaw rate with respect to the centre of gravity of the MAV. Nr and $N\delta r$ are the stability derivatives for yaw motion. The following equations can model the dynamics of roll angle:

$$\begin{cases} \dot{\phi} = p \\ \dot{p} = L_p p + L_{\delta_a} \delta_a \end{cases}, \tag{8.4}$$

where p is the roll rate, φ is the roll angle, and δ_a is the deviation of the ailerons. L_p and $L_{\delta a}$ are the stability derivatives of the roll motion. Figure 8.2(c) shows the roll motion variables.

8.1.2 Rotary-wing MAVs

RMAVs look like helicopters with rotary wings. RMAVs are functionally similar to helicopters or hovercraft. Spinning rotor blades generate lift and thrust. The necessity of these rotors to efficiently generate aerodynamic forces places a limit on how small they can be made. The rotors must have enough surface area to produce a sufficient magnitude of aerodynamic forces. The use of multiple sets of rotors (e.g., quadrotors, hexarotors, etc.) increases the lift and thrust they can generate. By spinning the rotors in opposite directions (in a balanced manner), the RMAV can be quickly stabilised, and the rotation of downwash air minimised [16]. This makes multiple rotor aircraft easy to remotely pilot (manoeuvre and hover).

One of the essential merits of RMAVs is their small dimensions, which allow them to fly in confined spaces, notably the RMAVs that hovers and has high manoeuvrability [17]. Unlike the FMAVs, RMAVs can fly in every direction, horizontally, vertically, and can hover in a fixed position [18]. These characteristics make them the perfect drones for surveying hard-to-reach areas, such as bridges, pipelines, among others [19]. RMAVs that are similar to helicopters generate lift from the constant rotation of the rotor blades [18]. Those comprising one motor and one blade are known as mono-copters, which got inspiration from whirling seeds that fall from some trees [3]. Currently, RMAVs can have from 2 to 12 motors [20], where the quadcopters and hexacopters are the most popular [21].

In the following, dynamics equations of motion for RMAVs are represented. Consider that O_i and CM are two points located on the rigid body (see Figure 8.3(b)), and $s_{oi,cm} \in \Re^3$ is the vector that joints the extreme border O_i with the rigid body's centre of mass, the angular and translational velocities (ω, v) and forces (f, τ), respectively, at any point on a body in $\Re^3$ are related as shown in Eq. (5). For kinematics transformation, 6×1 column vectors can be described for the physical quantities based on spatial algebra and Euler parameterisation. The system model is illustrated as [16]:

$$\begin{aligned} \dot{V}_{cm} &= \begin{bmatrix} \dot{\omega}_{cm} \\ \dot{v}_{cm} \end{bmatrix} = I_{cm,T}^{-1}[F_{cm,T} - \dot{I}_{cm,T} V_{cm}] \\ &= \begin{bmatrix} J_{cm,T} - \widetilde{s}_{oi,cm} & m_T \widetilde{s}_{oi,cm} \\ -m_T \widetilde{s}_{oi,cm} & m_T U \end{bmatrix}^{-1} \begin{bmatrix} \tau_{cm,T} - \widetilde{\omega}_{cm} J_{cm,T} \widetilde{\omega}_{cm} \\ f_{cm,T} \end{bmatrix}. \end{aligned} \tag{8.5}$$

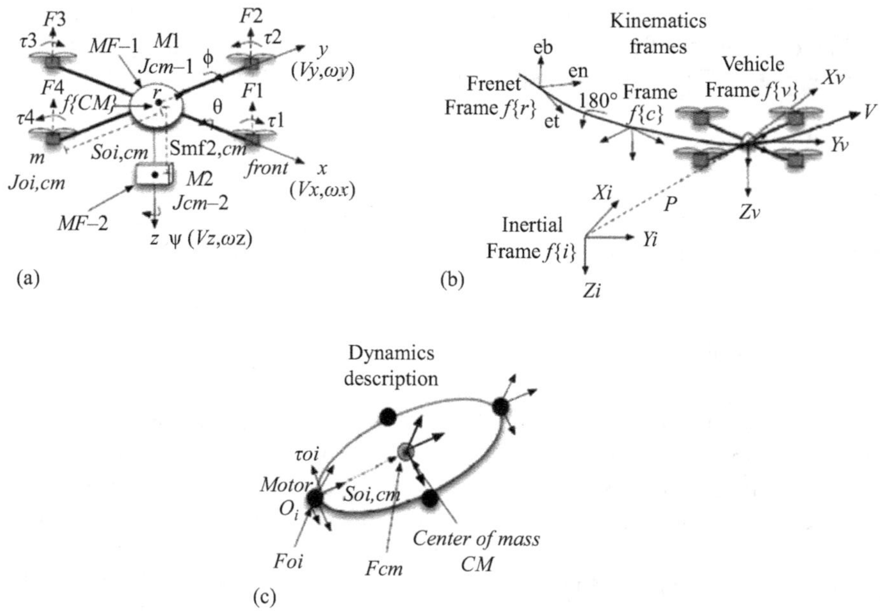

Figure 8.3 Modelling of the RMAVs [16]

Total inertia of the vehicle due to rotors and electronics can be calculated as follows:

$$J_{cm,T} = \frac{2}{5}M_1 r^2 + M_2\left(\frac{a^2+b^2}{12} + s^2_{mf2,cm}\right) + m\left(r^2_m + s^2_{oi,cm}\right) \tag{8.6}$$

For the RMAV, spherical-shape has been adopted from the main electronics with mass M_1 and radius r, whereas the rectangular shape with mass M_2 and cylindrical (m, r_m) shape for other electronics (battery + IMU) and the 4 rotors, respectively. The terms (a,b) refer to the lengths of the rectangular shape assumed for the second electronics (see Figure 8.3(a)).

$$\dot{V}_{cm} = \begin{bmatrix} J_{cm,T} - \tilde{s}_{oi,cm} m \tilde{s}_{oi,cm} & m\tilde{s}_{oi,cm} \\ -m\tilde{s}_{oi,cm} & mU \end{bmatrix}^{-1} \left(\begin{bmatrix} \tau_{cm,T} \\ f_{cm,T} \end{bmatrix} - \begin{bmatrix} \zeta_T \\ 0 \end{bmatrix} \right), \tag{8.7}$$

where $\tilde{s}_{oi,cm} \in \Re^{3\times 3}$ is the skew-symmetric matrix corresponding to the vector cross product operator of $\tilde{s}_{oi,cm}$ (see Figure 8.3 for operators description), and $\mathbf{U} \in \Re^{3\times 3}$ is the identity operator.

8.1.3 Flapping-wing or biomimetic MAVs

Bats, birds, and insects [22] inspired the design of a flapping-wing or BMAVs. BMAVs consist of the flexible and flapper wings, which use an actuation mechanism for their flapping motion [23–25]. Most of the flapping wings have flexible and light wings, as observed in birds and insects, which indicate that the flexibility and weight of wings are essential for aerodynamics and flight stability. The research on natural and human-made flapping wings showed that these types of air vehicles have more complexities as compared to FMAVs and RMAVs, mainly due to the aerodynamics complexity [3].

Biologic inspiration indicates that flying with flapping wings presents unique manoeuvrability advantages. There are fundamental challenges for tiny fixed and rotary wings to fly reliably. When the wing area is reduced, a flow transition to low Reynolds number occurs, which reduces the aerodynamic wing efficiency [26].

FMAVs and RMAVs that operate at low Reynolds numbers are prone to flow separation resulting in a drag increase and loss of efficiency.

A flapping wing can benefit from the advantages of other UAV types. The hovering ability of insects, coupled with the ability for a quick transition to forwarding flight, provides an ideal drone for search and rescue and other applications [27]. BMAVs can be designed and fabricated in three configurations, namely, monoplane, biplane, and tandem [3].

UAVs inspired by the flapping wings of the birds' or insects', or by the bats' flight model have been developed. Lift and thrust are achieved by flapping the wings. Flapping the wings generates a more significant force per wing surface area. Therefore, a BMAV has the potential to be much smaller than either an FMAV or an RMAV. The flapping frequency depends on the wing surface area. A dragonfly-inspired aircraft, for example, must flap at a lower frequency than a bee-inspired MAV since bees have much smaller wing-to-body ratios.

Some methods to design BMAVs use empirical formulae [3]. These formulae originated from allometric data extracted from avian flight [10] and involve design parameters of flapping wings, such as wing area, weight, and wing loading to the flapping frequency, flight speed, required power for flight, the geometry of the wing including the area and wingspan and the aircraft weight. Other methods based on statistical and experimental sizing and testing were applied.

Most of the sizing methods for MAVs depend on (allometric) expressions extracted from natural birds and insects to directly sizing artificial flapping wings without considering the impacts of other parameters such as the used materials for the wing membranes. The non-optimised MAVs result from the use of the empirical formulae derived from the observation of natural birds and insects and these mathematical models need revision and, probably, some correction factors [26].

After sizing and during the design process of BMAVs, different aerodynamic and structural analyses can be performed on them. Usually, in natural and human-made flapping wings, their aerodynamics, structure, and flight dynamics intersect with some fascinating problems like unsteady three-dimensional (3D) separation, transition in boundary layers and shear layers, uncertain flight environment, aeroelasticity, and anisotropic wing structure, and nonlinear and adaptive control [28]. Different theories are used to model the aerodynamic forces of the natural and human-made flapping wings, such as quasi-steady, strip theory, unsteady, and Navier–Stokes methods. It should be mentioned that the type of analysis is dependent on the type of flapping wing, its configuration, and flight modes. For instance, the complexity of aerodynamic analysis is increasing for flapping wings in tandem wing configurations. This wing configuration has been used by nature's flyers, such as dragonflies. Studying the unsteady flow interactions between the two wings is more complicated than the case of a single wing; however, two pairs of wings can provide increased lift and thrust and gust resistance [3].

Some BMAV design methods use empirical formulae established using quasi-steady aerodynamic models [22] centred on a slow wingtip speed relative to the overall velocity of the body based on the instantaneous velocity, wing geometry, and angle of attack when the steady-state aerodynamic model is used [29]. A quasi-steady model dramatically simplifies aerodynamics because it disregards the wing motion and flows history resulting from unsteady motion [22] while substantially reducing the model complexity. However, it falls short in accounting for the unsteady effects seen in flapping motion [29]. However, many animals and systems require other types of models to accurately approximate their flight lift due to very high flapping frequencies that produce unsteady flight aerodynamics conditions [22].

The strip theory is another model for the flapping motion of natural and human-made flapping wings [30]. The strip theory divides the wing into multiple sections, thus generating an integral function to account for the effects of each strip into an accurate aerodynamic model to determine the average lift and thrust through the cycle of the flapping motion of the system [30,31] to study the performance of flapping wings.

Aerodynamics of birds and insects during the flapping flight can also be modelled within the framework of unsteady Navier–Stokes equations. In this method, nonlinear physics with multiple variables, such as velocity and pressure, and time-varying geometries, are among the aspects of interest [22].

Particle image velocimetry (PIV) methods can also help to model the flow field because they combine different aerodynamic models with the experimental study to devise more realistic outcomes [28].

Next, the aerodynamic model of a BMAV is represented according to the pitching-flapping motion of a rigid wing that is a structured adaptation and simplification of the procedure adopted by [32]. A flapping angle has varied B as a sinusoidal function. The following equations represent β and its rate:

$$\begin{aligned} \beta(t) &= \beta_{\max} \cos(2\pi ft) \\ \dot{\beta}(t) &= 2\pi ft \sin(2\pi ft) \\ \theta(t) &= \frac{r(i)}{B}\theta_0 \cos(2\pi ft + \varphi) \end{aligned} \tag{8.8}$$

where θ_0 is the pitching motion, $\beta(t)$ is the flapping angle, and $\dot{\beta}(t)$ is the flapping rate.

The horizontal and vertical components of relative wind velocity are as follows:

$$V_x = U\cos(\delta) + (0.75 \times c \times \theta \times \sin(\theta)), \tag{8.9}$$

and

$$V_y = \sin(\delta) + (-r(i) \times \beta \times \cos(\beta)) + (0.75 \times c \times \theta \times \cos(\beta)) \tag{8.10}$$

The flight path angle γ for horizontal flight is zero. Furthermore, the relationship $0.75.c.\theta$ models the relative air effect of the pitching rate θ, which manifests at 75% of the chord length [33].

From the above equations, the relative velocity, the relative angle between the two velocity components ψ, and the effective AOA can be obtained:

$$\begin{aligned} V_{rel} &= \sqrt{V_x^2 + V_y^2}, \\ \psi &= \tan^{-1}\left(\frac{V_x}{V_y}\right), \\ \alpha_{eff} &= \psi + \theta \end{aligned} \tag{8.11}$$

By assuming the Kutta–Joukowski condition, the section lift due to circulation can be calculated as:

$$dL_c = \frac{1}{2} \times \rho \times V_{rel}^2 \times C_{l-c} \times c \times dr,$$

where c and dr are the chord length and the width of the element of the wing under consideration, respectively, and

$$C_{l-c} = 2\pi C(k)\sin(\alpha_{eff}). \tag{8.12}$$

The apparent mass effect (momentum transferred by accelerating the air to the wing) for the section is perpendicular to the wing and acts at mid-chord. Profile drag dD_p and induced drag dD_i forces can be calculated as follows:

$$\begin{aligned} dD_p &= \frac{1}{2} \times \rho \times V_{rel}^2 \times C_{dp} \times c \times dr \\ dD_i &= \frac{1}{2} \times \rho \times V_{rel}^2 \times C_{di} \times c \times dr \end{aligned} \tag{8.13}$$

Moreover, the total section drag is equal to:

$$dD_d = dD_p + dD_i. \tag{8.14}$$

The horizontal and vertical components of the forces can be defined as follows:

$$\begin{aligned} dF_{ver} &= dL_c \cos(\psi).\cos(\delta) + dN_{nc}\cos(-\theta).\cos(\beta).\cos(\delta) \\ dF_{hor} &= dL_c \sin(\psi).\cos(\delta) + dN_{nc}\sin(-\theta).\cos(\delta) - dD_d\cos(\psi).\cos(\delta) \end{aligned} \tag{8.15}$$

By collecting the whole vertical and horizontal components of forces, the lift and thrust of the BMAV can be calculated as follows:

$$dN_{nc} = -\frac{\rho\Pi c^2}{4}(\dot{\theta}U + r\ddot{\beta}\cos(\theta) - 0.5\ddot{\theta})dr. \tag{8.16}$$

Table 8.1 presents a brief overview of the advantages and disadvantages of the types of MAVs.

Table 8.1 Advantages and disadvantages of the several MAV types

Type of MAV	Advantages	Disadvantages
FMAVs	• Long endurance space • Large coverage • Fast flight speed • Heavy payload	• Launch-landing specific • No hover flight • Constant forward velocity to fly
RMAVs	• High speeds • Hover flight • VTOL[1] flight • Indoor flight and short-range reconnaissance	• Low efficiency of the rotor • Low thrust-to-weight ratio
BMAVs	Ornithopters: • Integrate lift and thrust	• Cannot hover • Need to obtain an initial airspeed before taking off
	Entomopters: • Large and rapid change of angle of incidence • VTOL flight • Hover flight	• Not yet widely available in commercial actuators.
Hybrid MAVs	• Long endurance • Large coverage • VTOL flight	• Under development • Transition between hovering and forward flight

[1]Vertical Takeoff and Landing Vehicle

8.1.4 Hybrid MAVs

Each of the four groups has its advantages and disadvantages. Aeroplane mode MAVs can fly in a long time, and they have higher efficiency than the others. Helicopters like MAVs have the ability to fly in each direction. The BMVs have the highest manoeuvrability, but they have still some works to do.

The flight agility (e.g., hovering and forward flight) of RMAV and BMAV allows using them in spatially compact areas or even indoors. There has been substantial research on swarm technology (especially with RMAV), which would enable multiple MAVs to work together using integrated navigation and search algorithms to conduct a variety of missions over broad areas. MAV (especially BMAV) made with low cost, and biodegradable materials are also being researched. The intention is to make them disposable and suitable for one-way missions. Not having to fly a return trajectory would effectively double their flight endurance and range, improving upon one of the primary limitations of MAV.

Research in low Reynolds number unsteady aerodynamics and BMAV propulsion has developed an alternative flapping-wing propelled MAV. Fixed/flapping-wing

MAVs are hybrid designs, which use fixed wings for lift and flapping wings for propulsion. In this type of MAVs, the drone usually consists of a low aspect ratio fixed-wing with a trailing pair of higher aspect ratio flapping wings, which flap in the counter phase. The flapping-wing part increases efficiency, provides a mechanically and aerodynamically balanced planform, and quenches stall over the fixed-wing by entraining flow [34]. This type of drone also can be seen in dragonfly with tandem wings, where they apply two pairs of wings to increase the lift and thrust forces.

The rest of this chapter is organised as follows. In Section 2, computer vision (CV) and its application in the biological-based MAVs are presented. Section 3 describes the role of sensing in MAVs based on different sensors. Section 4 presents the effects of illumination on the MAVs' vision. Section 5 presents the conception of navigation, pathfinding, and orientation on the MAVs. Section 6 explains about CCD cameras and their applications in machine vision. Section 7 presents future trends about MAVs. Finally, Section 8 concludes the chapter.

8.2 Computer vision as a biological inspiration

Nowadays, the inspiration from biological processes has been turned into an interesting category in science [35–40].

Most animals have a peculiar way of using visual sense and movement in nature [41]. This capacity allows them to improve their capability in finding food, escaping from threats, migration, and so forth. To do so, each group of animals has different methods based on their environmental conditions. For example, some animals can hunt preys in the darkness, such as bats and owls. Alternatively, some others can hunt their victims with fast diving to the land from the sky like eagles and some others with diving from the sky into the water, or another type of environmental change.

Nowadays, with developments on the robots, bio-inspired or biotypes robots can work on both the sky and the land, either by jumping or by walking [42–45]. Bio-inspired robots are a kind of robot that mimics the behaviour of the living entities. Scientists are trying to figure out how the muscular and bony structures of living things lead to all kinds of moves, such as crawling, slipping, swimming, walking on the wall, or even flying. Such robots have the potential to help humans in many applications. Bio-inspired robots can be used in tight spaces to scan destruction, slip into spirals and save survivors [46].

In this chapter, recent advances in the CV utilisation for the control of MAVs are discussed. The term MAV typically denotes a kind of minor version of UAVs, which has a size restriction [47].

In recent decades, the development of UAVs has grown, and different types of autonomous vehicles have been studied and developed throughout the world. In particular, drones may be used in many applications of emergencies that humans can often not act like earthquakes, floods, active volcanoes, or a nuclear disaster [48]. The initial researches about drones were focused on military applications.

However, the demand for drones, like airborne robots and bombers that can be used in emergencies and industrial applications, has just come into being. Recently, the MAVs, among the full range of drones, have been developed to fly in challenging conditions and have become popular UAVs.

MAVs have the ability to carry a variety of sensors; however, the small-scale types have some limitations in carrying the heavier objects. The most types of MAVs can carry a small digital camera to take aerial images for the preset goals.

The capability of carrying in MAVs makes them the ability to pilot. However, this process needs lots of training for piloting. This problem can be covered by considering the autopilot system to keep it in a definite spot.

For outdoor applications, the Global Positioning System (GPS) can be achieved; however, indoor applications cannot rely on GPS and requires an alternative sensor for the autopilot system. The best choice for autopilot is using on-board cameras.

These autopilot systems have been utilised on-board cameras successfully. The CV algorithms then evaluate the obtained ego-motion information from the camera and then fused with inertial measurement units (IMUs) to employ in the hover control loop in MAVs.

Image processing and CV algorithms can be employed for achieving the environment information for better controlling of the MAVs. In some applications, this information can be synthesised with measurements of IMUs to better controlling of the MAVs (Figure 8.4).

To reach this purpose, there are some challenges to overcome:

- The limitation of the on-board processors, particularly for image processing that demands a high computational capacity for the operations.
- For exact controlling, the highest frame rate is needed.
- The requirement of MAVs for robust operations in sensitive conditions.
- The metric scale cannot be measured by ego-motion estimation from a single camera.

Fusing camera-based measurements with IMU generates an efficient and robust ego-motion estimation algorithm [9,23–25,49–53]. Furthermore, we can employ the camera images for interpretation and environment sensing.

Using the connected cameras, MAVs can compute a 3D map of the environment, which can then utilised for autopilot applications [54].

Figure 8.4 A sample MAV: MikroKopter L4-ME [48]

Next, some of the applications of image processing and CV methods for controlling, autopilot, 3D mapping the MAVs are introduced and presented.

Yang *et al.* [55] introduced an autonomous search method for landing the MAVs on a pre-defined landing place using real-time monocular vision with only one single reference image of the landing place with an unknown size before initiating the task. Some other works in this area can be achieved in [56–58]. Different from these systems, systems with on-board sensors have different and harder methods for achieving the performance parameters [59,60].

There are several animals with different locomotion ways in the environment. Among these animals, birds and insects have efficient movements with their wings, which allow them to have more dynamics, more speed, and more manoeuvrability.

Bio-inspired robots can move on both the sky and the land, either by jumping or by walking [61]. The ability of the BMAVs to cope with wind turbulence and also their behaviour in front of the surrounding environmental conditions makes them fascinating.

For instance, consider a dragonfly-inspired robot. This robot has the following features:

- The presence of four wings in this robot makes it possible to have the maximum ability for lifting extra payloads like sensors and more processing power.
- The existence of four wings provides high manoeuvrability to these robots while maintaining robust and stable flight performance to environmental disturbances.
- The existence of one actuator per wing makes the system control simpler.
- This type of wings makes MAV sustain long flight times at minuscule energy requirements.

The dynamics and kinematics of MAVs can be modelled using CAD software tools. A mathematical model of the system is then can be verified by simulation software. The design of the Autopilot system can be performed in parallel with the mechanical testing and construction. The aerodynamic analysis is finally employed as leading to design and fabrication of the final wing for the bio-inspired MAV (Figure 8.5).

8.3 The role of sensing in MAVs

MAVs use various types of sensors as inputs (1) to estimate the pose (position and attitude), and (2) to observe the surroundings (in path planning, surveillance, mapping, and so on).

Environmental awareness (EA) can be done using acoustic sensors such as the sonar, LIDAR, IR, and mono/stereo vision.

Pose estimation can rely on the output of devices to perceive translation/rotation (e.g., magnetometer and GNSS), inertial sensors, optical flow (OF), and flow/pressure detectors. Some IR range sensors are limited to indoor use due to interference with electromagnetic energy emitted from the sun.

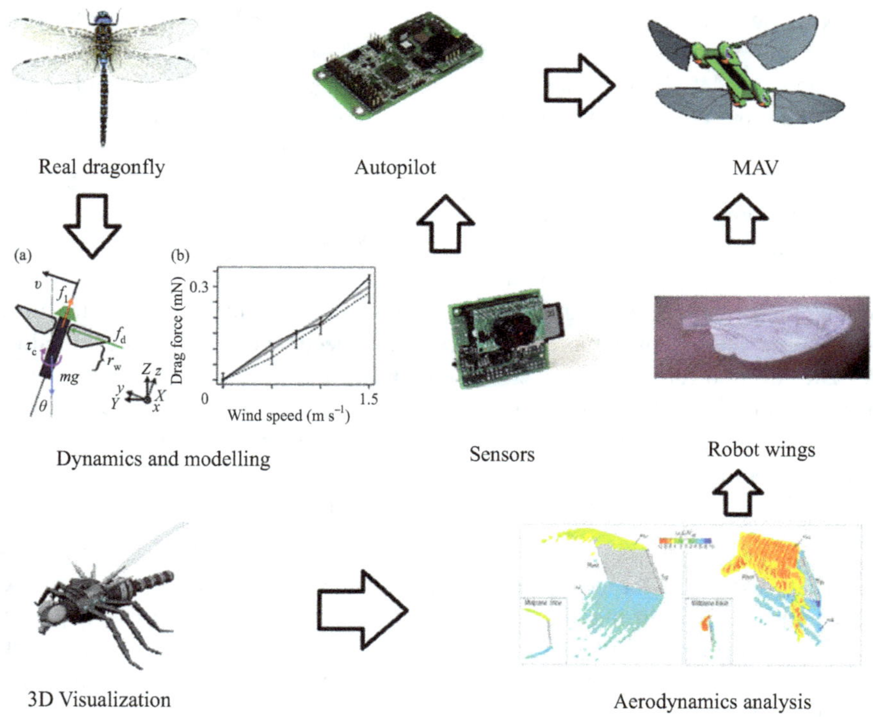

Figure 8.5 Design and implementation steps of a dragonfly inspired MAV

8.3.1 Pose-estimation sensors

In general, MAVs operate in turbulent indoor and outdoor environments. Current MAVs, therefore, face operational constraints due to severe turbulence at low altitudes. Hence, increasing the operational spectrum of MAVs in turbulent conditions is essential to enable them to conduct autonomous missions in urban environments safely. The large turbulence scales relevant to the miniature size of the vehicle, in addition to the fast perturbation rates, can degrade the pose estimate significantly, thus degrading flight-path and attitude tracking performance. This degradation can be hazardous for low-level flight in complex environments and is particularly significant for under-actuated vehicles, which can undergo significant flight-path deviation before returning to a planned path.

MAV pose estimation is considered critical due to significant stability issues in turbulent flow environments. Pose estimation involves localisation and attitude sensors, which, respectively, can be used to determine the position and orientation of the vehicle. Inertial navigation units (INUs), which employ a combination of inertial, rotational, and translational sensors, are commonly used for pose estimation. Inertial-based sensors can often lack the update rate, resolution, accuracy, precision, robustness, and reliability needed for MAVs. Sensor fusion, in combination with efficient and effective algorithms, can lead to substantial pose estimate enhancement in addition to bounding accumulated errors.

The utility of optical sensors for pose estimation is critical for the automatic guidance of aerial vehicles that require some degree of decision-making autonomy [62]. Through a bio-inspired image-processing approach for visual guidance known as OF, translation, and rotation in a series of consecutive images can be detected. OF can be described as the apparent visual motion in a scene as seen by a moving observer. Advancements in solid-state sensors and digital signal processors have helped to implement OF in MAVs.

There have also been recent attempts to utilise multiple sensors for OF performance improvement [63,64]. A major limitation to OF and all other electro-optical (EO) approaches is that their performance is highly dependent on the visual scene. Lighting conditions, lack of discernible or distinct features, and weather conditions (e.g., rain or fog) can degrade their performance. This limits a MAV utility in certain conditions and reduces its operational spectrum. Despite significant advances in computer processing hardware, OF approaches are computationally intensive, and this may introduce undesirable latencies [49–53]. The additional computational hardware required to process optical measurement-based approaches means that they have higher relative power consumption than many other sensing options. OF may also be constrained by the bandwidth of the visual data input. The video sampling rate must account for aliasing because the update rate of the computed pose estimate must at least match the perturbation rate.

The fusion of the output of flow sensors to the INU to improve pose estimation is essential [8]. Flow sensors capable of sensing flow disturbances and control systems capable of counteracting them before they induce a perturbation are desired. Such systems have been shown to improve pose estimation and thus to enhance attitude stability [65]. The minimal processing required to produce an output makes this sensory technique ideal for enhancing the pose estimates of INUs. This emerging area of research may provide the sensory needs for MAVs to fly in severe turbulence.

8.3.2 Environmental awareness sensors

The assessment of EA sensors is made on an ordinal measurement scale, which has the qualitative values of Unacceptable, Marginal, and Acceptable. Next, some MAV EA sensors are discussed below.

8.3.3 Sonar ranging sensor

Sonar sensors rely on emitting pressure waves (referred to as probing pulses) and measuring the time taken for the wave to return to a receiver. This concept is often referred to as time-of-flight (ToF) sensing.

There are two types of sonar-transducer technologies: piezoelectric and electrostatic [66]. Electrostatic sonars function by vibrating two voltage-biased plates and changing the capacitance between them. Piezoelectric materials can also be used, as they vibrate when a voltage is applied across the material and vice versa. The operational bands of piezo-based sonars are limited to the piezoelectric material oscillation frequency.

The operational frequencies of MEMS transducers can, therefore, be tailored to match the acoustic impedance of air to offer higher performance.

Sonars are suited for MAV operations because they are resilient to routine operational impacts and are relatively affordable. Their power and range are directly proportional and entirely hardware-dependent.

They are not sensitive to the colour or reflectivity of the detected object. Nonetheless, they are generally suited for sensing obstacles normal to the direction of pulse propagation. Angled, smooth objects may lead to the redirection of the beam resulting in false readings of an incident object. The reflected signal can also be particularly sensitive to the target area, relative object attitude, and the ability of the material to absorb sound energy (such as foam). The reliance on acoustics is particularly limited in sensing.

Ultrasonic sensors are generally immune to background sounds. This is not the case for the high-frequency noise that may be induced, for instance, from propeller motion. Thus, sensor placement may also present an issue. Environmental changes such as pressure, altitude, and temperature variations will alter the speed of sound and, consequently, the readings. Additionally, operating in turbulent or particle dense (scattering) atmospheres will also influence the sonar readings. Sonars have high latency compared with other range sensors. The first cause is that the transducers must be stopped before receiving a signal to avoid interference, which is known as the blanking time. This is in addition to the reliance on sound, which is physically limited compared to electromagnetic waves. The combination of limited range and high latency of sonars may render them unreliable for the safe navigation of the MAV.

These sensors are often used for obstacle detection, where the distance of a measured obstacle can be inferred from the ToF measurement. A sensor array or a rotating sensor for a 360° scan can help to map the nearby obstacles. Scan patterns detect features like corners and specific geometries. It is possible to use a dual sonar receiver structure for indoor Simultaneous Localisation and Mapping (SLAM) [67].

8.3.4 Infrared-range sensors

Infrared (IR) range (proximity) sensors may be constructed using pairs of LEDs and phototransistors. The LED is programmed to emit an IR signal at a certain wavelength (typically in the near-IR band), and the phototransistor sends an electric signal when it is exposed to the light of that frequency. The operation of IR proximity sensors is conceptually similar to the ToF sensing used by sonar and Light Detection and Ranging (LIDAR) sensors employing different media.

The benefits of IR range sensors are their low SWaP, low latency, suitable cost-effectiveness, and robustness. In contrast, they have a limited detection range and are high-powered. Their FoV is usually restricted to a few degrees.

As the detection distance grows, the signal-to-noise ratio (SNR) decreases, and the readings become less trustworthy. Their outdoor use is restricted to low light conditions, thanks to the interference of solar IR radiation. This, in turn, limits the use of increasingly popular RGB-depth imaging sensors, which have been used for indoor-mapping and navigation [68]. They are also highly sensitive to the reflectivity and direction of motion of detected objects.

IR proximity sensors are more oriented towards obstacle detection. Sensor arrays could be arranged around a MAV structure due to their beneficial SWaP characteristics. They are often used for obstacle avoidance and tasks such as GNSS-denied altitude estimation. For MAVs, they are better suited for tasks in which visibility is limited, such as in the presence of smoke and low-lighting/night operation.

8.3.5 Thermal imaging

Thermal imaging cameras detect objects radiating heat from their surroundings. They consist of a lens, an IR sensor array, and a signal-processing unit. Often a cooling unit is required to reduce noise in the resulting images. They offer more robust sensing in low light than the traditional EO sensors because they do not use visible light. All cameras need some light and contrast to permit the detection of different objects. Thermal imagers have been traditionally employed as payloads for use in addition to imaging and mapping applications. Thermal imaging also aids fire detection and pattern recognition for SLAM.

Nevertheless, their size limits their use on MAVs. Despite the miniature size of thermal-radiation sensors, the size of the optics (lenses) is significantly larger. The focal length, which is the distance between the lens and the sensor, is more significant than that of typical EO cameras [69].

8.3.6 LIDAR

LIDAR was initially developed for atmospheric sensing and topographic mapping but now can be found in GNSS and INU. LIDAR sensors used for range measurements characteristically operate similarly to IR-range and sonar sensors, i.e., as ToF sensors. Hence, unlike IR proximity, there is a limited divergence from the source that enables ToF sensing using the phase shift between the emitted and returned signals.

A large number of pulses are emitted per unit time for 2D range scans where the sensor captures the reflected waves. In many cases, one or more sources are used to accomplish each scan. For 3D scanning with compound sources, angled rotating mirrors, and rotating housings are used. LIDAR results in large data sets and a very high processing computer load. The intensity of the returned wave depends on the reflectivity of the detected object and may be affected by ambient illumination. However, LIDARs are active devices less sensitive than EO sensors to environmental factors and can also relay on depth information. EO sensors rely on passively sensing ambient light. LIDAR ranging requires more power, particularly as the measurement range and scanning frequency increase. Another aspect that must be considered with the use of lasers is safety, as MAV will operate in urban environments. There are regulations for the various classes of lasers when humans are exposed that are eye-safe [69].

LIDARs using SwaP are suited for MAV operations [70,71] although they jeopardise other operational necessities and suffer solar interference in outdoor operations such as cost, robustness, accuracy, sensitivity, and range.

8.3.7 Cameras

CV is an essential aspect of MAVs since images captured by one or more onboard cameras can provide both EA and pose estimation [7,69]. These techniques are used for image segmentation, target detection, target tracking, feature extraction, and mapping. Motion estimation entails the comparison of frames adjacent in time to detect the displacement between objects [72–77]. Visual SLAM algorithms perform feature extraction for landmark detection and must deal with all types of uncertainties and cooperative mapping. CV is appropriate for MAVs since visual sensors are compact and consume less power than active ranging sensors do.

Nonetheless, there are associated challenges that must be addressed with CV solutions. First, the computational load of image processing and map building is demanding limited on-board resources. As with any computationally intensive task, this is often handled by relying on powerful off-board computers, which leads to data latency and increases power consumption. MAVs must follow the Johnson criteria (JC) for the quality of video transmission [69,78]. JC establishes the resolution thresholds for object detection, orientation, recognition, and identification. Compared with LIDAR scanners, cameras have a limited FoV, which limits the use of solutions relying solely on vision, principally in active urban environments. Combining multiple EO sensors and curved lenses to improve the FoV can increase the role of CV-based systems.

Optical sensing can suffer from sensitivity to external changes in environmental conditions and disturbances. A single-camera is not suitable for MAV environmental-awareness applications, such as mapping, because instantaneous depth data are not captured. A complementary sensor can provide depth information such as using two cameras (stereovision) [69] and monocular cameras with INUs and/or range sensors [79]. Kinect and other RGB-D sensors provide single-camera images with depth information, but they are limited in range, and their use is compromised by strong sunlight. The use of a single camera with an INU is more practical for MAVs because most onboard flight control systems are equipped with INUs. Consequently, this leads to a reduction in payload mass and energy consumption. On the contrary, it has been shown the INUs are typically ill-suited for MAVs operations, because of their relatively high levels of uncertainty.

8.4 Illumination

Illumination is one of the most significant components for configuring the hardware. To achieve the most efficient system, the light source should be chosen based

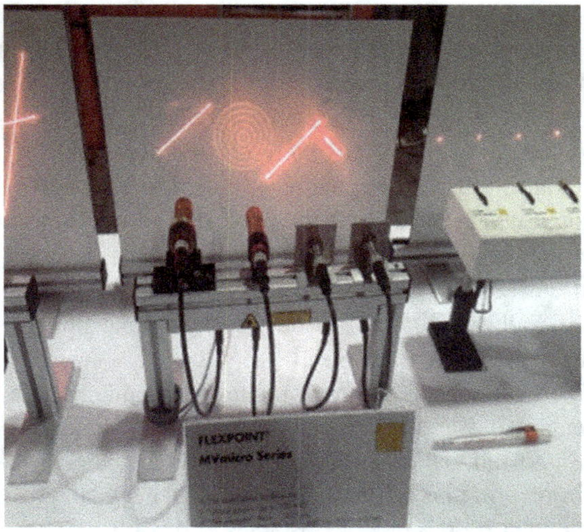

Figure 8.6 Structural lighting

on its applications, like the required stand-off distances, the environmental problems and the amount of light [80].

The main requirement of the system for achieving proper monitoring is to provide adjusted contrast between the object and the background regions. We should also adjust the illumination angle and the intensity source to emphasise the tool region of interest. Usually, lasers, incandescent lamps, and sometimes the natural light are used as light sources for condition monitoring.

In addition to appropriate light source selection, it is also essential to consider proper techniques to give the optimum results. There are three main techniques which are used extensively for different machine vision applications [81] (Figure 8.6):

1. Backlighting: This method illuminates the subject from the back and involves a proper contrast, but it is limited to silhouette information.
2. Front lighting: This method provides direct illumination of the object; this method has been extensively used in image processing applications.
3. Structured lighting: This method projects a known pattern (often grids or horizontal bars) on to a scene. This method is usually utilised to estimate the depth of the region.

8.5 Navigation, pathfinding, and orientation

Path planning algorithms for UAVs, in many reconnaissance applications, count on GPS data for localisation. Nevertheless, disruption of GPS signals, intentionally or otherwise, can make these plans and procedures useless.

Many insects use the sky polarisation patterns for navigation. For instance, honeybees use celestial polarisation to move between the foraging locations and the hive [82]. The salmon fishes to orient underwater light fields [83] use a similar ability. Another example is the light reflection on the water surface, which helps the insects to find their path and orientation [84].

Some research studies have focused on Guidance, Navigation, and Control (GNC) for UAVs, resulting in various methods and systems [85]. There are three main categories: control, navigation, and guidance. For each category, methods are grouped at the highest level based on the autonomy level they provide, and then according to the algorithmic approach used, which in most cases depends on the type of sensors used [85].

UAV GNC is traditionally carried out through three methods, namely radio control, video-based, and autopilot [86].

The use of UAVs in an indoor environment is a challenging task, and it needs higher requirements than the outdoor environment. It is challenging to use GPS to avoid collisions in an indoor environment. Usually, an indoor MAV deals with a GPS-denied environment, thus inhibiting the use of RF signals. RF signals could be reflected, and degraded by the indoor obstacles and walls.

Stationary landmarks can aid localisation in GPS-disrupted or GPS-denied environments. In particular, given the vehicle's path, we formulate a landmark-placement problem and present algorithms to place the minimum number of landmarks while satisfying the localisation, sensing, and collision-avoidance constraints.

In a radio-control system, the UAV navigation system uses a transmitter and a receiver [86]. Remote control (RC) equipment consists of a radio transmitter that includes several radio channels. The pilot transmits instructions to the UAV using any of these channels [3]. In RC systems, the transmitter range is different and usually covers a range of about five kilometre.

A UAV radio transmitter must have at least 4 to 6 channels to control their different flight levels. Additional channels can be used for camera control, where the receiver transmits instructions to the servomotors and speed controller [86] (Figure 8.7).

For video-based navigation systems, a camera installed on a MAV takes videos and photos when passing regions and sending them to the ground station by a video transmitter. Small size, low weight, and high visibility and clarity are considered as the essential features of a video system [86]. In a video-based system, the images sent from the video-transmitter and the images received by the antenna are displayed on a screen at the ground station. Analysis of the output waves can be used to evaluate antennas. In some cases, amplifiers are combined with the antenna, which makes it much easier to receive images. Currently, ultrasonic sensors, colour, thermal, or IR cameras obtain information about the environment. MAVs often use colour cameras, which are more useful only in the daytime and cannot provide scale and depth information for the observed environment. In a video-based navigation system, CV plays a vital role in the MAV automation because they extract the required data from images for navigation, stabilisation, and additional data collection.

Ordinarily, video transmitters can broadcast signals over a limited distance that is not suitable for long-distance flights. When out of range, drones signal displaying a dead zone and decrease the flight radius [86]. The best type of GNC for MAVs is the autopilot, which is a set of software and hardware tools to enable their flight

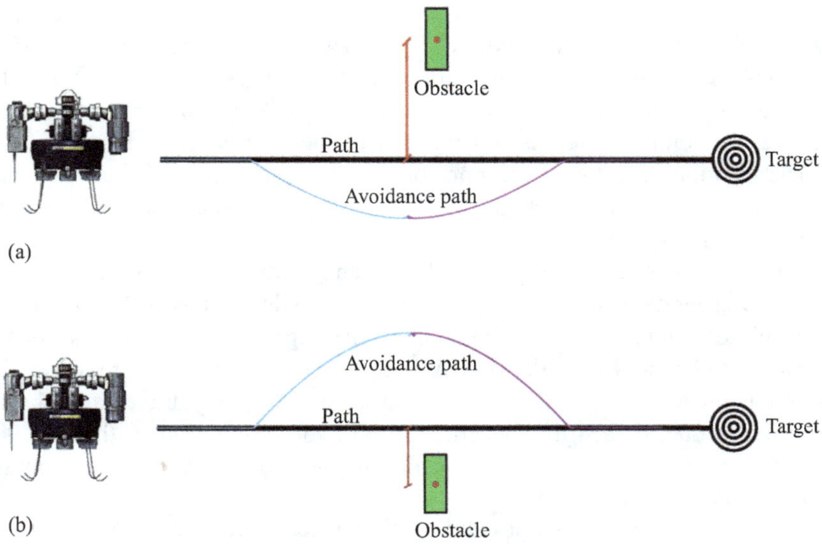

Figure 8.7 Pathfinding robot with an obstacle in the (a) bottom and (b) top of the avoidance path [13]

operations automatically. For example, by defining flight plans, direction, and speed in different parts of the flight, then the drone automatically conforms to this flight plan to perform its assignment with minimal errors [86].

Nowadays, several types of autopilots exist, for example, Micropilot [87], Barton [88], and Paparazzi. In an autopilot system, the flight plan is uploaded on-board before the flight, and at any moment, the drone can contact the ground station and transmit altitude, and velocity data, for instance. The ground station can send different instructions through RF modems to the UAV. After receiving instructions, the autopilot sends the corresponding commands to the servo to perform the desired operation [86].

New navigation approaches can be applied to MAVs in the future. Head movements can be used to control a quadrotor by transforming head nodes into flying instructions and sending them over to the MAVs, which can be appropriate in the case of limited flight ranges and to perform the hovering flight, such as RMAVs.

A brain–computer interface (BCI) controlled the movement (turn, rise, dip, and fly through a ring) of a quadcopter in [89]. Electroencephalography (EEG) is a non-invasive technique, which can record the electrical activity of the subject's brain with a cap with electrodes to send the signals to a quadrotor through Wi-Fi [90]. In BCI, when there is a movement or the thought about it, neurons in the motor cortex send small currents. Thinking about a different movement activates a new set of neurons. Smartphones can control and navigate MAVs.

The key differentiating factor in UAV control and navigation is the quality of communication expressed regarding the lag of the control loop, control bandwidth, and communication loss. Generally, UAV control over short distance results in a negligible lag and high bandwidth with minimal losses, while control over thousands of miles results in severe control lag, low bandwidth, and substantial losses. Hence, UAVs for long-distance and endurance flights typically have augmentation autopilots to stabilise flights if there is a loss of the command and control link. Likewise, the control system is structured differently for the same reason.

The positioning system is responsible for the central part of the MAV navigation. There are different methods for UAV positioning, such as GPS and inertial navigation system (INS) [3].

In drones, to detect the position, velocity, and altitude, GPS is usually used. To provide an accurate position of the drone, GPS should be in contact with at least four satellites simultaneously [86]. Since GPS signals are easily disturbed by external noise or interference [18], MAVs equipped only with GPS may lose their connection for some time. Such a situation leads to landing drones and aborting their mission due to security concerns. So, to avoid this problem, the INS is an appropriate method to estimate the location of the drones when they temporarily lose their GPS connection [3].

The INS includes gyroscopes and accelerometers, which are used to calculate the position and orientation of the UAVs. Nowadays, GPS and INS are combined to avoid errors in positioning and to produce accurate navigational evidence. The Kalman filter is a typical fusion algorithm and, the extended Kalman filter (EKF) can estimate the MAVs location when they lose their GPS info momentarily [3].

MAVs navigation and direction can use wireless networks and the Internet to exchange instructions [91–93].

8.6 Communication and polarisation-inspired machine vision applications

Some animals use polarisation patterns to signal with each other. Among the signals, some are controlled by linear polarised light reflection. By employing the polarisation-dependent reflectivity, forest butterflies use their wings to identifying markers [94,95] whereas males diagnose females based on that way [94–96].

Polarised patterns in some animals have a significant effect on the image formation as they comprise important image characteristics by the animals' visual fields. The described important polarised characteristics can be used in computer applications like bio-inspired applications in image formation, camera systems, robotics, and CV.

A large number of research works have addressed natural polarisation characteristics that inspired artificial vision techniques. Currently, there are also some applications, which are inspired by the natural behaviour of the animals to utilise in robot navigation and orientation, communication, camouflage breaking, and areal vision.

8.6.1 Robot orientation and navigation

There are lots of theoretical and experimental research works about the skylight polarisation for navigation and autonomous orientation [97]. Scattering of sunlight by particles polarised the sky and air molecules in the atmosphere. Polarisation information, which can be achieved by the sky polarisation patterns, can be used as an external compass. Skylight polarisation patterns can be affected by sun location [98].

The main advantage of employing the skylight polarisation pattern toward directly employing the sun is that only patches of sky are sufficient for orientation tasks. Photodiodes use orientation techniques to read polarisation data. Moreover, by getting cheaper and advanced camera technology, new methods using CCD, CMOS, and/or Fisheye lens are introduced to achieve the polarisation information.

8.6.2 Polarisation-opponent sensors

There is a kind of sensors which has inspired by the polarisation (POL)-neurons found in the insects. This system is a biological neuronal circuit, whose input devices include the POL-opponent (POL-OP) units [99,100].

POL-OP units include a pair of polarised light sensors (POL-sensors) and have the characteristics below (Figure 8.8):

- Each pair of POL-sensors fed its signals into a log-ratio amplifier.
- Polarised light sensors consist of a blue transmitting filter and photodiodes with a linear polariser. The polarising axis of each POL-sensor unit is adapted on 90° for the polarising axis of the other sensor, thus inspiring the crossed-analyser configuration in the POL-area of insect eyes.
- The mobile robot consists of three pairs of POL-sensors that are mounted on it and tuned as the polarising axis of the positive channel was similar to insect layout (0°, 60°, and 120°) regard to the body axis of the robot.

Achieving the compass direction from the POL-OP responses can be done twofold: simultaneous models and scanning models.

In scanning models, the solar meridian should be found to utilise as reference direction 0° for its activated system.

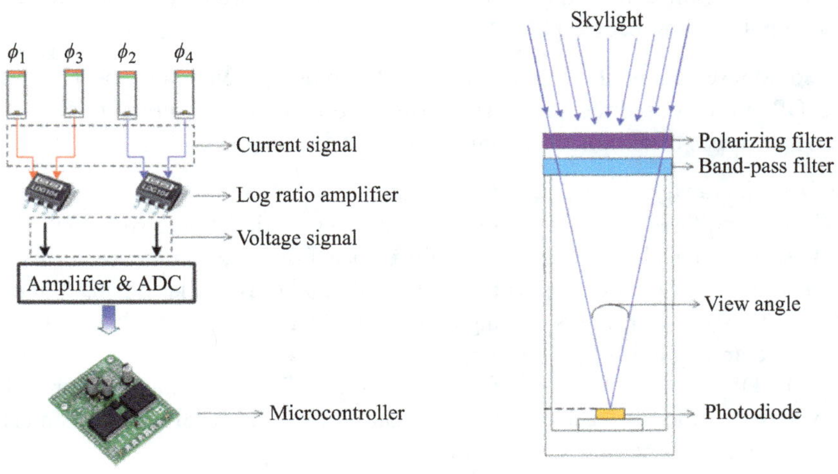

Figure 8.8 Schematic of POL sensors

1. To do this, the sky should be scanned actively by swinging it around the vertical body axis. Likewise, when reaching the maximum value on the output signal of each POL-OP units (or a combination of multiple POL-OP units), the robot swings itself into the solar meridian direction.
2. The method has employed current information from the solar meridian to find the heading direction.
3. By comparing the corresponding direction of the robot with the POL-OP unit's current output values with a lookup table, the compass direction can be achieved.
4. Before experiments by a single 360° rotation on the robot, we should record the lookup table.

Instead, in the simultaneous model, there is no need to continue scanning movements for determining the heading direction continuously.

In [101], a simultaneous model is employed, which does not need the lookup table; instead, an analytical procedure is utilised to derive compass information from the POL-OP units values.

They also consider the polarisation pattern changes during the day, which can be done by either normalising the outputs of the POL-OP units or regularly updating the lookup table by the following considerations.

- The sigmoid function is applied to the POL-OP signals for de-logarithmized
- Two candidate orientations can be defined as follows [101]:

$$\varphi = \frac{1}{2}\arctan\frac{v_1(\varphi) + 2v_2(\varphi) - \frac{3}{2}}{\sqrt{3}(v_1(\varphi) - \frac{1}{2})}, \tag{8.17}$$

  where $v_1(\varphi)$ is the output of the sensor.
- A set of ambient-light sensors can be implemented on the robot to solve the ambiguity.
- Transform the achieved data for POL-OP to signals which are independent of the polarisation degree [101].

Some researchers like [102] worked on the bionic orientation sensor with the POL-OP unit and used the same mathematical formulation (Figure 8.9).

These types of design have some definite characteristics:

1. Direction analysers are placed on a plate.
2. They used three polarisation direction analysers with the polarising axis of the positive channel adjusted to a 60° difference from one to one. Each of them includes two POL-sensors with the shape of regular triangular prisms. For each direction analyser, the polarising axis of one POL-sensor was adjusted 90° for the polarising axis of the sensors.
3. POL-OP unit is composed of a POL-sensor following a log-ratio amplifier where the amplifier recovers the input data from POL-sensors and transmits the difference logarithmic signal of them.
4. Ambient-light sensors are placed and arranged as a ring in the metal cylinder of the POL-sensors.

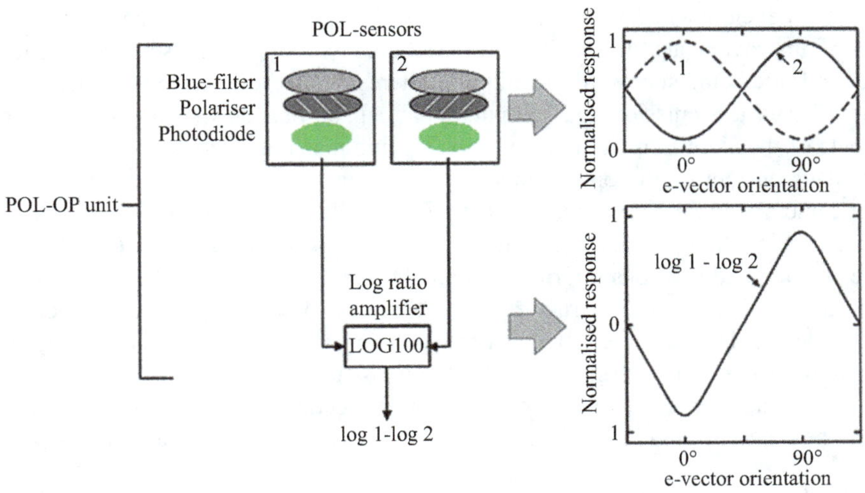

Figure 8.9 Schematic of the POL-OP unit

There are many types of research like [99,103–105], which illustrated that achieving the directional information from polarisation has high accuracy, and the travelling distance shows the value of the error.

Schmolke *et al.* [106] compared the performance using a polarisation compass with the performance of the path integration without external reference. In this research, results showed that the e-vector compass has the highest accuracy.

Chahl and Mizutani [107] improved two bio-mimetic sensors, and then they performed flight tests for orientation and stabilisation of an aircraft using the polarisation, spatial, and spectral distribution of the environment light. The main attempt in their work was to imitate the head of a kind of dragonfly.

8.7 CCD cameras and applications in machine vision

With the advancement of CCD sensors in image understanding and processing speed, they have been turned into developed camera-based systems, which can be used for different purposes like UAVs. There are different works about using CCDs in UAVs [108] and MAVs [109]. For example, the Delft University has introduced a MAV with a flapping wing, which depends on a flapping configuration bi-plane wing for thrust and lifts [110].

Recently, because of the profits of CCD sensors like improved solid-state technology and the emergence of cost-effective, they are now accessible for different purposes of machine vision applications.

CCD consists of an analogue shift register with a series of capacitors that are closely spaced. Ordinary CCDs have a pixel resolution of 768 × 493 with imaging rates of 30 images per set, and high-resolution CCDs have sized up to 2048 × 2048 pixels, but with a lower temporal resolution with a very high cost.

By development and miniaturisation of the actuators (like lasers) and especially improvements in the detectors technology (e.g., CCD sensors), optical methods are being successful. In addition, there are also some developments in the rate of data transmission and computational capabilities, as well as the improvements of the algorithms in image processing and machine vision.

By considering the optical indoor positioning systems, a mobile sensor (i.e., the camera) should be placed, and static sensors locate the moving objects in the images. Since the primary observations in CCD sensors are 2D positions, to find the position and rotations information in a 3D world, we should measure the image coordinates for all of the structures on the camera-based system; system coordinates define the exclusively built on triangulation and angular information by the angle of arrival (AoA) principle [111] (Figure 8.10).

We can achieve the transformation of the projection centre of a camera (X_0, Y_0, Z_0), the object space (X, Y, Z), and the image coordinates (x'', y'') as follows:

$$\begin{pmatrix} X \\ Y \\ Z \end{pmatrix} = \begin{pmatrix} X_0 \\ Y_0 \\ Z_0 \end{pmatrix} + \lambda \Re \begin{pmatrix} x'' \\ y'' \\ -c \end{pmatrix} \tag{8.18}$$

where λ is a coefficient (also called distance or depth), R is a rotation matrix that defines the orientation angles on the camera, and c is the camera constant. Typically, several images are taken from multiple cameras or multiple views of a single camera.

From the above formula, object coordinates (camera orientation parameters) from known 3D object coordinates can be determined by using a resection procedure [111,112].

It is essential to know that if one single image is available, the scale factor λ differs for each pixel and is unknown; so, additional depth information is needed to transform from the image space into the object space.

Figure 8.10 CCD-based MAVs

By moving the camera in different orientations, we can achieve depth information of the images. For example, in synthetic stereo vision, the view can be observed sequentially from different places by the same camera.

Furthermore, a supplementary method is required for determining the baseline between sequential images. So, the system coefficient λ needs a separate solution in addition to determining from the images.

Hence, using a stereo camera system with an accurate baseline helps to determine the coefficient from the stereoscopic images.

One of the crucial problems in the stereo camera system is that the performance directly depends on the length of the stereo baseline. By using a short baseline, the geometry has become unfavourable, especially for the forward intersection. Hence, it is not performable to use miniaturisation for handheld devices applications.

We can alternatively sense the distances using additional sensors like laser-scanners or range imaging cameras. The latter returns a distance value for every pixel of an image. The position of the autofocus can be utilised to describe the scale.

The method of how reference information can be achieved is an essential characteristic of the system architecture (Figure 8.11).

Table 8.2 shows an overview of the mentioned system's key characteristics [111].

A new overview of the video tracking systems can be found in [135]. The principle methods include different approaches to rigid object tracking, window tracking, deformable contour tracking, feature tracking, and visual learning.

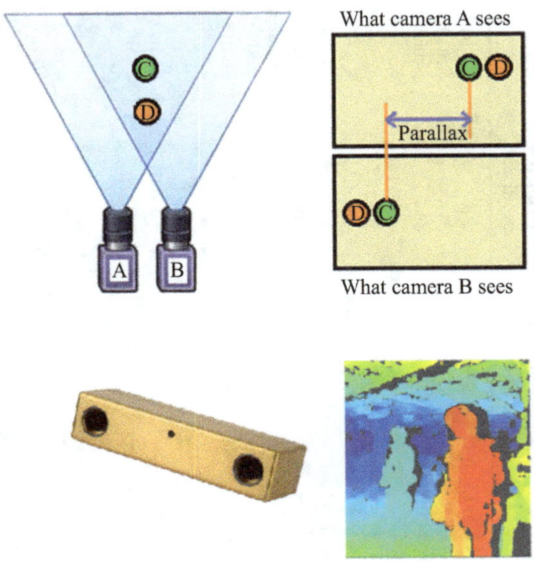

Figure 8.11 Performance and the output of a stereo camera system

Table 8.2 Optical indoor positioning systems

Name	Coordinate reference	Reported accuracy	Coverage	CCD size (pixel)	Frame rate	Object/camera positioning	Camera costs	Market maturity
Kohoutek [113]	CityGML	dm	Scalable	176×144	54 Hz	Cam, SR 4000	9000 $	Suggestion
Hile [114]	Floor plan	30 cm	Scalable	640 × 480	0.1 Hz	Obj., Cell Phone	100 $	Development
Kitanov [115]	Vector model	dm	Scalable	752 × 585	10 Hz	Cam., EVI-D31	£ 245	Development
Schlaile [116]	Edges segments	1 dm/min	Scalable	752 × 582	50 Hz	Cam., VC-PCC48P	175 €	Development
Ido [117]	Images	30 cm	Scalable	320 × 240 × 4	30 Hz	Cam., IEEE1394	–	Development
Sjö [118]	Images/scans	sub m	Scalable	320 × 240	30 Hz	Cam., VC-C4R	700 €	Development
Muffert [119]	Images	0.15 gon/min	Room	1616 × 1232 × 6	15 Hz	Cam., Ladybug3	>10000$	Development
Maye [120]	Images/landm.	1%	Scalable	16 × 16	2300 Hz	Cam., ADNS-2051	1.35 €	Development
Mulloni [121]	Coded markers	cm – dm	Scalable	176 × 144	15 Hz	Cam., Cell Phone	Low	Product
StarGazer [122]	Coded markers	cm – dm	Scalable	–	20 Hz	Cam.	980 $	Product
Lee [123]	Coded markers	dm	36 m^2	1280 × 1024	30 Hz	Cam.,VX-6000	40 $	Development
naviSCAN3D	Projection	50 μm	1.5–10 m	2448 × 2048 × 2	1 Hz	50 μm Obj., steroSCAN	High	Product
TrackSense [124]	Projection	4 cm	25 m^2	640 × 480	15 Hz	Obj.,CamPro4000	200 $	Development
CLIPS [125]	Projection	0.5 mm	36 m^2	1032 × 778	30 Hz	Cam., GuppyF80	1000 €	Development
Habbecke [126]	Projection	mm	25 m^2	1280 × 960	–	Obj.	1000 €	Development
Popescu [127]	Projection	cm	25 m^2	720 × 480	15 Hz	Cam.	1500 $	Development
DEADALUS [128]	–	0.04 mm	m – km	1024 × 768	30 Hz	Obj., GuppyF80	High	Development
Boochs [129]	–	0.05 mm	4 m^3	2000 × 2000 × 4	–	Obj.,	High	Development
Tappero [130]	–	dm – m	30 m^2	356 × 292	3 Hz	Obj., OV6620	20 $ US	Suggestion
Soloviev [131]	GNSS	cm	Scalable	1240 × 1024	–	Obj.	–	Suggestion
Aufderheide [132]	Image features	–	Scalable	–	5–30 Hz	Cam.	–	Suggestion
Liu T. [133]	Scanner, image	1%	Scalable	1338 × 987 × 3	10 Hz	Obj.	Low	Development
Liu W. [134]	Magnetic field	1 mm	1 m^3	768 × 576 ×	4 25 Hz	Obj., Sony ICX	800 €	Development

8.8 Error modelling of environments with uncertainties

Here, the development of MAVs vision in the presence of uncertainties is studied. Generally, controlling of MAV for obstacle avoidance and handling uncertain urban environment is known as the most complicated part of the autonomous flight control problem.

Generally, all of the MAV handling methods, especially vision-based methods, need high-performance microprocessors, which is a significant limitation of their usage in real MAV [136].

The primary purpose of the autonomous flight control is to handle MAVs such that they have independence during the flight in an unknown and uncertain environment as much as it is possible.

One of the hard environment, which accounts as a complicated and uncertain region for MAVs is the uncertain urban environment [137].

Indeed, the uncertainty of an urban environment is the city's unknown streets, buildings, and other obstacles location and orientation, which their places are sometimes changes. Therefore, having a warranty for rapid response against environment obstacles is a significant problem in the MAVs.

In this case, the algorithm should also navigate MAV through an urban environment in a safe and optimal path; i.e., the performance of the autonomous flight control is utterly dependent on the navigation and control algorithms.

There are several types of research about improving the effectiveness of autonomous flight control of MAVs in an uncertain environment, which most of them are focused on MAVS based on digital image processing and machine vision [79,138–142].

The importance of vision-based systems is that they allow analysing a considerable area around MAV, and it makes the possibility to create a map of the environment used next by path planner for the environments by uncertainties [140].

The main shortcoming of these MAVs is their need for complex imaging processing systems, which needs efficient DSP processors with high power consumption, i.e., its effects on the available amount of power stored in on-board batteries. This shortcoming can be covered by increasing the capacity of batteries, but this approach has a big problem: it increases the total weight of MAV.

In vision-based MAVs, image stabilisation is an additional problem. Changing in shadows place, weather changes, and other uncertainties make the vision system to face a complicated target. Generally, vision-based can be considered as the future of the autonomous MAVs. There are still more works to able to use them in the real-world systems and the size of MAVs.

8.9 Further work and future trends

The design of MAVs involves conceptual design, preliminary design, and detailed design [143]. Each step requires increasingly sophisticated sizing, aerodynamic, aeroelastic, structural, propulsion, stability, control, electronic, and fabrication analysis [11,26,27].

Size impacts the optimum values of their dimensions and weights and has usually five steps: (1) defining the mission, (2) setting the flight mode based on the type, (3) determining the wing shape (planform) and aspect ratio, (4) constraint analysis, and (5) weight estimation [11,26,27]. The analysis of the route happens during the mission definition, and it helps to determine the flight time, cruise speed, turning speed, flight modes, the wing shape, and its aspect ratio for the type of mission. Then, to determine the appropriate UAV wing and thrust loadings, a constraint analysis is carried out to simulate the kinematic and dynamic equations of the flight. Along with the steps above, different methods for weight estimation can be employed resulting in the determination of the UAV geometry and dimensions besides the calculation of some aerodynamic parameters for each type [27].

MAVs are more than scaled-down versions of larger aircraft [10] in a small volume. The challenges and complexity in their design and fabrication of insect-sized drones grow significantly due to the corresponding low Reynolds numbers, which results from their low speed and small sizes [144]. Generally, the difficulties during the flight in this regime of flow led to the study of the flight of insects [145]. Next, some challenges for designing some types of MAVs are discussed.

8.9.1 MAV challenges

FMAVs are the most advanced MAVs and the easiest ones to design and fabricate because there are different methods for larger fixed-wing aeroplanes, which can be applied with some modifications in aerodynamic and geometric characteristics [146]. There is a wide variety of FMAVs across the world with different flight speeds, altitudes, and endurance, depending on the mission [26,27]. These FMAVs necessitate relatively higher speeds for flight in comparison to rotary wings or flapping wings, cannot hover or fly slowly, and the indoor flight is very challenging for them [7,8].

FMAVs usually require a thrust loading less than one and less power to fly than a helicopter with the same weight in hovering mode. When the velocity and dimensions decrease, the operating Reynolds number is reduced, and thus, the efficiency of the drone is also decreased [22].

RMAVs rely on the number and positions of their motors, can fly at high speeds, and can perform the vertical take-off, landing, and hovering flight [3,10]. RMAVs can fly in indoor spaces and are perfect for surveillance with limited endurance due to the required higher power for the hovering flight mode. As the size and weight of RMAVs decrease [3], many design challenges appear. Despite these shortcomings, RMAVs can fly with high and low speeds and can perform hovering flight based on the defined undertaking [72].

8.9.2 Proposed solutions for MAV design challenges

To overcome the mentioned challenges for different types of MAVs, developers and designers of drones should consider various parameters in the design process, which can result in developing optimised drones. As discussed in the previous sections, each type of drones and their design methods have advantages and disadvantages. Therefore, by using theoretical, statistical, revised allometric, and bio-inspiration methods, a comprehensive methodology can be proposed, which finds solutions for the drawbacks of previous methods. Various types of MAVs can be introduced by taking inspiration from nature [73]. Currently, the ability to transform and change the configuration is a new field of research.

The MAV design process should consider the aircraft configuration and its design methodology. Recently, there are some efforts to project UAVs with alternative configurations inspired by nature, e.g., birds, insects, marine organisms, and others.

The main problem with MAVs is their low flight endurance when compared to larger UAVs due to their power consumption. Different geometrical and physical parameters, for example, wing shape, wingspan, airfoil, speed, meteorological conditions, among others, can decrease the drag and consequently the power consumption of MAVs. Solar panels and piezoelectric energy harvesters can be used as renewable energy sources to augment MAV flight endurance or to operate extra sensors and cameras [74,75,147].

Challenging issues are their high power consumption and limited power capacity attributable to their weight limitation that causes little flight endurance [74]. Now, mounting solar panels on MAVs can be a conventional method to increase flight endurance, and, commonly, the battery is used as a backup when the solar cells cannot generate enough power and flying in or under clouds or in the dark. In other words, a hybrid source, which is a combination of the solar cells and battery, is usually used for powering UAVs [74]. Solar cells, which are thin, flexible, low weight, and efficient, are applied to the wings of different types of drones.

The solar cells must be light, flexible, and highly efficient. Thin-film solar cells (TFSCs) on the drone wing surfaces do not substantially affect the aerodynamic efficiency [74]. Significant limitations of the solar cells can be their high costs, low efficiency, and their temperature sensitivity. Increased temperatures reduce the power output of solar cells [76]. One of the parameters that have an enormous impact on the maximum power output of solar cells is the amount of solar radiation absorbed by the solar cell [77]. Series and parallel connections of solar cells are used to achieve the required voltage and current to improve their performance.

Since the solar-powered UAVs cannot produce enough power to fly in or under clouds or in the dark, their use at night is limited. One answer to this problem is to use laser light from a conventional power source, such as a portable generator or an electrical grid. For example, a laser beam is directed to a photovoltaic receiver, which is installed under the drone [148] to provide unlimited flight endurance to overcome the limitations of most UAVs [149]. When it comes to wireless power systems, having the energy source on the ground simplifies design and makes power generation inexpensive [150]. Laser systems do not need to turn off at night and can continuously charge the battery [148]. Even though this system can solve the endurance issues, it has some problems in the flight range. For instance, this system cannot be part of high altitude UAVs, but it can be the right choice for RMAVs.

Harvesting energy can increase MAV endurance from flapping motion and morphing, where there is an optimum electrical load resistance that optimises the harvested power. The piezoelectric energy harvested from the morphing of wings can also power many sensors and cameras using wasted mechanical energy [75].

8.9.3 New frontiers in sensors

Safe navigation and path planning require a combination of multiple sensors to provide sufficient information for the MAV [151]. Hence, sensor fusion of signals of several different modalities and filtering techniques are needed for MAV perception. Characteristics such as payload, the on-board computer capacity, and aircraft energy limitations can impact technology solutions for a specific mission.

For a set of MAV characteristics, the selection of suitable sensors with a given duration, distance, and environment requirements must consider all the inherent limitations. In real applications, dynamic obstacles can appear at any distance. RMAVs and recently introduced hybrid BMAVs can hover, while a path is being calculated around the dynamic obstacle. MAV Sensors require further research to address some unique constraints, especially in urban environments, to cut power consumption, computational complexity, bandwidth restrictions, and allow miniaturisation [69,152,153].

Light-Emitting Diode Detection and Ranging (LEDDAR) technology that utilises separate LED beams to diminish computational complexity and power consumption and to increase the range of ToF measurements. The beam size reduces the LEDDAR sensor resolution at long distances.

The radar can be implemented on a single Complementary Metal-Oxide Semiconductor (CMOS) chip to reduce power consumption and costs while maintaining an acceptable target-detection range; and Single-Photon Avalanche Diodes (SPAD) can fuse images with range data for individual photons, to generate 3D images. Multi-channel solid-state SPAD sensors can improve image and timing resolution, but keep a compromise between spatial and temporal resolution for the SPAD sensors.

8.10 Conclusion

This chapter provides a survey of bio-inspired UAV application for imagery acquisition in different environments, design challenges along with possible solutions. Moreover, the used manufacturing methods and challenges, propulsion systems and actuators, power supply and endurance, control and navigation of drones are reviewed with proposing new ideas to get rid of the existing limitations. The importance of swarm flight and separation of drones is also discussed. Using only one drone for a given mission can be risky in view of some technical or other problems. Hence, various missions can be performed more efficiently with multiple UAVs or drone swarms. A swarm flight of drones has an advantage. If one drone of the swarm strays in flight, the rest of the drones can carry out the mission. These air vehicles push the flight limitations in the low Reynolds number regime, requiring new research in designing optimisation procedures; lightweight structures and materials; microelectronic devices; and aerodynamic modelling tools. The robust behaviour of living animals against nature problems makes a strong inspiration for the researchers to work on the robots which can do the hard things like the animals in nature. These inspirations make the researchers produce robots with high robustness and flexibility. There are different inspiration from the animals like their behaviours, their motions, their nervous system, and others. The main focus of this chapter is about the role of CV in MAVs.

References

[1] V. V. Estrela, O. Saotome, H. J. Loschi, *et al.*, "Emergency response cyber-physical framework for landslide avoidance with sustainable electronics," *Technologies*, 2018;6(2):42.

[2] V. V. Estrela, J. Hemanth, O. Saotome, E. G. H. Grata, and D. R. F. Izario, "Emergency response cyber-physical system for flood prevention with sustainable electronics," in *Proceedings of the 3rd Brazilian Technology Symposium (BTSym 2017)*, Springer, Cham, 2019. DOI: 10.1007/978-3-319-93112-8_33.

[3] M. Hassanalian and A. Abdelkefi, "Classifications, applications, and design challenges of drones: a review," *Progress in Aerospace Sciences*, vol. 91, pp. 99–131, 2017.

[4] T. A. Ward, C. J. Fearday, E. Salami, and N. Binti Soin, "A bibliometric review of progress in micro air vehicle research," *International Journal of Micro Air Vehicles*, vol. 9, pp. 146–165, 2017.

[5] P. N. Sivasankaran and T. A. Ward, "Spatial network analysis to construct simplified wing structural models for biomimetic micro air vehicles," *Aerospace Science and Technology*, vol. 49, pp. 259–268, 2016.

[6] T. Gorjiara and C. Baldock, "Nanoscience and nanotechnology research publications: a comparison between Australia and the rest of the world," *Scientometrics*, vol. 100, pp. 121–148, 2014.

[7] A. Mohamed, R. Clothier, S. Watkins, R. Sabatini, and M. Abdulrahim, "Fixed-wing MAV attitude stability in atmospheric turbulence, part 1: suitability of conventional sensors," *Progress in Aerospace Sciences*, vol. 70, pp. 69–82, 2014.

[8] A. Mohamed, S. Watkins, R. Clothier, M. Abdulrahim, K. Massey, and R. Sabatini, "Fixed-wing MAV attitude stability in atmospheric turbulence—Part 2: investigating biologically-inspired sensors," *Progress in Aerospace Sciences*, vol. 71, pp. 1–13, 2014.

[9] A. M. Coelho and V. V. Estrela, "EM-based mixture models applied to video event detection," *Principal Component Analysis – Engineering Applications*, IntechOpen, pp. 101–124, 2012. DOI: 10.5772/38129

[10] M. Hassanalian and A. Abdelkefi, "Design, manufacturing, and flight testing of a fixed wing micro air vehicle with Zimmerman planform," *Meccanica*, vol. 52, pp. 1265–1282, 2017.

[11] M. Hassanalian, H. Khaki, and M. Khosravi, "A new method for design of fixed wing micro air vehicle," *Proceedings of the Institution of Mechanical Engineers, Part G: Journal of Aerospace Engineering*, vol. 229, pp. 837–850, 2015.

[12] K. P. Valavanis, *Advances in Unmanned Aerial Vehicles: State of the Art and the Road to Autonomy*, vol. 33: Springer Science & Business Media, Amsterdam, Netherlands, 2008.

[13] A. E. Fraire, R. P. Morado, A. D. López, and R. L. Leal, "Design and implementation of fixed-wing MAV controllers," in *Proceedings of 2015 Workshop on Research, Education and Development of Unmanned Aerial Systems (RED-UAS)*, Cancun, Mexico, 2015, pp. 172–179.

[14] D. McLean, *Automatic Flight Control Systems*, vol. 16: Prentice Hall, New York, 1990.

[15] J. Guerrero and R. Lozano, *Flight Formation Control*, John Wiley & Sons, New Jersey, USA, 2012.

[16] A. Barrientos Cruz, J. Colorado Montaño, A. Martínez Álvarez, and J. R. Pereira Valente, "Rotary-wing MAV modeling & control for indoor scenarios," In Proc. 2010 IEEE International Conference on Industrial Technology, Vina del Mar, Chile, pp. 1475–1480, 2010.

[17] P. M. Joshi, "Wing analysis of a flapping wing unmanned aerial vehicle using CFD," *Development*, vol. 2, 216–22, 2015.

[18] R. Austin, *Unmanned Aircraft Systems: UAVS Design, Development and Deployment*, vol. 54: John Wiley & Sons, New Jersey, USA, 2011.

[19] W.-J. Han, Y.-H. Lei, and X.-W. Zhou, "Application of unmanned aerial vehicle survey in power grid engineering construction," *Electric Power Survey & Design*, vol. 3, p. 019, 2010.

[20] E. Altug, J. P. Ostrowski, and R. Mahony, "Control of a quadrotor helicopter using visual feedback," in *Proceedings of IEEE International Conference on Robotics and Automation, 2002. ICRA'02. IEEE*, 2002, Washington D.C., USA, pp. 72–77.

[21] G. Cai, J. Dias, and L. Seneviratne, "A survey of small-scale unmanned aerial vehicles: recent advances and future development trends," *Unmanned Systems*, vol. 2, pp. 175–199, 2014.

[22] W. Shyy, Y. Lian, J. Tang, D. Viieru, and H. Liu, *Aerodynamics of Low Reynolds Number Flyers*, vol. 22: Cambridge University Press, Cambridge, USA, 2007.

[23] S. R. Fernandes, V. V. Estrela, and O. Saotome, "On improving sub-pixel accuracy by means of B-spline," in *Proceedings of 2014 IEEE International Conference on Imaging Systems and Techniques (IST)*, Santorini, Greece, 2014, pp. 68–72. DOI: 10.1109/IST.2014.6958448.

[24] A. M. Coelho, V. V. Estrela, F. P. do Carmo, and S. R. Fernandes, "Error concealment by means of motion refinement and regularized Bregman divergence," in *International Conference on Intelligent Data Engineering and Automated Learning*, Natal, RN, Brazil, 2012, pp. 650–657. DOI: 10.1007/978-3-642-32639-4_78.

[25] M. A. de Jesus, V. V. Estrela, O. Saotome, and D. Stutz, "Super-resolution via particle swarm optimization variants," in *Biologically Rationalized Computing Techniques for Image Processing Applications*, Springer, Zurich, Switzerland, 2018, pp. 317–337.

[26] M. Hassanalian, A. Abdelkefi, M. Wei, and S. Ziaei-Rad, "A novel methodology for wing sizing of bio-inspired flapping wing micro air vehicles: theory and prototype," *Acta Mechanica*, vol. 228, pp. 1097–1113, 2017.

[27] C. T. Orlowski and A. R. Girard, "Dynamics, stability, and control analyses of flapping wing micro-air vehicles," *Progress in Aerospace Sciences*, vol. 51, pp. 18–30, 2012.

[28] G. Abate, M. Ol, and W. Shyy, "Introduction: biologically inspired aerodynamics," *AIAA Journal*, vol. 46, pp. 2113–2114, 2008.

[29] G. Throneberry, M. Hassanalian, and A. Abdelkefi, "Optimal design of insect wing shape for hovering nano air vehicles," in *Proceedings of 58th AIAA/ASCE/AHS/ASC Structures, Structural Dynamics, and Materials Conference*, 2017, p. 1071.

[30] M. Hassanalian, G. Throneberry, and A. Abdelkefi, "Forward flight capabilities and performances of bio-inspired flapping wing nano air vehicles," in *55th AIAA Aerospace Science Meeting*, Grapevine, Texas, USA, 2017, p. 0499.

[31] S. Ho, H. Nassef, N. Pornsinsirirak, Y.-C. Tai, and C.-M. Ho, "Unsteady aerodynamics and flow control for flapping wing flyers," *Progress in Aerospace Sciences*, vol. 39, pp. 635–681, 2003.

[32] J. D. DeLaurier, "An aerodynamic model for flapping-wing flight," *The Aeronautical Journal*, vol. 97, pp. 125–130, 1993.

[33] J. Whitney and R. Wood, "Conceptual design of flapping-wing micro air vehicles," *Bioinspiration & Biomimetics*, vol. 7, p. 036001, 2012.

[34] K. Jones, C. Bradshaw, J. Papadopoulos, and M. Platzer, "Bio-inspired design of flapping-wing micro air vehicles," *The Aeronautical Journal*, vol. 109, pp. 385–393, 2005.

[35] N. Razmjooy, F. R. Sheykhahmad, and N. Ghadimi, "A hybrid neural network – world cup optimization algorithm for melanoma detection," *Open Medicine*, vol. 13, pp. 9–16, 2018.

[36] N. Razmjooy, M. Ramezani, and A. Namadchian, "A new LQR optimal control for a single-link flexible joint robot manipulator based on grey wolf optimizer," *Majlesi Journal of Electrical Engineering.*, vol. 10, p. 53, 2016.

[37] N. Razmjooy, M. Ramezani, and N. Ghadimi, "Imperialist competitive algorithm-based optimization of neuro-fuzzy system parameters for automatic red-eye removal," *International Journal of Fuzzy Systems*, vol. 19, pp. 1144–1156, 2017.

[38] N. Razmjooy and M. Ramezani, "An improved quantum evolutionary algorithm based on invasive weed optimization," *Indian Journal of Scientific Research*, vol. 4, pp. 413–422, 2014.

[39] M. Khalilpuor, N. Razmjooy, H. Hosseini, and P. Moallem, "Optimal control of DC motor using invasive weed optimization (IWO) algorithm," in *the Majlesi Conference on Electrical Engineering, Majlesi town, Isfahan, Iran*, 2011.

[40] N. Razmjooy, V. V. Estrela, and H. J. Loschi, "A study on metaheuristic-based neural networks for image segmentation purposes," in Q. A. Memon, and S. A. Khoja (eds), *Data Science Theory, Analysis and Applications*, Taylor & Francis, Abingdon, UK, 2019.

[41] P. Moallem, N. Razmjooy, and M. Ashourian, "Computer vision-based potato defect detection using neural networks and support vector machine," *International Journal of Robotics and Automation*, vol. 28, pp. 137–145, 2013.

[42] D. Chiarella, M. Bibuli, G. Bruzzone, *et al.*, "Gesture-based language for diver-robot underwater interaction," in *Proceedings of the Conference Oceans 2015*, Genova, 2015, pp. 1–9.

[43] H. Albitar, "Enabling a robot for underwater surface cleaning," Örebro University, 2017.

[44] Y. Mulgaonkar, A. Makineni, L. Guerrero-Bonilla, and V. Kumar, "Robust aerial robot swarms without collision avoidance," *IEEE Robotics and Automation Letters*, vol. 3, pp. 596–603, 2018.

[45] D. J. Hemanth and V. V. Estrela, "Deep learning for image processing applications, Advances in Parallel Computing Series" Vol. 31, IOS Press, Netherlands, 2017, ISBN 978-1-61499-821-1 (print), ISBN 978-1-61499-822-8 (online).

[46] E. Garea Llano, D. E. Osorio Roig, and Y. Chacon Cabrera, Unsupervised segmentation of agricultural crops In UAV RGB Images. Rev Cubana Cienc Informat [online]. 2018, vol. 12, n. 4, pp.17–28. ISSN 2227-1899.

[47] M. Kamel, M. Burri, and R. Siegwart, "Linear vs nonlinear MPC for trajectory tracking applied to rotary wing micro aerial vehicles," *IFAC-PapersOnLine*, vol. 50, pp. 3463–3469, 2017.

[48] M. Saska, T. Baca, J. Thomas, *et al.*, "System for deployment of groups of unmanned micro aerial vehicles in GPS-denied environments using onboard visual relative localization," *Autonomous Robots*, vol. 41, pp. 919–944, 2017.

[49] V. V. Estrela and A. M. Coelho, "State-of-the art motion estimation in the context of 3D TV," in *Multimedia Networking and Coding*, ed: IGI Global, Hershey, PA, USA, 2013, pp. 148–173.

[50] H. R. Marins and V. V. Estrela, "On the use of motion vectors for 2D and 3D error concealment in H. 264/AVC video," in *Feature Detectors and Motion Detection in Video Processing*, ed: IGI Global, Hershey, PA, USA, 2017, pp. 164–186. DOI: 10.4018/978-1-5225-1025-3.ch008.

[51] V. V. Estrela, L. A. Rivera, P. C. Beggio, and R. T. Lopes, "Regularized pel-recursive motion estimation using generalized cross-validation and spatial adaptation," in *Proceedings of the 16th Brazilian Symposium on Computer Graphics and Image Processing (SIBGRAPI 2003), Sao Carlos, SP, Brazil.* DOI: 10.1109/SIBGRA.2003.1241027.

[52] V. V. Estrela, H. A. Magalhaes, and O. Saotome, "Total variation applications in computer vision," in *Handbook of Research on Emerging Perspectives in Intelligent Pattern Recognition, Analysis, and Image Processing*, IGI Global, Hershey, PA, USA, pp. 41–64, 2016.

[53] A. M. Coelho, J. T. de Assis, and V. V. Estrela, "Error concealment by means of clustered blockwise PCA," in *Proceedings of the 2009 IEEE*

Picture Coding Symposium, Chicago, IL, USA. DOI: 10.1109/PCS.2009.5167442.

[54] F. Fraundorfer, A. M. López, A. Imiya, T. Pajdla, and J. M. Álvarez, "Computer vision for MAVs," *Computer Vision in Vehicle Technology: Land, Sea & Air*, 1, pp. 55–74, 2017.

[55] S. Yang, S. A. Scherer, K. Schauwecker, and A. Zell, "Autonomous landing of MAVs on an arbitrarily textured landing site using onboard monocular vision," *Journal of Intelligent & Robotic Systems*, vol. 74, pp. 27–43, 2014.

[56] M. Müller, V. Casser, J. Lahoud, N. Smith, and B. Ghanem, "Sim4CV: a photo-realistic simulator for computer vision applications," *International Journal of Computer Vision*, 1, pp. 1–18, 2018.

[57] C. Kanellakis and G. Nikolakopoulos, "Survey on computer vision for UAVs: Current developments and trends," *Journal of Intelligent & Robotic Systems*, vol. 87, pp. 141–168, 2017.

[58] A. M. López, A. Imiya, T. Pajdla, and J. M. Álvarez, *Computer Vision in Vehicle Technology: Land, Sea, and Air*, John Wiley & Sons, New Jersey, USA, 2017.

[59] A. S. Huang, A. Bachrach, P. Henry, *et al.*, "Visual odometry and mapping for autonomous flight using an RGB-D camera," in *Robotics Research*, Springer, Zurich, Switzerland, 2017, pp. 235–252.

[60] E. Price, G. Lawless, H. H. Bülthoff, M. Black, and A. Ahmad, "Deep neural network-based cooperative visual tracking through multiple micro aerial vehicles," IEEE Robotics and Automation Letters, 2018;3(4):3193–3200.

[61] T. Jitsukawa, H. Adachi, T. Abe, H. Yamakawa, and S. Umezu, "Bio-inspired wing-folding mechanism of micro air vehicle (MAV)," *Artificial Life and Robotics*, vol. 22, pp. 203–208, 2017.

[62] F. Kendoul, K. Nonami, I. Fantoni, and R. Lozano, "An adaptive vision-based autopilot for mini flying machines guidance, navigation and control," *Autonomous Robots*, vol. 27, p. 165, 2009.

[63] J. Kim and G. Brambley, "Dual optic-flow integrated navigation for small-scale flying robots," in *Proceedings of Australasian Conference on Robotics and Automation*, Brisbane, Australia, 2007.

[64] J. Chahl, K. Rosser, and A. Mizutani, "Bioinspired optical sensors for unmanned aerial systems," *Bioinspiration, Biomimetics, and Bioreplication*, Vol. 1, p. 797503, 2011.

[65] A. Mohamed, S. Watkins, A. Fisher, M. Marino, K. Massey, and R. Clothier, "A feasibility study of bio-inspired wing-surface pressure sensing for attitude control of MAVs," *Journal of Aircraft*, vol. 152, pp. 827–838, 2014.

[66] L. Kleeman and R. Kuc, "Sonar sensing," *Handbook of Robotics*, Springer, Zurich, Switzerland, 2016, pp. 753–782.

[67] J. Steckel and H. Peremans, "BatSLAM: simultaneous localization and mapping using biomimetic sonar," *PLoS One*, vol. 8, p. e54076, 2013.

[68] F. Endres, J. Hess, J. Sturm, D. Cremers, and W. Burgard, "3-D mapping with an RGB-D camera," *IEEE Transactions on Robotics*, vol. 30, pp. 177–187, 2014.

[69] M. Elbanhawi, A. Mohamed, R. Clothier, J. Palmer, M. Simic, and S. Watkins, "Enabling technologies for autonomous MAV operations," *Progress in Aerospace Sciences*, vol. 91, pp. 27–52, 2017.

[70] M. Dekan, D. František, B. Andrej, R. Jozef, R. Dávid, and M. Josip, "Moving obstacles detection based on laser range finder measurements," *International Journal of Advanced Robotics Systems*, vol. 15, p. 1729881417748132, 2018.

[71] J. Morales, V. Plaza-Leiva, A. Mandow, J. A. Gomez-Ruiz, J. Serón, and A. García-Cerezo, "Analysis of 3D scan measurement distribution with application to a multi-beam LIDAR on a rotating platform," *Sensors*, vol. 18, p. 395, 2018.

[72] S. Hanford, L. Long, and J. Horn, "A small semi-autonomous rotary-wing unmanned air vehicle (UAV)," in *Infotech@ Aerospace*, ed. 2005, p. 7077.

[73] M. Hassanalian, H. Abdelmoula, S. B. Ayed, and A. Abdelkefi, "Thermal impact of migrating birds' wing color on their flight performance: Possibility of new generation of biologically inspired drones," *Journal of Thermal Biology*, vol. 66, pp. 27–32, 2017.

[74] M. Hassanalian, M. Radmanesh, and A. Sedaghat, "Increasing flight endurance of MAVs using multiple quantum well solar cells," *International Journal of Aeronautical and Space Science*, vol. 15, pp. 212–217, 2014.

[75] A. Abdelkefi and M. Ghommem, "Piezoelectric energy harvesting from morphing wing motions for micro air vehicles," *Theoretical and Applied Mechanics Letters*, 2013;3(5), https://doi.org/10.1063/2.1305204.

[76] P. Singh and N. M. Ravindra, "Temperature dependence of solar cell performance—an analysis," *Solar Energy Materials and Solar Cells*, vol. 101, pp. 36–45, 2012.

[77] M. K. Islam, T. Ahammad, E. H. Pathan, A. Mushfiqul, and M. R. H. Khandokar, "Analysis of maximum possible utilization of solar radiation on a solar photovoltaic cell with a proposed model," *International Journal of Modeling and Optimization*, vol. 1, p. 66, 2011.

[78] *Long-Range Surveillance Cameras & Johnson's Criteria*. Available from: http://www.aissecuritysolutions.com/white-paper-on-long-range-surveillance-cameras.pdf

[79] M. W. Achtelik, S. Lynen, S. Weiss, M. Chli, and R. Siegwart, "Motion-and uncertainty-aware path planning for micro aerial vehicles," *Journal of Field Robotics*, vol. 31, pp. 676–698, 2014.

[80] J. C. Russ, *The Image Processing Handbook*, CRC Press, Boca Raton, Florida, United States, 2016.

[81] V. Krátký, V. Spurny, T. Báca, and M. Saska, "Documentation of large historical buildings by UAV formations-scene perception-driven motion planning and predictive control." In Proc. ICRA 2017 Workshop on Multi-robot Perception-Driven Control and Planning, Marina Bay Sands in Singapore, 2017

[82] T. W. Cronin, N. Shashar, R. L. Caldwell, J. Marshall, A. G. Cheroske, and T.-H. Chiou, "Polarization vision and its role in biological signaling," *Integrative and Comparative Biology*, vol. 43, pp. 549–558, 2003.

[83] I. N. Flamarique and C. W. Hawryshyn, "Is the use of underwater polarized light by fish restricted to crepuscular time periods?" *Vision Research*, vol. 37, pp. 975–989, 1997.

[84] R. Schwind, "Polarization vision in water insects and insects living on a moist substrate," *Journal of Comparative Physiology A*, vol. 169, pp. 531–540, 1991.

[85] F. Kendoul, "Survey of advances in guidance, navigation, and control of unmanned rotorcraft systems," *Journal of Field Robotics*, vol. 29, pp. 315–378, 2012.

[86] M. Hassanalian, M. Radmanesh, and S. Ziaei-Rad, "Sending instructions and receiving the data from MAVs using telecommunication networks," in *Proceedings of International Micro Air Vehicle 2012 Conference (IMAV2012)*, pp. 3–6, 2012.

[87] S. Trites, "Miniature autopilots for Unmanned Aerial Vehicles," *MicroPilot,* Available from: http://www.micropilot.com.

[88] J. D. Barton, "Fundamentals of small unmanned aircraft flight," *Johns Hopkins APL Technical Digest*, vol. 31, pp. 132–149, 2012.

[89] *Fly mind controlled plane scientist pilots drone using just thoughts technology one day used commercial aircraft.* Available from: http://www.dailymail.co.uk/sciencetech/article-2970073/Would-fly-mind-controlled-plane-Scientist-pilots-drone-using-just-thoughts-technology-one-day-used-commercial-aircraft.html

[90] *University Minnesota researchers control flying robot only the mind.* Available from: https://cse.umn.edu/college/news/university-minnesota-researchers-control-flying-robot-only-mind

[91] V. V. Estrela, A. C. B. Monteiro, R. P. França, Y. Iano, A. Khelassi, and N. Razmjooy, "Health 4.0: Applications, management, technologies and review," *Medical Technologies Journal*, vol. 2, no. 4, pp. 262–276, 2019. DOI:10.26415/2572-004X-vol2iss4p262-276.

[92] V. V. Estrela, A. Khelassi, A. C. B. Monteiro, *et al.*, "Why software-defined radio (SDR) matters in healthcare?" *Medical Technologies Journal*, vol. 3, no. 3, pp. 421–9, 2019. DOI:10.26415/2572-004X-vol3iss3p421-429.

[93] R. P. Franca, Y. Iano, A. C. B. Monteiro, R. Arthur, and V. V. Estrela, "Betterment proposal to multipath fading channels potential to MIMO systems," in Y. Iano *et al.* (eds) *Proceedings of the 16th Brazilian Symposium on Computer Graphics and Image Processing (BTSym'18)*, Smart Innovation, Systems and Technologies, vol. 140. Springer, Zurich, Switzerland, 2019. DOI: 10.1007/978-3-030-16053-1_11.

[94] A. Sweeney, C. Jiggins, and S. Johnsen, "Insect communication: polarized light as a butterfly mating signal," *Nature*, vol. 423, p. 31, 2003.

[95] J. M. Douglas, T. W. Cronin, T.-H. Chiou, and N. J. Dominy, "Light habitats and the role of polarized iridescence in the sensory ecology of neotropical nymphalid butterflies (Lepidoptera: Nymphalidae)," *Journal of Experimental Biology*, vol. 210, pp. 788–799, 2007.

[96] T. Cronin, N. Shashar, and L. Wolff, "Portable imaging polarimeters," in *Proceedings of 12th IAPR International Conference Pattern Recognition, Vol. 1.*, Jerusalem, Israel, Israel, pp. 606–609, 1994.

[97] D. Konings, B. Parr, C. Waddell, F. Alam, K. M. Arif, and E. M. Lai, "HVLP: Hybrid visible light positioning of a mobile robot," in *Proceedings of the International Conference on Mechatronics and Machine Vision in Practice (M2VIP)*, pp. 1–6, 2017.

[98] B. Suhai and G. Horváth, "How well does the Rayleigh model describe the E-vector distribution of skylight in clear and cloudy conditions? A full-sky polarimetric study," *Journal of the Optical Society of America A*, vol. 21, pp. 1669–1676, 2004.

[99] J. Dupeyroux, J. Diperi, M. Boyron, S. Viollet, and J. Serres, "A novel insect-inspired optical compass sensor for a hexapod walking robot," in *the 2017 IEEE/RSJ International Conference on Intelligent Robots and Systems (IROS 2017)*, 2017.

[100] T. Ma, X. Hu, J. Lian, and L. Zhang, "A novel calibration model of polarization navigation sensor," *IEEE Sensors Journal*, vol. 15, pp. 4241–4248, 2015.

[101] D. Lambrinos, R. Möller, T. Labhart, R. Pfeifer, and R. Wehner, "A mobile robot employing insect strategies for navigation," *Robotics and Autonomous Systems*, vol. 30, pp. 39–64, 2000.

[102] J. Chu, K. Zhao, Q. Zhang, and T. Wang, "Design of a novel polarization sensor for navigation," in *International Conference on Mechatronics and Automation, 2007 (ICMA 2007)*, pp. 3161–3166, 2007.

[103] J. Dupeyroux, J. Diperi, M. Boyron, S. Viollet, and J. Serres, "A bio-inspired celestial compass applied to an ant-inspired robot for autonomous navigation," in *European Conference on Mobile Robotics (ECMR)*, 2017.

[104] Z. Xian, X. Hu, J. Lian, *et al.*, "A novel angle computation and calibration algorithm of bio-inspired sky-light polarization navigation sensor," *Sensors*, vol. 14, pp. 17068–17088, 2014.

[105] J. Wood, "Visual analytic design for detecting airborne pollution sources," 2017. In Proc. 2017 IEEE Conference on Visual Analytics Science and Technology (VAST), Award: Comprehensive Mini-Challenge 2 Answer. 201–202

[106] A. Schmolke, H. Mallot, and K. Neurowissenschaft, "Polarization compass for robot navigation," in *The Fifth German Workshop on Artificial Life*, 2002, pp. 163–167.

[107] J. Chahl and A. Mizutani, "Biomimetic attitude and orientation sensors," *IEEE Sensors Journal*, vol. 12, pp. 289–297, 2012.

[108] J.-H. Lee, S.-S. Yoon, I.-H. Kim, and H.-J. Jung, "Diagnosis of crack damage on structures based on image processing techniques and R-CNN using unmanned aerial vehicle (UAV)," in *Sensors and Smart Structures Technologies for Civil, Mechanical, and Aerospace Systems*, vol. 2018, p. 1059811, 2018.

[109] J. Register, D. M. Callahan, C. Segura, *et al.*, "Advances in flexible optrode hardware for use in cybernetic insects," In Proceedings of the SPIE NANOSCIENCE + ENGINEERING, San Diego, CA, USA, 2017.

[110] M. Groen, B. Bruggeman, B. Remes, R. Ruijsink, B. Van Oudheusden, and H. Bijl, "Improving flight performance of the flapping wing MAV DelFly

II," in *International Micro Air Vehicle Conference and Competition (IMAV 2010)*, 2010.

[111] R. Mautz and S. Tilch, "Survey of optical indoor positioning systems," in *International Conference on Indoor Positioning and Indoor Navigation (IPIN 2011)*, pp. 1–7, 2011.

[112] T. Luhmann, S. Robson, S. Kyle, and I. Harley, *Close Range Photogrammetry*, Wiley, New Jersey, PA, USA, 2007.

[113] T. K. Kohoutek, R. Mautz, and A. Donaubauer, "Real-time indoor positioning using range imaging sensors," in *Real-Time Image and Video Processing 2010*, p. 77240K, 2010.

[114] H. Hile and G. Borriello, "Positioning and orientation in indoor environments using camera phones," *IEEE Computer Graphics and Applications*, 2008;28(4):32–39

[115] A. Kitanov, S. Bisevac, and I. Petrovic, "Mobile robot self-localization in complex indoor environments using monocular vision and 3D model," in *Paper presented at IEEE/ ASME International Conference on Advanced Intelligent Mechatronics*, pp. 1–6, 2007.

[116] C. Schlaile, O. Meister, N. Frietsch, C. Keßler, J. Wendel, and G. F. Trommer, "Using natural features for vision based navigation of an indoor VTOL MAV," *Aerospace Science and Technology*, vol. 13, pp. 349–357, 2009.

[117] J. Ido, Y. Shimizu, Y. Matsumoto, and T. Ogasawara, "Indoor navigation for a humanoid robot using a view sequence," *The International Journal of Robotics Research*, vol. 28, pp. 315–325, 2009.

[118] K. Sjö, D. G. López, C. Paul, P. Jensfelt, and D. Kragic, "Object search and localization for an indoor mobile robot," *Journal of Computing and Information Technology*, vol. 17, pp. 67–80, 2009.

[119] M. Muffert, J. Siegemund, and W. Förstner, "The estimation of spatial positions by using an omnidirectional camera system," in *Proceeding of the 2nd International Conference on Machine Control & Guidance*, 2010.

[120] O. Maye, J. Schaeffner, and M. Maaser, "An optical indoor positioning system for the mass market," in *Proceedings of the 3rd Workshop on Positioning, Navigation and Communication*, pp. 111–116, 2006.

[121] A. Mulloni, D. Wagner, I. Barakonyi, and D. Schmalstieg, "Indoor positioning and navigation with camera phones," *IEEE Pervasive Computing*, 2009;8:22–31.

[122] N. Fallah, I. Apostolopoulos, K. Bekris, and E. Folmer, "The user as a sensor: navigating users with visual impairments in indoor spaces using tactile landmarks," in *Proceedings of the SIGCHI Conference on Human Factors in Computing Systems*, 2012, pp. 425–432.

[123] S. Lee and J.-B. Song, "Mobile robot localization using infrared light reflecting landmarks," in *International Conference on Control, Automation and Systems, 2007 (ICCAS'07)*, pp. 674–677, 2007.

[124] M. Köhler, S. N. Patel, J. W. Summet, E. P. Stuntebeck, and G. D. Abowd, "Tracksense: infrastructure free precise indoor positioning using projected patterns," in *International Conference on Perv. Comp.*, 2007, pp. 334–350.

[125] S. Tilch and R. Mautz, "Development of a new laser-based, optical indoor positioning system," *The International Archives of the Photogrammetry, Remote Sensing and Spatial Information Sciences Commission*, vol. 1501, pp. 575–580, 2010.

[126] M. Habbecke and L. Kobbelt, "Laser brush: a flexible device for 3D reconstruction of indoor scenes," in *Proceedings of 2008 ACM Symposium on Solid and Physical Modeling*, pp. 231–239, 2008.

[127] V. Popescu, E. Sacks, and G. Bahmotov, "Interactive modeling from dense color and sparse depth," In Proceedings 2nd International Symposium on 3D Data Processing, Visualization and Transmission, 3DPVT 2004. Thessaloniki, Greece, Greece, 2004, p. 430–437. DOI: 10.1109/TDPVT.2004.1335270.

[128] B. Bürki, S. Guillaume, P. Sorber, and H.-P. Oesch, "DAEDALUS: a versatile usable digital clip-on measuring system for Total Stations," in *2010 International Conference on Indoor Positioning and Indoor Navigation (IPIN)*, pp. 1–10, 2010.

[129] F. Boochs, R. Schütze, C. Simon, F. Marzani, H. Wirth, and J. Meier, "Increasing the accuracy of untaught robot positions by means of a multi-camera system," in *2010 International Conference on Indoor Positioning and Indoor Navigation (IPIN)*, pp. 1–9, 2010.

[130] F. Tappero, "Low-cost optical-based indoor tracking device for detection and mitigation of NLOS effects," *Procedia Chemistry*, vol. 1, pp. 497–500, 2009.

[131] A. Soloviev and D. Venable, "When GNSS goes blind-integrating vision measurements for navigation in signal-challenged environments," *GNSS Inside*, 1, pp. 18–29, 2010.

[132] D. Aufderheide and W. Krybus, "Towards real-time camera egomotion estimation and three-dimensional scene acquisition from monocular image streams," in *2010 International Conference on Indoor Positioning and Indoor Navigation (IPIN)*, pp. 1–10, 2010.

[133] T. Liu, M. Carlberg, G. Chen, J. Chen, J. Kua, and A. Zakhor, "Indoor localization and visualization using a human-operated backpack system," in *2010 International Conference on Indoor Positioning and Indoor Navigation (IPIN)*, pp. 1–10, 2010.

[134] W. Liu, C. Hu, Q. He, M. Q.-H. Meng, and L. Liu, "An hybrid localization system based on optics and magnetics," in *2010 IEEE International Conference on Robotics and Biomimetics (ROBIO)*, pp. 1165–1169, 2010.

[135] E. Trucco and K. Plakas, "Video tracking: a concise survey," *IEEE Journal of Oceanic Engineering*, vol. 31, pp. 520–529, 2006.

[136] C. Kownacki, "Guidance and obstacle avoidance of MAV in uncertain urban environment," in *European Micro Aerial Vehicle Conference and Flight Competition*, Delft, Netherlands, 2009.

[137] V. Baiocchi, D. Dominici, and M. Mormile, "UAV application in post-seismic environment," *International Archives of the Photogrammetry, Remote Sensing and Spatial Information Sciences, XL-1 W*, vol. 2, pp. 21–25, 2013.

[138] R. He, A. Bachrach, and N. Roy, "Efficient planning under uncertainty for a target-tracking micro-aerial vehicle," in *2010 IEEE International Conference on Robotics and Automation (ICRA)*, pp. 1–8, 2010.

[139] M. W. Achtelik, S. Weiss, M. Chli, and R. Siegwart, "Path planning for motion dependent state estimation on micro aerial vehicles," in *2013 IEEE International Conference on Robotics and Automation (ICRA)*, pp. 3926–3932, 2013.

[140] C. Forster, M. Faessler, F. Fontana, M. Werlberger, and D. Scaramuzza, "Continuous on-board monocular-vision-based elevation mapping applied to autonomous landing of micro aerial vehicles," in *2015 IEEE International Conference on Robotics and Automation (ICRA)*, pp. 111–118, 2015.

[141] N. Razmjooy, M. Ramezani, and V. V. Estrela, "A solution for Dubins path problem with uncertainties using world cup optimization and Chebyshev polynomials," in Y. Iano *et al.* (eds), *Proceedings of the 4th Brazilian Technology Symposium (BTSym'18)*, Smart Innovation, Systems and Technologies, vol. 140. Springer, Zurich, Switzerland, 2019. DOI: 10.1007/978-3-030-16053-1_5.

[142] N. Razmjooy, M. Ramezani, V. V. Estrela, H. J. Loschi, and D. A. do Nascimento, "Stability analysis of the interval systems based on linear matrix inequalities," in Y. Iano *et al.* (eds), *Proceedings of the 16th Brazilian Symposium on Computer Graphics and Image Processing* (BTSym'18), Smart Innovation, Systems and Technologies, vol. 140, Springer, 2019. DOI: 10.1007/978-3-030-16053-1_36.

[143] M. Sadraey, "A systems engineering approach to unmanned aerial vehicle design," in *10th AIAA Aviation Technology, Integration, and Operations (ATIO) Conference*, p. 9302, 2010.

[144] W. Shyy, Y. Lian, J. Tang, *et al.*, "Computational aerodynamics of low Reynolds number plunging, pitching and flexible wings for MAV applications," *Acta Mechanica Sinica*, vol. 24, pp. 351–373, 2008.

[145] T. Nguyen, D. S. Sundar, K. S. Yeo, and T. T. Lim, "Modeling and analysis of insect-like flexible wings at low Reynolds number," *Journal of Fluids and Structures*, vol. 62, pp. 294–317, 2016.

[146] I. M. Al-Qadi and A. M. Al-Bahi, "Micro aerial vehicles design challenges: State of the art review," in *SSAS UAV Scientific Meeting & Exhibition*, 2006, pp. 29–45.

[147] N. Razmjooy, M. Khalilpour, and V. V. Estrela, and H. J. Loschi, "World Cup optimization algorithm: An application for optimal control of pitch angle in hybrid renewable PV/wind energy system," in M. Quiroz, A. Lara, Y. Maldonado, and L. Trujillo and O. Schuetze. (eds), *Proceedings of the NEO 2018: Numerical and Evolutionary Optimization*, 2019.

[148] P. A. T.P., R. Pandiarajan, and P. Raju, "Wireless power transmission to UAV using LASER beaming," *International Journal of Mechanical Engineering and Research*, vol. 5, no.1, 137–142, 2015.

[149] M. C. Achtelik, J. Stumpf, D. Gurdan, and K.-M. Doth, "Design of a flexible high performance quadcopter platform breaking the MAV endurance record with laser power beaming," in *Proceedings of 2011 IEEE/RSJ*

International Conference on Intelligent Robots and Systems (IROS), pp. 5166–5172, 2011.

[150] S. S. Mohammed, K. Ramasamy, and T. Shanmuganantham, "Wireless power transmission – a next generation power transmission system," *International Journal of Computer Applications*, vol. 1, pp. 100–103, 2010.

[151] D. Scaramuzza, M. C. Achtelik, L. Doitsidis, *et al.*, "Vision-controlled micro flying robots: From system design to autonomous navigation and mapping in GPS-denied environments," *IEEE Robotics & Automation Magazine*, vol. 21, pp. 26–40, 2014.

[152] P. Tripicchio, M. Satler, M. Unetti, and C. A. Avizzano, "Confined spaces industrial inspection with micro aerial vehicles and laser range finder localization," *International Journal of Micro Air Vehicles*, vol. 10, pp. 207–224, 2018.

[153] M. Li, R. J. Evans, E. Skafidas, and B. Moran, "Radar-on-a-chip (ROACH)," in *Proceedings of 2010 IEEE Radar Conference*, pp. 1224–1228, 2010.

Chapter 9

Computer vision in UAV using ROS

Gustavo de Carvalho Bertoli[1], Osamu Saotome[1] and Vania V. Estrela[2]

This chapter presents the concepts, advantages, and practical examples of computer vision on robot operating system (ROS), applied explicitly on unmanned aerial vehicle (UAV) domain. ROS was built with the concept of abstraction (using messages interfaces) to allow re-use of software. This concept allows the integration with already built source code for computer vision tasks with the ROS environment.

In conjunction with the available simulators for ROS, it is possible to design UAV solutions in a model-based concept, anticipating many operational and technical constraints prior to operational release.

9.1 Introduction

The robot operating system (ROS) is not a real operating system (OS), but a framework and set of tools that offer the functionality of an OS on a heterogeneous computer arrangement or cyber–physical system (CPS). The ROS ecosystem consists of robotics middleware that is to say an assortment of software tools to develop robots for example hardware abstraction, low-level device control, implementation of commonly used functionality, message passing between processes, and package management. Running sets of ROS-based processes are represented in a graph architecture where processing takes place in nodes that may receive, post and multiplex sensor, control, state, planning, actuator, and other messages.

Computer vision (CV) is a field of engineering, which enables computers to comprehend what is being shown in an image, identifying its attributes and symbolic representing it. It has synergy with artificial intelligence in applications such as object detection and recognition, navigation, tracking, and image-based decision systems allowing better and efficient systems deployment. It faces many challenges to be accomplished such as sensors (e.g., camera) quality, an environment that is applied, sometimes facing light restriction or obstacles, target dynamics, and so on.

[1]DCTA-ITA-IEEA, Sao Jose dos Campos, SP, Brazil
[2]Universidade Federal Fluminense (UFF), RJ, Brazil

These challenges directly relate to real-time processing and typical usage for CV in robotics or industrial processes.

9.2 Computer vision on ROS

The great benefit of doing CV in ROS is the possibility of reusing code. There are several third-party packages available to develop new CV programs for real UAVs or to simulate them. Such packages implement some exciting CV tools for the most common visual tasks:

- data fusion [1];
- multimodality perception [2];
- object identification [3];
- scene segmentation [4];
- object recognition [5];
- face recognition [6];
- gesture recognition [7];
- crowd recognition/understanding [8,9];
- motion understanding/tracking [10,11];
- structure from motion (SFM) [12];
- stereo vision [13];
- visual odometry [14];
- navigation [15];
- flight control [16]; and
- flight planning/reconfiguration [17,18] among others.

Also, using simulation [19,20], it is possible to test these algorithms integrated with robots prior to physical deployments, which can anticipate integration and development issues that support systems development – anticipating issues before deployment, reducing development costs, directly supporting model-based design, and verification and validation tasks.

This section will introduce applications considering the UAV domain, figures detection through examples using OpenCV libraries integrated with ROS infrastructure [21].

The CV UAV navigation will be presented and accomplished with a custom ROS package (tum ardrone) [22].

9.3 Applications

9.3.1 OpenCV in ROS

As previously stated, the smooth integration of ROS with other packages allows new functionalities to be available on ROS deployments. OpenCV, which is a widely used open-source CV library [23], will be used for the following examples.

OpenCV has algorithms for image manipulations such as translation, rotation, crop, arithmetic, and bitwise operations, convolution, blurring, de-noising, sharpening, threshold operation, dilatation, erosion, edge detection, and segmentation techniques like contours.

For these manipulations, OpenCV translates images into a matrix representation where matrix cells containing colour space values represent each pixel. On this text, colour space will be RGB (red–green–blue), which OpenCV uses as BGR. It is important to mention that other colour spaces are available, but they are out of the scope, such as HSV (hue–saturation–value) and CMYK (cyan–magenta–yellow–key). Based on this matrix representation, algorithms access, manipulate, and analyse the images (Figures 9.1–9.3).

Figure 9.1 Grayscale example (200 × 150 pixels)

Figure 9.2 Cropped from grayscale example (20 × 20 pixels)

107	110	114	113	112	113	113	117	117	113	112	112	113	114	116	116	109	105	106	104
106	108	113	114	114	116	114	117	123	118	116	116	114	115	117	117	113	109	106	105
108	108	114	116	115	121	119	118	125	122	118	117	112	113	116	116	116	112	103	104
113	114	116	115	115	122	124	121	119	120	122	116	111	114	114	112	113	112	106	101
115	117	116	115	116	121	123	121	117	118	123	117	115	115	114	109	109	112	114	105
116	118	118	117	118	121	122	126	123	119	122	120	114	117	119	113	107	110	116	110
114	116	119	119	121	121	124	128	131	124	117	120	109	106	111	115	110	105	112	112
114	116	119	119	120	120	127	126	132	129	114	119	113	99	97	108	111	102	110	115
117	120	122	122	121	122	124	121	124	130	121	118	115	108	103	107	108	106	113	114
119	123	125	126	122	120	122	120	117	129	129	118	110	111	110	111	113	114	115	111
120	123	126	127	122	121	124	123	115	123	123	113	107	109	110	111	114	118	113	108
120	121	125	130	128	121	115	118	114	118	116	110	107	105	105	105	113	119	110	104
118	118	125	135	136	126	110	114	115	118	112	107	107	100	102	104	115	121	109	104
119	120	127	131	131	129	114	117	114	108	109	103	106	98	99	107	111	118	107	102
117	120	127	128	129	132	121	115	115	99	101	106	109	102	96	99	109	119	105	100
116	117	125	130	130	136	131	115	119	97	89	102	110	105	96	101	113	116	105	103
116	117	125	130	132	138	136	124	125	104	85	91	109	102	98	117	117	108	108	100
116	117	124	128	133	140	134	126	120	109	89	88	108	106	105	120	112	106	107	95
116	118	121	124	130	140	132	119	116	106	91	88	102	103	104	112	107	108	106	96
116	119	120	122	130	137	130	112	115	103	90	88	93	93	99	106	103	109	105	98

Figure 9.3 Grayscale pixel values between 0 and 255 from the cropped image

To move forward and consider the Ubuntu Linux as ROS host, it is required the installation of Ubuntu Linux packages as follows:

Listing 9.1: OpenCV packages installation

```
$ sudo apt-get install libopencv-dev

# For ROS Kinetic Distribution

$ sudo apt-get install ros-kinetic-opencv3
```

A typical Python code for ROS node that consumes and processes ROS images requires the specific libraries:

- rospy – ROS library for Python
- cv2 – OpenCV library
- cv bridge – Converts between ROS images and OpenCV images on Python

The following example is a node that subscribes to UAV camera data topic and then performs a conversion from ROS image data type to OpenCV image data type through a library called *cv bridge* (Figures 9.4 and 9.5). This Python ROS node is as follows:

Listing 9.2: Python code for image conversion between ROS and OpenCV

```
#!/usr/bin/env python

import rospy
import cv2
from cv_bridge import CvBridge
```

```

# to handle ROS camera image
from sensor_msgs.msg import Image

bridge = CvBridge()

def callback(data):
cv_image = bridge.imgmsg_to_cv2(data,"bgr8")
cv2.imshow("Image from ROS", cv_image)
cv2.waitKey(0)

def main():
image_subscribe =
rospy.Subscriber("/ardrone/image_raw",
Image, callback)

rospy.init_node("ROSImgToCV")
rospy.spin()
cv2.destroyAllWindows()

if __name__ == '__main__':
main()
```

The Python code presented creates an ROS topic that subscribes to */ardrone/image raw* topic (Figure 9.6). This topic outputs the UAV camera image. Once on ROSImgToCV node, this image is converted using the *cv bridge* library.

After this conversion, all OpenCV processing can be conducted through available algorithms or reused codes. Its processing outputs can be published containing commands and/or feedback for the ROS system.

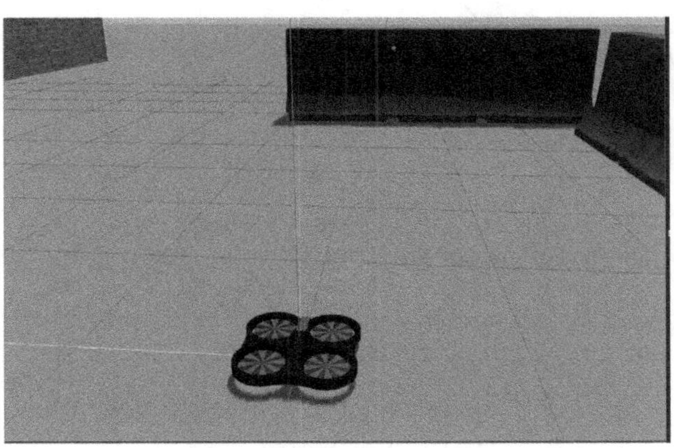

Figure 9.4 Gazebo simulation view

Figure 9.5 Converted ROS image to OpenCV

9.3.1.1 Object detection

A simple application to demonstrate CV in UAV using ROS is the landing command based on image processing.

In this case, when the UAV detects a specific object on the image, it will publish (trigger) to UAV a landing command.

Based on */ROSImgToCV* Node code, post-processing with OpenCV built-in algorithms will be introduced to determine if the condition is met and publishing the landing command to UAV.

For landing and takeoff, the command to UAV shall be sent through ROS Topics */ardrone/land* and */ardrone/takeoff*.

Another possibility is the ROS topic */cmd_vel* which is of type *geometry msgs/ Twist*, according to ROS documentation it expresses velocity in free space broken into its linear and angular parts. Through */cmd vel* it is possible to command UAV navigation.

The prototype of function to directly command takeoff, landing, and moves can be summarised as follows:

Listing 9.3: Examples of UAV control on ROS

```
#!/usr/bin/env python

import rospy
import time

# Empty type required by Takeoff/Land
from std_msgs.msg import Empty

# Twist type required by cmd_vel
from geometry_msgs.msg import Twist

# Create Node
rospy.init_node('UAV_Command')
```

```

# publishers creation
cmd_vel = rospy.Publisher('/cmd_vel', Twist, queue_size=1)
takeoff = rospy.Publisher('/ardrone/takeoff', Empty,
queue_size=1)
land = rospy.Publisher('/ardrone/land', Empty,
queue_size=1)

# message type definition
move_msg = Twist()
takeoff_msg = Empty()
land_msg = Empty()

# it is required to pay attention between ROS commands,
# to overcome any issue, it is being used timers

time.sleep(5)

# example of Takeoff
takeoff.publish(takeoff_msg)
time.sleep(1)

# example of stopping
move_msg.linear.x = 0.0
move_msg.angular.z = 0.0
cmd_vel.publish(move_msg)
time.sleep(1)

# example of turning 90 degrees
move_msg.linear.x = 0.0
move_msg.angular.z = 1.0
cmd_vel.publish(move_msg)
time.sleep(1)

# example of moving forward
move_msg.linear.x = 1.0
move_msg.angular.z = 0.0
cmd_vel.publish(move_msg)
time.sleep(1)

# example of Landing
land.publish(land_msg)
time.sleep(1)
```

To simulate this CV application, let us consider the UAV is navigating, and when detecting a person, it will be commanded to land (Figures 9.7 and 9.8).

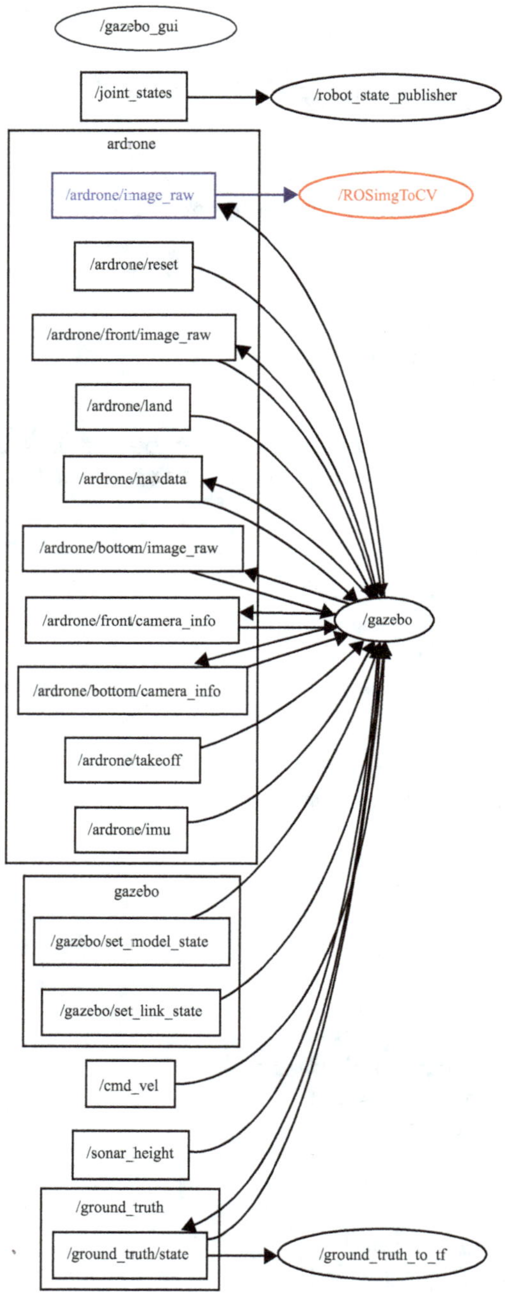

Figure 9.6 ROS graph with node subscribed to UAV camera

Figure 9.7 Gazebo simulation view

Figure 9.8 UAV view

OpenCV has a built-in algorithm to detect a person, and it is the HOG with SVM approach [24]. HOG is *Histograms of Oriented Gradients* and is a 'feature descriptor' to generalise objects. SVM means *Support Vector Machine* [25], which is a machine-learning algorithm for classification.

The arguments of *hog.detectMultiScale* present the following impacts on HOGSVM algorithm [26]:

- *winStride*
  Determine the step size in both x and y for the sliding window. Smaller winStride will require more evaluations by the algorithm.

Listing 9.4: OpenCV people detection algorithm

```
import cv2
import imutils

image = cv2.imread('UAV_Output.png')

# initialize HOG detector
hog = cv2.HOGDescriptor()

# set SVM detector as default for person detection
hog.setSVMDetector(
cv2.HOGDescriptor_getDefaultPeopleDetector())

# resize to a maximum width of 400 pixels
image = imutils.resize(image,
width=min(400, image.shape[21]))

# parameters winStride, padding and Scale are configurable
# to increase algorithm precision

(rects, weights) = hog.detectMultiScale(image,
winStride=(4,4), padding=(16,16), scale=1.05,
useMeanshiftGrouping=False)

for (x, y, w, h) in rects:
cv2.rectangle(image,(x,y), (x+w, y+h), (0, 255,0),
2)

cv2.imshow('Detections', image)

cv2.waitKey(0)
cv2.destroyAllWindows()
```

- *padding*
  Indicates the number of pixels in both x and y in which the sliding window is padded.
- *scale*
  Controls the factor that image is resized at each layer of the image pyramid.
- *useMeanshiftGrouping*
  Boolean value that indicates if mean-shift grouping will be performed to handle potential overlapping bounding boxes.

The algorithm output is illustrated in Figure 9.9.

Finally, an integrated algorithm that takes advantage of the ROS easy integration with OpenCV is presented. This algorithm consumes UAV camera images through ROS nodes and topics infrastructure and according to CV evaluation, if there is a person on the image, the UAV is sent to land.

Figure 9.9 People detected with OpenCV (HOG/SVM)

Listing 9.5: ROS and OpenCV integrated node (/LandingCV)

```

#!/usr/bin/env python

import rospy
import cv2
import imutils
from cv_bridge import CvBridge

from sensor_msgs.msg import Image
from std_msgs.msg import Empty

bridge = CvBridge()
hog = cv2.HOGDescriptor()
hog.setSVMDetector(
cv2.HOGDescriptor_getDefaultPeopleDetector())
land = rospy.Publisher('/ardrone/land', Empty,
queue_size=1)
land_msg = Empty()

def callback(data):
cv_image = bridge.imgmsg_to_cv2(data,"bgr8")

image = imutils.resize(cv_image,
width=min(400, cv_image.shape[21]))

(rects, weights) = hog.detectMultiScale(image,
winStride=(4,4), padding=(16,16),
scale=1.05, useMeanshiftGrouping=False)
```

```python

if len(rects):
land.publish(land_msg)

def main():
image_subscribe = rospy.Subscriber(
"/ardrone/image_raw", Image, callback)
rospy.init_node("LandingCV")
rospy.spin()


if __name__ == '__main__':
main()
```

Moreover, the modified ROS graph including the new */LandingCV* node is presented by Figure 9.10. It can be confirmed that this node subscribes on */ardrone/image raw* topic and publish on */ardrone/land* topic.

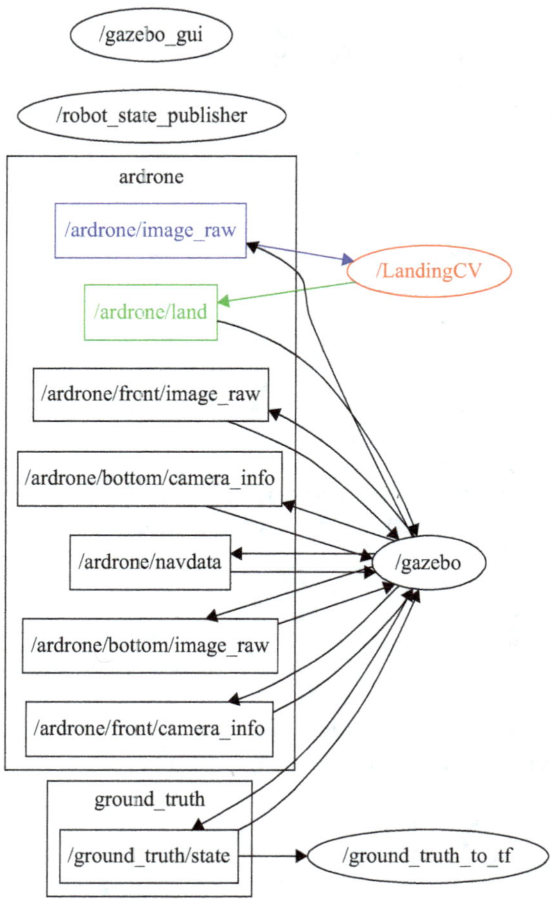

Figure 9.10 ROS graph with final node (/LandingCV)

9.3.2 Visual navigation

A concern for UAV operation is related to its dependence on position awareness. This can be accomplished using a Global Positioning System (GPS) but considering indoor or restricted environments, GPS cannot overcome these restrictions. A way to enable position awareness in restricted environments is through visual navigation, which is an enabling technology for autonomous navigation.

9.3.2.1 Parallel tracking and mapping (PTAM)

Before understanding PTAM, let us talk about Simultaneous Localization and Mapping (SLAM), which allows the robot to infer its location. Based on available sensors data, a map of its environment is created; this map is then utilised to determine robot new position in regular time intervals. It enables the continuously physical environment awareness and localisation estimation [27]. Parallel Tracking and Mapping (PTAM) [28] [29] is based on two tasks, tracking and mapping, processed in parallel threads. The tracking thread is responsible for the camera pose estimation, and the mapping thread is responsible for producing a 3D map of point features based on the previous camera frames.

9.3.2.2 ROS package – autonomous flight

The ROS package tum ardrone provides autonomous flight through PTAM-based visual navigation and it is made for Parrot AR.Drone. tum ardrone is composed of three components for UAV application [30]:

- Monocular SLAM
  Monocular SLAM uses only a single visual camera, which becomes a simpler, smaller, and cheaper implementation – as compared to stereo SLAM. On the other hand, it requires a more complex algorithm and more data to support depth inference using a single image, doing it through multiples frames of video. This deployment of monocular SLAM uses PTAM that, after map initialisation, rotates the visual map such that the *XY*-plane corresponds to the horizontal plane. Using the accelerometer data, scaling is done such that the average keypoint depth is 1. The scale of the map is estimated using metric sensors like ultrasound altimeter and estimations from an extended Kalman filter (EKF) to determine if a frame should be rejected or not [31].
- EKF
  The UAV is a dynamic system, change its states over time. EKF is used to filter and fuse available UAV data from sensors measurements such as attitude, height, and horizontal velocity in conjunction with UAV physical dynamics to determine its pose and serve as input for control laws.
- PID control
  Consuming position and velocity estimations from EKF, a proportional-integral derivative (PID) controller is a control loop feedback mechanism responsible for controlling the UAV dynamics behaviour like steering UAV to desired positions aiming to minimise the error between its inputs (reference) and measured outputs. For each of four degrees of freedom, a separate PID controller is trimmed.

Installation and initialisation instructions are as follows [22]:

Listing 9.6: tum ardrone ROS package installation

```
$ cd catkin_ws/src
$ git clone https://github.com/tum-vision/tum_ardrone.
git
-b hydro-devel
$ cd ..
$ source devel/setup.bash
$ rosdep install tum_ardrone
$ catkin_make
```

Listing 9.7: Launching ROS nodes

```
$ roslaunch tum_ardrone ardrone_driver.launch
$ roslaunch tum_ardrone tum_ardrone.launch
```

After tum ardrone initialisation, it is possible to inspect ROS infrastructure created through ROS graph (Figure 9.11) showing all the nodes as ellipses and all of the topics as rectangles.

Can be noted topics providing camera data (/ardrone/bottom/image raw and /ardrone/front/image raw), IMU data (/ardrone/imu) and important nodes for State Estimation (/drone stateestimation) and Autopilot (/drone autopilot) and interfaces with Simulation Suit Node (/gazebo).

Important to detail the following topics – that can be used later for reproducing flights (e.g., using ROS bag files):

- /cmd_vel – Data about the movements and speeds of the UAV
- /ardrone/takeoff – Data related to the takeoff of UAV
- /ardrone/land – Data related to the landing of UAV
- /ardrone/image raw – UAV's camera data
- /ardrone/navdata – Navigation-related data

It is important to mention that these topics are specific to Parrot AR.Drone.

After launching tum ardrone ROS package following windows will be available:

- PTAM UAV View
  A Map based on the different data provided by the UAV.
- PTAM UAV Camera Feed
  A video stream of what the UAV is seeing. One will also use this screen for sending 'goals' to the UAV.
- tum ardrone graphical user interface (GUI)
  The main window where one can interact with the different applications of the package (Figure 9.12).

9.3.2.3 tum ardrone GUI

tum ardrone GUI (Figure 9.13) offers a simple way to control the drone autopilot node, drone state estimation node and fly the UAV.

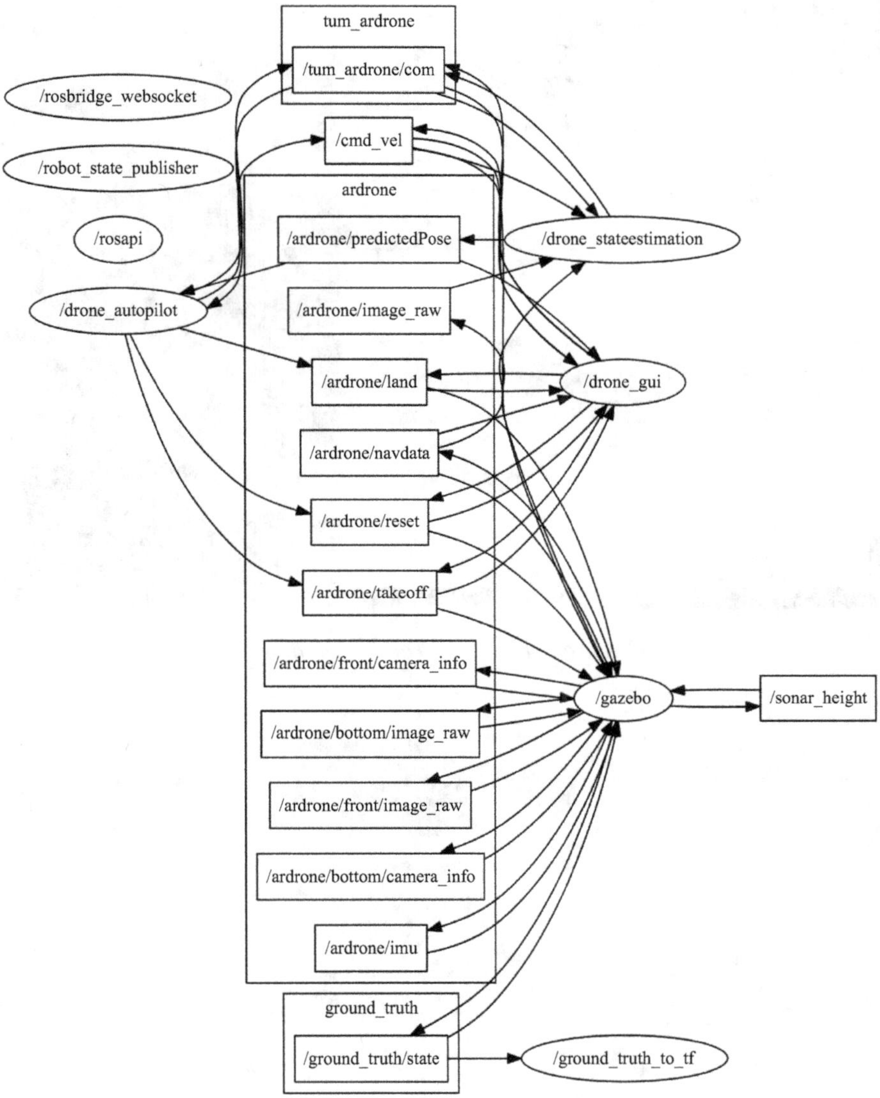

Figure 9.11 ROS graph for tum ardrone deployment (nodes and topics)

On the top-right corner of the GUI (refer to the Node Communication Status), information related to communication between nodes and UAV is provided like the navigation data (Navdata) frequency, Control frequency, Pose Estimation frequency and motors statuses.

For reference, Navdata shall be higher than 100 Hz and Pose Estimate 33 Hz to indicate correct operation.

There is 'Autopilot Status' that shows the status of Autopilot node. This node allows UAV navigation through waypoint sent via camera feed screen.

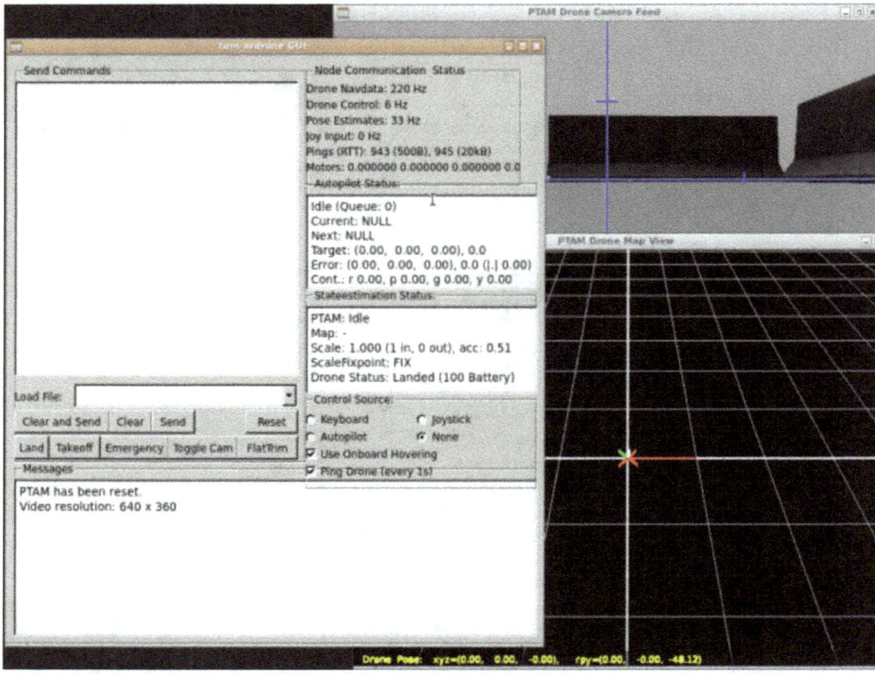

Figure 9.12 tum ardrone windows

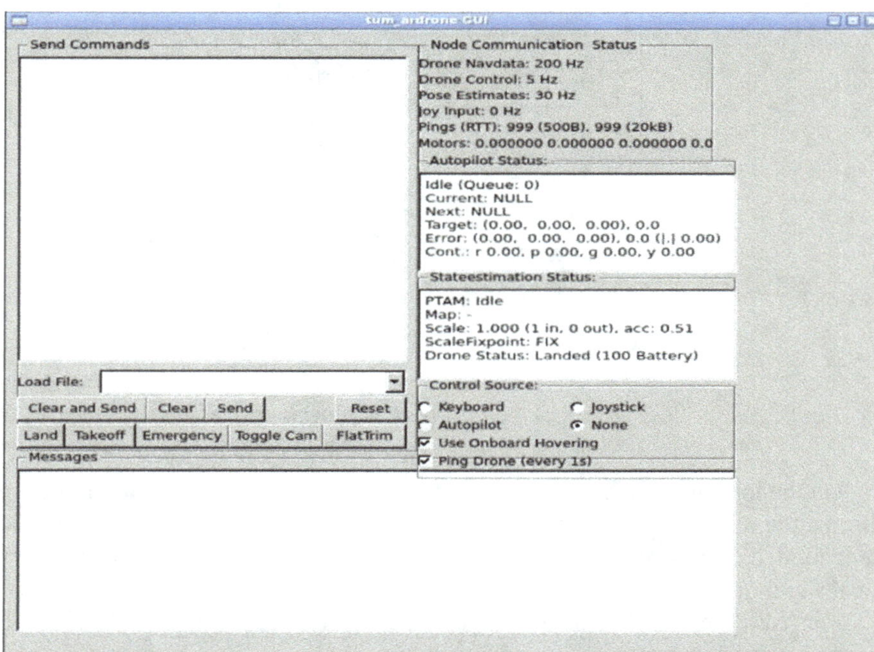

Figure 9.13 TUM parrot AR.Drone GUI

The status of StateEstimation node is presented by 'Stateestimation Status', this estimation is based on data sent by navdata, and this node is also responsible for sending control commands and PTAM.

Control Source section is responsible for switching between multiples control modes such as Keyboard, Autopilot, Joystick or none.

On the 'Send Commands' section, direct commands to be sent to the node are allowed in order to perform specific actions. Some ready commands are available through buttons such as Landing and Takeoff.

Finally, the 'Messages' sections output all logs generated by the application.

To perform an autopilot initialisation do as follows (Figure 9.14):

- On the 'Send Commands' section: autoInit 500 800
- Click on the button 'Clear and Send'
- Messages will show up on 'Messages' section indicating that it is successfully initialised.

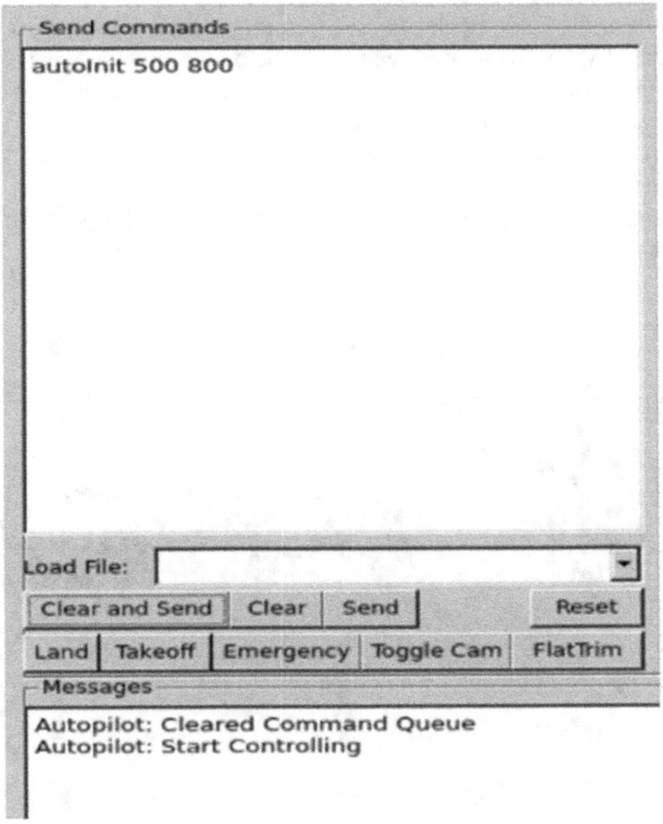

Figure 9.14 Autopilot Init

9.3.2.4 PTAM UAV camera feed and navigation

Once the Autopilot mode is activated, the PTAM UAV camera feed is the way to interact with the UAV navigation subsystem by clicking on the window (Figure 9.15).

The window is divided by two axes, *x* and *y*, clicking on the video window will generate waypoints that are sent to drone autopilot node. On tum ardrone implementation, there are differences between left and right mouse click on the window.

- left-click: fly (*x*,*y*,0)m relative to the current position, image-center is (0,0), borders are 2m, respectively
- right-click: fly (0,0,*y*)m and rotate yaw by *x* degree, image-centre is (0,0), borders are at 2 m and 90°, respectively.

These waypoints will control the UAV after its take-off.

9.3.3 Setting the drone state estimation node

The drone_stateestimation will estimate the UAV's position based on the sent navdata, control commands, and PTAM. In order to have it running, first of all, it is required to have the UAV flying. So, the first thing required is the UAV take-off.

Once the UAV is flying, one needs to focus on the camera window and in the bottom of the window presents a yellow message that states that Point the camera at a planar scene and press spacebar to start tracking for the initial map (Figure 9.15).

Make sure that the UAV is looking to the scene, and press the spacebar button in the keyboard.

Automatically after pressing spacebar, some points will appear in the image (Figure 9.16).

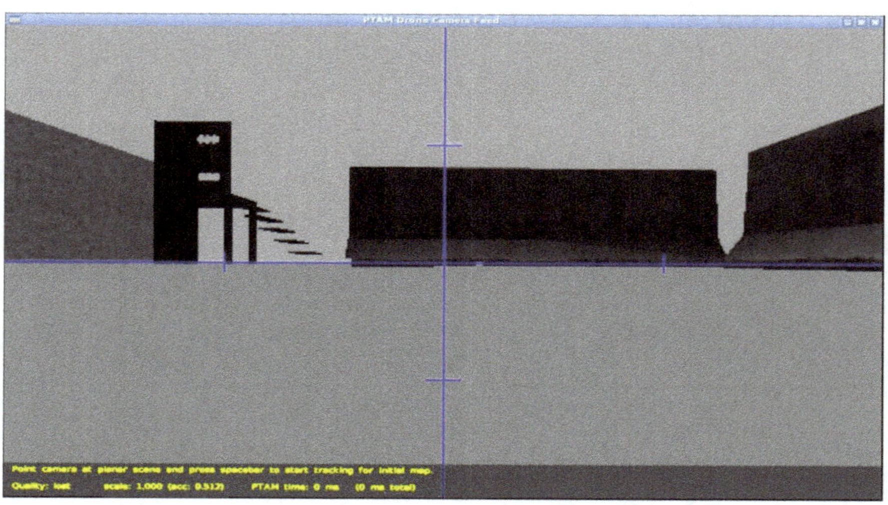

Figure 9.15 PTAM UAV camera feed

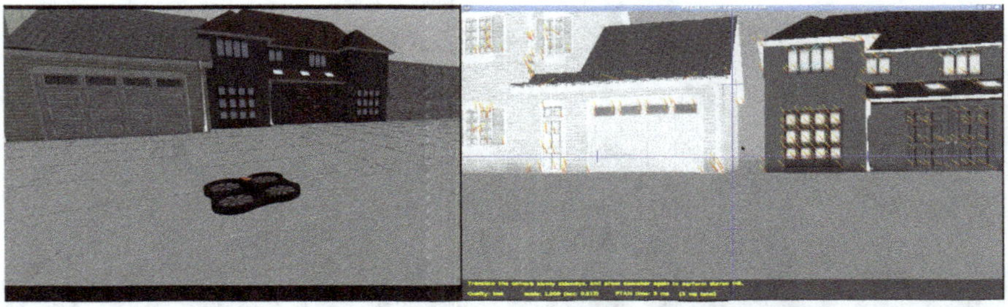

Figure 9.16 Tracking (initial map)

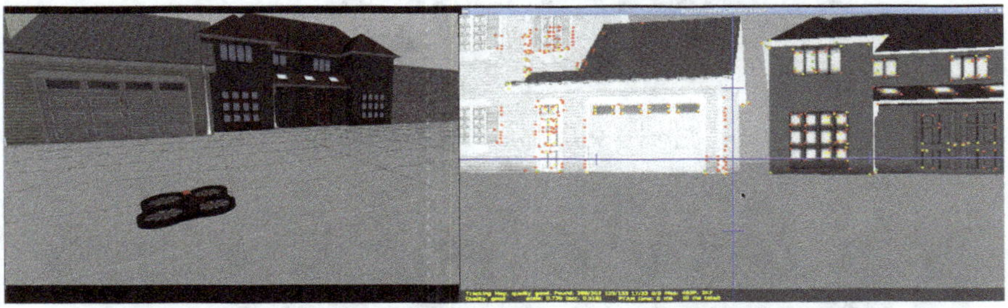

Figure 9.17 Tracking map

At the bottom of the window, the text has changed. It says:

Translate the camera slowly sideways, and press spacebar to perform stereo init.

Now, move the UAV sideways a little bit. This way, the camera will also move sideways a little bit. Then, immediately after this, press the spacebar again.

Automatically, after pressing the spacebar, points now appear much clearer in the image (Figure 9.17).

NOTE: To properly estimate PTAM's scale, it is best to fly up and down a little bit (for instance, 1m up and 1m down) immediately after initialisation.

This is the way the drone stateestimation node is configured. So now, the objects captured by the camera are being position tracked about the UAV position.

Now at the PTAM Map window, the points captured in the camera are placed as well in the map (Figure 9.18). The points that represent the house are placed on the map with an estimated distance about the UAV.

9.3.3.1 Simple navigation

The tum ardrone GUI allows loading pre-defined flights using PTAM estimating UAV's position each time. It can be through 'Load File' option on GUI or using the 'Send Commands' textbox.

After the PTAM calibration, the following commands can be tested (Figure 9.19):

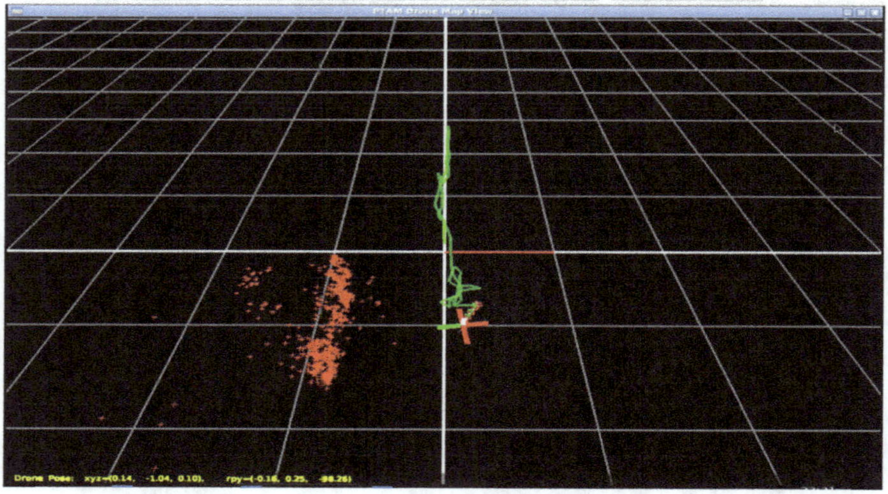

Figure 9.18 tum ardrone map

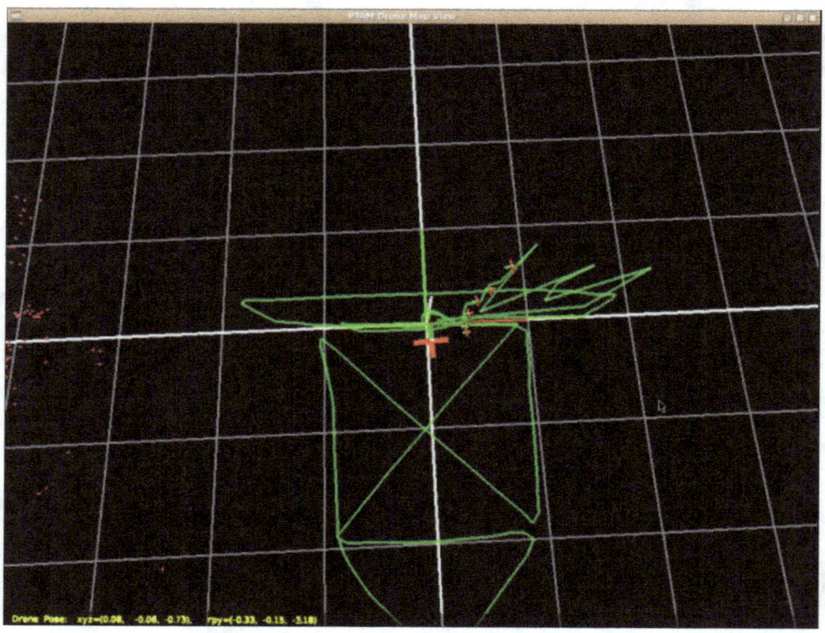

Figure 9.19 Navigation example

Listing 9.8: tum ardrone simple navigation example

```

autoInit 500 800

```

```
setMaxControl 1
setInitialReachDist 0.2
setStayWithinDist 0.5
setStayTime 3

# rectangle
goto 0 0 1 0
goto 0 0 0 0
goto -1.5 0 0 0
goto -1.5 0 1 0
goto 1.5 0 1 0
goto 1.5 0 0 0
goto 0 0 0 0

# house
goto -1 0 0 0
goto 1 0 0 0
goto -1 -2 0 0
goto 1 -2 0 0
goto 0 -3 0 0
goto -1 -2 0 0
goto -1 0 0 0
goto 1 -2 0 0
goto 1 0 0 0
goto 0 0 0 0

land
```

9.4 Future developments and trends in ROS

ROS integration with other frameworks, libraries, and already built code creates an excellent platform for robotics systems development using CV tasks through an agile development workflow from prototype to production.

ROS and CV applications have been used in research areas of self-driving cars, industrial robots (manufacturing), smart home robots, surgical robotics, flying robots, space robots, underwater vehicles, inspection robots, mining, military applications, among others.

These integration aspects are highly tied with new technology trends like digital twin and Internet of Things (IoT) allowing tasks like augmented reality, navigation, motion planning, perception – based on CV – to be simulated and evolved during all system life cycle phases.

With the release of ROS 2.0 more focus on real-time control and distributed processing was provided which allow applications of deep learning – high throughput – due to distributed processing and embedded application with real-time requirements and also providing real-time operating systems (RTOS) interface allowing compliance with the strong real-time response that can lead to edge embedded applications.

9.5 Conclusion

ROS is a very useful tool for robot development and simulation [19,20]. In this chapter, it was shown that deployment of CV tasks on ROS is easy mainly by its integration with the available libraries and tools for development, such as those demonstrated as Gazebo simulator, ready to integrate ROS packages (*tum ardrone*) and OpenCV libraries.

It was demonstrated that with simulation, it is possible to train and test models producing a more mature solution before physical or real-world deployment, anticipating issues and reducing costs.

Despite the importance of reactivity and low latency in UAV control, ROS, itself, is not a RTOS, but it allows some real-time code to be used.

The lack of real-time support is being addressed by the ROS 2.0 including security aspects and distributed processing [32,33].

CPSs have restricted computational and storage facilities owing to the fact they generally possess small size and are embedded into or interact with larger systems. With the advent of cloud computing and the IoT [34,35], UAV-CPSs encounter several new prospects to lengthen their capacities via the cloud assets in different forms. There is a need to develop efforts towards tools that will facilitate the integration of UAV-CPSs with cloud computing, classify and extend them to areas like (1) remote intelligence [34,36–39], (2) big data handling [40], (3) virtualisation [17,41], (4) mobile robots [32,42], (5) wireless sensor networks [32,42], and (6) vehicular networks [32,42,43].

References

[1] Koubaa A Robot Operating System (ROS) The Complete Reference (Volume 2). Studies in Computational Intelligence book series (SCI, volume 707), Springer Cham, Zurich, Switzerland, 2017.

[2] Sinapov J, Schenck C, and Stoytchev A. Learning relational object categories using behavioral exploration and multimodal perception. 2014 IEEE International Conference on Robotics and Automation (ICRA). 2014, p. 5691–5698.

[3] Gupta S, Girshick RB, Arbeláez PA, and Malik J. Learning rich features from RGB-D images for object detection and segmentation. Proceedings of 2014 ECCV. Zurich, Switzerland, 2014.

[4] Razmjooy N, Mousavi BS, and Soleymani F. A hybrid neural network imperialist competitive algorithm for skin color segmentation. Mathematical and Computer Modelling, Vol. 57, 2013, p. 848–856.

[5] Khalilpour M, and Razmjooy N. A hybrid method for gesture recognition. Journal of World's Electrical Engineering and Technology, Vol. 2, no. 3, 2013, p. 47–53.

[6] Firoze A, and Deb T. Face recognition time reduction based on partitioned faces without compromising accuracy and a review of state-of-the-art face recognition approaches. Proceedings of ICIGP 2018, Hong Kong, China.

[7] Panella M, and Altilio R. A smartphone-based application using machine learning for gesture recognition: Using feature extraction and template matching via hu image moments to recognize gestures. IEEE Consumer Electronics Magazine, Vol. 8, 2019, p. 25–29.

[8] Miraftabzadeh SA, Rad P, Choo KKR, and Jamshidi MM. A privacy-aware architecture at the edge for autonomous real-time identity reidentification in crowds. IEEE Internet of Things Journal, Vol. 5, 2018, p. 2936–2946.

[9] Yildirim F, Meyer V, and Cornelissen FW. Eyes on crowding: Crowding is preserved when responding by eye and similarly affects identity and position accuracy. Journal of Vision, Vol. 15, 2015, p. 2.

[10] Hart S, Dinh P, and Hambuchen KA. The affordance template ROS package for robot task programming. 2015 IEEE International Conference on Robotics and Automation (ICRA) 2015, Montreal, QC, Canada, p. 6227–6234.

[11] Coelho AM, and Estrela VV. A study on the effect of regularization matrices in motion estimation. International Journal of Computer Applications, Vol. 51, no. 19, 2012, p. 17–24. doi: 10.5120/8151-1886.

[12] Estrela VV, and Coelho AM. State-of-the art motion estimation in the context of 3D TV. In: Multimedia Networking and Coding. IGI Global, 2013, p. 148–173. doi:10.4018/978-1-4666-2660-7.ch006.

[13] Brzozowski B, and Szymanek N. Stereo vision module for UAV navigation system. In Proceedings of the 5th IEEE International Workshop on Metrology for AeroSpace (MetroAeroSpace) 2018, Rome, Italy, p. 2422–2425.

[14] Zhou Y, Li H, and Kneip L. Canny-VO: Visual odometry with RGB-D cameras based on geometric 3-D–2-D edge alignment. IEEE Transactions on Robotics, Vol. 35, 2019, p. 184–199.

[15] Potena C, Nardi D, and Pretto A. Effective target aware visual navigation for UAVs. Proceedings of 2017 European Conference on Mobile Robots (ECMR), Paris, France, 2017, p. 1–7.

[16] Nogar SM, and Kroninger CM. Development of a hybrid micro air vehicle capable of controlled transition. IEEE Robotics and Automation Letters, Vol. 3, 2018, p. 2269–2276.

[17] Kosak O, Wanninger C, Hoffmann A, Ponsar H, and Reif W. Multipotent systems: Combining planning, self-organization, and reconfiguration in modular robot ensembles. Sensors, 2019, 19(1), 17.

[18] Salla LML, Odubela A, Espada G, Correa MCB, Lewis LS, and Wood A. The EDNA public safety drone: Bullet-stopping lifesaving. In: Proceedings of 2018 IEEE Global Humanitarian Technology Conference (GHTC), San Jose, California, USA, 2018, p. 1–8.

[19] Hailong Q, Meng Z, Meng W, *et al.* Autonomous exploration and mapping system using heterogeneous UAVs and UGVs in GPS-denied environments. IEEE Transactions on Vehicular Technology, Vol. 68, 2019, p. 1339–1350.

[20] Valente J, Almeida R, and Kooistra L. A comprehensive study of the potential application of flying ethylene-sensitive sensors for ripeness detection in apple orchards. Sensors, 2019, 19(2), 372.

[21] Martinez A, and Fernandez E. Learning ROS for robotics programming. Birmingham: Packt Publishing Ltd; 2013.

[22] tum ardrone package; 2014. Available from: https://github.com/tum-vision/tumardrone.

[23] Mahtani A, Sanchez L, Fernandez E, *et al.* Effective robotics programming with ROS. Birmingham: Packt Publishing Ltd; 2016.

[24] Dalal N, and Triggs B. Histograms of oriented gradients for human detection. In: IEEE Computer Society Conference on Computer Vision and Pattern Recognition, 2005 (CVPR 2005). Vol. 1. IEEE; 2005, p. 886–893.

[25] Razmjooy N, Mousavi BS, and Soleymani F. A real-time mathematical computer method for potato inspection using machine vision. Computers & Mathematics with Applications, Vol. 63, 2012, p. 268–279.

[26] Rosebrock A. HOG detectMultiScale parameters explained – PyImageSearch. PyImageSearch; 2018. Available from: https://www.pyimagesearch.com/2015/11/16/hog-detectmultiscale-parameters-explained/.

[27] Riisgaard S, and Blas MR. SLAM for dummies. Vol. 22, 2003.

[28] Klein G, and Murray D. Parallel tracking and mapping for small AR Workspaces. In: Proceedings of the Sixth IEEE and ACM International Symposium on Mixed and Augmented Reality (ISMAR'07). Nara, Japan; 2007.

[29] Sturm J. TUM VISNAV 2013, Lecture Notes: Visual navigation for flying robots; 2013. Available from: http://vision.in.tum.de/teaching/ss2013/visnav2013. Last visited on 2018/07/01.

[30] Academy TCRI. Programming drones in ROS; 2018. Available from: http://www.theconstructsim.com/construct-learn-develop-robots-using-ros/robotignite academylearnros/ros-courses-library/ros-courses-programming-drones-ros/.

[31] Engel J, Sturm J, and Cremers D. Camera-based navigation of a low-cost quadrocopter. In: 2012 IEEE/RSJ International Conference on Intelligent Robots and Systems (IROS). IEEE; 2012, p. 2815–2821.

[32] Chaari R, Ellouze F, Koubaa A, *et al.* Cyber-physical systems clouds: A survey. Computer Networks, Vol. 108, 2016, p. 260–278.

[33] D'Emilia G, Gaspari A, and Natale E. Measurements for smart manufacturing in an industry 4.0 scenario a case-study on a mechatronic system. Proceedings of 2018 Workshop on Metrology for Industry 4.0 and IoT, Brescia, Italy, 2018, p. 1–5.

[34] Estrela VV, Monteiro ACB, França RP, Iano Y, Khelassi A, and Razmjooy N. Health 4.0: Applications, management, technologies and review. Medical Technologies Journal, Vol. 2, no. 4, 2019, p. 262–276, http://medtech.ichsmt.org/index.php/MTJ/article/view/205.

[35] Motlagh NH, Taleb T, and Osama O. Low-altitude unmanned aerial vehicles-based internet of things services: Comprehensive survey and future perspectives. IEEE Internet of Things Journal, Vol. 3, 2016, p. 899–922.

[36] Brahmbhatt S, Amor HB, and Christensen HI. Occlusion-aware object localization, segmentation and pose estimation. Proceedings of 2015 BMVC, Swansea, UK, 2015.

[37] Bachmann D, Weichert F, and Rinkenauer G. Review of three-dimensional human-computer interaction with focus on the leap motion controller. Sensors, 2018, 18(7), 2194.

[38] Hemanth DJ, and Estrela VV. Deep learning for image processing applications. Advances in Parallel Computing, Vol. 31, 2017, pp. 27–270.

[39] Zheng S, Lin Z, Zeng Q, Zheng R, Liu C, and Xiong H. IAPcloud: A cloud control platform for heterogeneous robots. IEEE Access, Vol. 6, 2018, p. 30577–30591.

[40] Atat R, Liu L, Wu J, Li G, Ye C, and Yi Y. Big data meet cyber-physical systems: A panoramic survey. IEEE Access, Vol. 6, 2018, p. 73603–73636.

[41] Liu W, Anguelov D, Erhan D, *et al.* SSD: Single shot multibox detector. Proceedings of 2016 ECCV, Amsterdam, Netherlands, 2016.

[42] Koubaa A, and Qureshi B. DroneTrack: Cloud-based real-time object tracking using unmanned aerial vehicles over the internet. IEEE Access, Vol. 6, 2018, p. 13810–13824.

[43] Peng HX, Liang L, Shen X, and Li GY. Vehicular communications: A network layer perspective. IEEE Transactions on Vehicular Technology, Vol. 68, 2018, p. 1064–1078.

Chapter 10
Security aspects of UAV and robot operating system

Gustavo de Carvalho Bertoli[1] *and Osamu Saotome*[1]

Unmanned aerial vehicles (UAVs) are an enabling technology being used to improve typical tasks such as delivery, agriculture, inspection, rescue, catastrophe response, among others. Initially used on military applications, nowadays they are increasing its presence as the commercial platform and consumer electronics. Despite its benefits and being an evolving technology, it embraces aspects such as complex architecture and mission-critical applications that require a rigorous development process to guarantee operational success during its whole life cycle. News about security-related events is becoming commonplace, compromising systems aspects such as confidentiality, integrity, and availability (i.e., security pillars), resulting in operational disruption, financial losses and safety incidents. This chapter preliminary presents typical, published and ongoing researches on security flaws on UAV domain and possible scenarios that can arise from them. The following security strategies to the resilient operation is presented to support new UAV designs based on typical security approaches (e.g., authentication, cryptography), current researches, and insights derived from aircraft design guidelines, which relates to intentional unauthorized interaction (cyber-threat) and its effects to safety. Finally, deployment of the robot operating system on consumer UAV (parrot AR.Drone 2) is performed, and a cybersecurity assessment is presented containing its findings, mitigations, and proposals to strengthen its operation (resiliency).

10.1 Introduction

Unmanned aerial vehicle (UAV) is a technology that improves typical tasks such as photography, delivery, agriculture, inspection, rescue, catastrophe response, among others. Initially used in military applications, nowadays they are increasing its presence as a commercial platform and consumer electronics. There are advantages such as low maintenance costs, the stability of navigation, reliability and compact size.

Despite its benefits and evolving technology aspects, it embraces traits such as complex architecture and mission-critical applications that require a rigorous development process to guarantee operational success during its whole life cycle.

[1]Instituto Tecnolgico de Aeronutica (ITA)

News about security-related events is becoming commonplace, compromising systems aspects such as confidentiality, integrity, and availability (i.e., security pillars), resulting in operational disruption, financial losses and safety incidents [1].

Currently, robot operating system (ROS) has been used on UAVs, and strategies for security-driven design decisions were not given enough attention aiming to cover its whole life-cycle phases and the vulnerabilities that it can be exposed.

This security-driven design deals with architectural issues and arises as another trade-off component during design decisions that results in a resilient system when considering intentional actions against it (malicious).

This chapter presents typical, published, and ongoing research works on security flaws on UAV and ROS domains and possible scenarios that can arise from them.

Next, security strategies to the resilient operation are presented to support ROS deployment on new UAV designs based on typical security approaches, current investigations information, and insights derived from design guidelines, which relate intentional unauthorised interaction (aka, cyber-threat) and its effects to safety.

10.2 Unmanned aerial vehicles

Multi-rotors, fixed-wing and rotorcrafts configurations compose the most used physical configurations of UAVs. Specifically, quadcopters (four rotors) have the flying principle of rotors with gears and propellers responsible for generating thrust. These rotors can be configured in many configurations such as six and eight rotors, respectively, hexacopters and octocopters. The dynamics of multi-rotors is controlled by changing the rotation speed of rotors. From now and on, it will be used UAV as synonymous with the quadcopter (Figure 10.1).

Regarding control of Autonomous Flight, it can be separated into two categories:

- Low level: It maintains attitude and stabilization, and compensates disturbance
- High level: It compensates drift, avoid obstacles, localisation and mapping, navigation, tracking

Figure 10.1 Parrot AR.Drone (quadcopter)

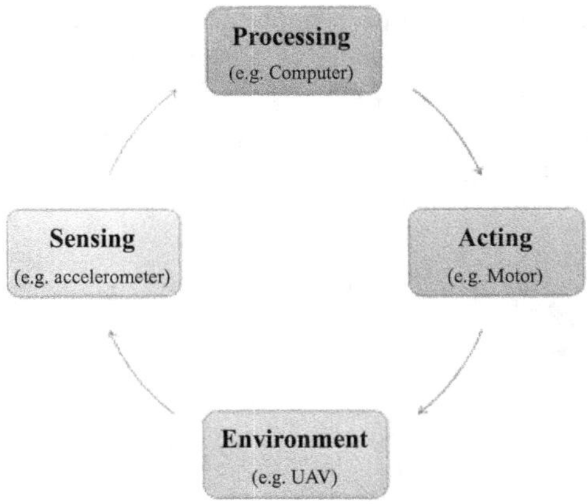

Figure 10.2 UAV system process

There are still some challenges regarding the UAV, such as limited payload, computational power and sensors, battery life, fast dynamics, and electronic control and safety, which tightly relate with results from security events or can be impaired or triggered by security issues. The UAV system consists of three main components, sensors such as camera, accelerometer, gyroscope; actuators such as motors, gimbals, and grippers; and control unit/software that is implemented by computers or embedded board computers. These components are responsible for performing a state machine consisting of sensing the environment, processing sensing data and acting on the environment (Figure 10.2).

10.3 ROS basic concepts

In a more connected world, illustrated by some buzzwords such as the Internet of Things (IoT) and the Industry 4.0, that takes advantage of technical evolutions on areas such as digitalisation, autonomous systems, cloud computing, data science, machine learning among others enabling technologies. On these applications, security concerns are pointed out as a major aspect to enable these expected enhancements provided by connecting things.

Specifically, on the robotics domain, it is common to see deployments of ROS on cyber–physical systems (CPS) that are engineered systems that are built from, and depend upon, the seamless integration of computational algorithms and physical components (NSF) [2]. Despite ROS increasing usage, its security characterisation is always a concern before widespread adoption by industry and governments.

Considering previous Control Categories presented, most ROS applications are related to High-Level Control Category, with ROS operating as middleware to support the processing phase. The most advantages of ROS are for development and operational phases through an infrastructure that is modular, robust, de-centralised, facilitates software re-use, allows hardware and software abstraction, supports debugging, data logging, tools for visualisation and easy to learn and extend.

ROS is released as distributions, where each distribution is a versioned set of packages like GNU/Linux distributions to support developers to work with stable codebases (Figures 10.3–10.5).

ROS is responsible for message-passing between process, hardware, and software, consisting of an ROS Master responsible by a namespace service in charge of

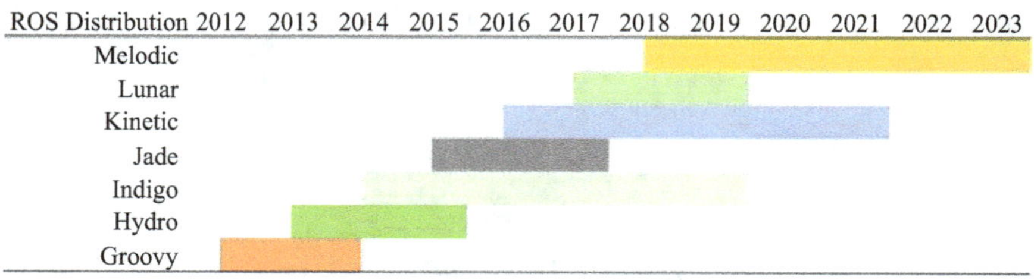

Figure 10.3 List of ROS distribution release and end-of-life date

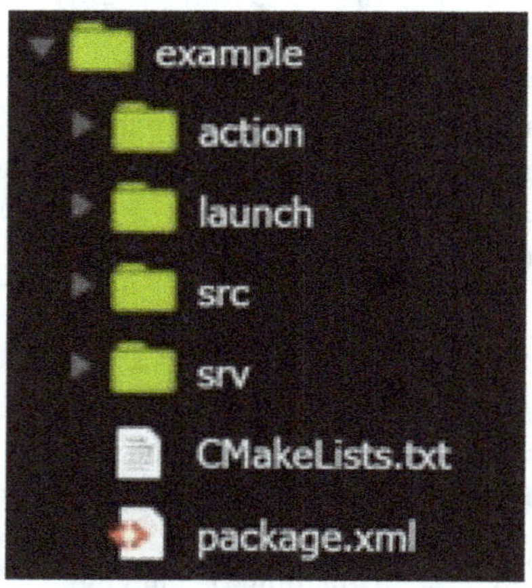

Figure 10.4 ROS package structure

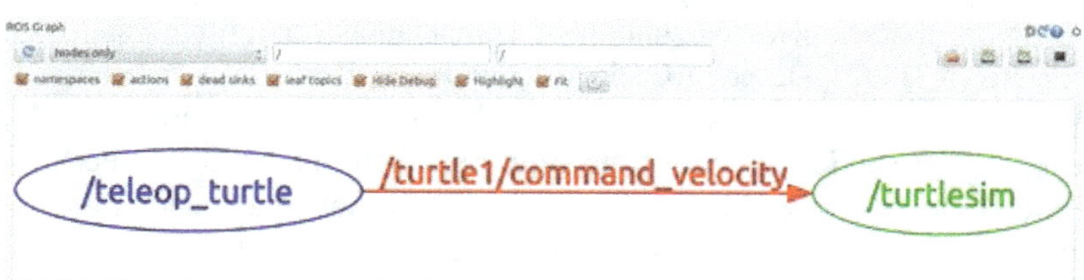

Figure 10.5 ROS graph structure

managing all components interactions. These components can be decomposed as follows:

- ROS package
  ROS is organised as packages that contain ROS Nodes and aims to support reuse and modularity. It is simply a directory tree, which is the way to release new ROS applications. Current ROS versions use the catkin build system for package management. Up to ROS distribution Fuerte, a build system was performed by rosbuild.
  As presented above, the ROS Package named example presents an action folder for custom made ROS Actions, a launch folder where the ROS launch scripts are saved, an src folder where Nodes source codes reside, and an srv folder where custom Services are implemented. CMakeLists.txt is the CMake build file and package.xml that defined project properties, like dependencies, used by catkin during the build process.
- ROS Nodes
  ROS Nodes are the processes, consist of software created on ROS and normally located at subfolder src of the Package and can be built in C++ or Python using the libraries roscpp and rospy. Nodes are combined to exchange messages (though topics) to perform a variety of functionalities. It is the central accountable part to support modularisation once in a project. Each node can be implemented to act in specific parts such as motor control, localisation and sensor readings. It is worth mentioning that ROS nodes can be running on a local machine or remotely.
- ROS topics
  Topics are like pipes or buses, where data exchanges between nodes. It consists of a publish (write) and subscribe (read) concept unidirectionally in a many-to-many one-way transport. An example is an accelerometer that will use a Topic to publish. It reads on a specific rate (continuous), and this information will be available to other ROS nodes, like the ones in charge of navigation and control.
  As presented by ROS Graph (connections between nodes created by topics), there are two nodes, blue and green, responsible to teleoperate the Turtle Robot (/teleop turtle), a demonstration Robot like a hello world in ROS available through turtlesim package and turtle simulation (/turtlesim) connected by the red arrow that is a topic. Blue Node publishes a command velocity (/command velocity) and green node acting as a subscriber to consume this data.
- ROS services
  Services allow specialisation of nodes; it functions like the Topic, but once it is based on response to a request like a procedure call, it is not the same as Topics.
  A sub-system that performs image recognition will be implemented as a Service, once it requires to be called and to wait until its recognition feedback (synchronous behaviour, cannot continue until it receives the result from service) [3].
- ROS actions
  Actions are the same as services. When one calls an action, this amounts to calling functionality that another node is providing just the same thing as for services. The difference is that calling service implies one must wait until the

service finishes. When one calls an action, this one does not necessarily have to wait for the action to complete.

A navigation system will be considered as an action once ROS will call it to perform some movement, and during this move, other tasks can be performed (asynchronous behaviour, like a thread).

10.4 Security UAV review

Security relates to events with intentionally misuse instead of well-known concepts of failure using the probability of components failures and its impacts, leading to redundancy and architectural changes.

Security analysis must think like a malicious attacker raising possibilities of misuse of available functions (scenarios) and its impacts on the whole system. After its findings, a trade-off shall be conducted to analyse the impact of these scenarios (for example, catastrophic) and its likelihood of aiming to support design decisions.

Security aspects or goals can be classified into three major categories (three pillars):

- Confidentiality: Only those with rights/privileges shall be capable of accessing information/data (e.g., Authentication)
- Integrity: Confidence on information/data content and sender/receiver (e.g., hashes)
- Availability: Information, data or system shall be available when required (for instance, redundancy)

All these aspects have many corresponding security threats or attacks; some examples appear below:

- Confidentiality: Traffic analysis
- Integrity: Data injection
- Availability: Denial of Service attack, flood attacks, and malware

Considering consumer UAV that uses IEEE 802.11 Wireless Communication (Wi-Fi-based) security findings are Denial of Service, Wireless Injection, and Hijacking Video Capturing [4–7].

Regarding the Parrot AR.Drone 2.0, this denial of service attack compromises flight resulting in a catastrophic impact. A common approach for these platforms is hardening propositions on cryptography and authentication schemes through a Wi-Fi communication protocol change.

Security findings relating to small UAV is more common in academia due to favourable environment for research, access platforms (e.g., Parrot AR.Drone 2.0) and publishing.

This does not allow the conclusion that military or commercial applications have a greater design concern relating to Security. Available news claims that cases of hijacking military UAV and data breaching were already evidenced, which leads to more powerful agents involved (governments) that can lead to what is called cyber warfare [8]. As news states and based on publications related to sensors reading spoofing (e.g., GPS), jamming, or even physical access and data management flaws allow concluding that critical application and platforms, commercial and military, could be vulnerable to security risks.

Establishing a parallel to the aircraft industry show that security concerns are rising, and guidelines are still too young to be on currently available aircraft. It is the case of RTCA DO-356 Airworthiness Security Methods and Considerations that was released in 2014. Currently, certification agencies, e.g., the Federal Aviation Administration (FAA) has no specific regulations to address cybersecurity. They are addressing these "gaps of analysis" during design and certification through special conditions to guarantee an equivalent level of safety as existing airworthiness standards focusing on aircraft designs.

Research on FAA Special Conditions database shows the history of issuances of special conditions for both transport category (commercial aircraft) and general aviation (business jets).

The multi-UAV operation has been used to provide network infrastructure on remote areas or during temporary network disruption in specific areas. In these applications, UAVs are the nodes of the network (aerial nodes) [9,10]. Networks available through multi-UAV networks are susceptible to attacks like unauthorised access, signal jamming, signal spoofing, and data injection. This type of UAV operation is an emerging area of research [11].

Attacks (e.g., the Coagulation Attack) were already studied on UAVs networks, that affects UAVs physical configuration and manoeuvres. These attacks have as impact network failures, crashes and UAV hijacking [12].

10.5 Security ROS review

In 2012, during DEFCON 20 (Security Conference) an ROS deployment was available for security assessment as a contest [13]. It was based on a small-scale car-like robot with conference attendees invited to exploit this CPS deployment hoping to discover and address security vulnerabilities on ROS.

The pointed vulnerabilities are plain-text communication, where a malicious agent can read to identify its structure and then perform injections of fake communication packages; unprotected TCP ports; no access control to a namespace, which stores the locations of ROS computational graphs and unencrypted data storage (Figure 10.6).

The work presented by [14] proposed a security architecture based on changes to the Application Layer (top layer of the seven-layer OSI Model) and presented a case of application and validation of this architecture. Highlighted the limited native security on ROS, this proposed architecture aggregate hardening and cryptographic aspects to increase ROS resiliency to Security Threatens, giving special attention to not introduce overhead to ROS.

The work presented by [15] proposed a cryptographic structure aiming to secure communication between ROS Nodes, introducing an encryption stage on the publisher and a decryption stage on subscriber presenting an evaluation of this

Figure 10.6 ROS security assessment on DEFCON 20

implementation, considering a Triple Data Encryption Standard (3DES) algorithm, on ROS performance (CPU usage and time delays) which can be critical when considering real-time applications.

SROS (secure ROS) is presented as an addition to ROS to support cryptography and other security measures to address existing vulnerabilities already mentioned, like plain-text communication and integrity checking of messages exchange. Introduces Transport Layer Security (TLS) using a Public-Key Infrastructure (PKI) additions accomplished in a library level allowing current deployments to support SROS also mentions that security logging functionality to support auditing is on the scope. The SROS is under massive development to be deployed as part of ROS 2.

The work presented by [16] performed an IPv4 internet-wide scan to identify exposed ROS deployments. Performing TCP SYN scan to ROS master port 11311, followed by an HTTP GET / request. These scans presented deployments around the world open to external access presenting security flaws that allowed data leakage and unauthorized robot actuation.

10.6 UAV security scenarios

From the information systems domain, it is a normal approach to take a service down when an attack arises. This system mode or approach is not feasible when considering CPS applications and more specific UAV operations that, besides its information aspects must take account of its physical interaction with the environment, for example, during a flight, requiring a mitigation or design decision to guarantee safe operation. The Security Scenarios considers the published ROS security flaws [17] and its possible impacts on the consumer UAV under test with ROS:

- Data Capturing and Interception (sniffing)
- Data Interruption aka Denial of Service (DoS)
- Data Fabrication (spoofing – non-repudiation aspect)

For UAV applications, some threat scenarios can be created. Some interesting examples can be found below:

- A malicious agent forcing a landing procedure during the flight to hijack the UAV through a fake topic publishing (injection);
- False sensor readings through node corruption causing misbehaviours or UAV crash;
- Fake topic publisher flooding with a large number of data to cause a subscriber denial of service due to high processing;
- Data breaching to capture image transmission compromising privacy.

10.7 Security assessment on consumer UAV operation with ROS

ROS security assessment will be based on a simulated Parrot AR.Drone 2.0 in a virtual environment using ROS Kinetic on Ubuntu 16 LTS in conjunction with the following packages Gazebo [18], tum simulator and rviz.

The Gazebo is a free 3D dynamic simulator that enables fast simulation, tests, and development in a realistic environment. The tum simulator is a package that contains the Gazebo data for Parrot AR.Drone 2.0 with most of the sensors modelled providing also the cameras, IMU and navigation outputs. rviz is an ROS visualisation tool that is used to plot a UAV flight path to determine the impacts of security interventions on UAV behaviour.

The following flow graph can summarise the infrastructure implementation to perform security assessment:

From this flow graph, it is essential to detail the node /move square, which is responsible for controlling Parrot AR.Drone 2.0 in a fixed square path continuously while the following security cases are performed. This node interfaces with inner nodes from the Parrot AR.Drone 2.0 like /ardrone/takeoff and /ardrone/land, which are responsible as their names show, for taking off and landing. All linear and angular movements of the UAV happen through the node /cmd vel (Figures 10.7 and 10.8).

Other nodes like those related to transformation (tf) and visualisation (/visualization marker) are used to support Marker Plot on rviz. The /gazebo node handles interfaces with most UAV nodes, and its behaviour influences the simulation at Gazebo Interface (Figure 10.9).

Right after ROS initialisation through roscore command line, it is possible to evidence the first Security Findings from ROS architecture, the Master on port 11311. According to ROS documentation, the ROS Master is implemented via the XMLRPC HTTP-based protocol with available APIs that can be easily handled.

The first mitigation for this finding could be via security through obscurity, which relies on secrecy to provide security, choosing a port for ROS Master different from the typical 11311. This can be accomplished by argument -p desired port in conjunction with the roscore command line. Other ports listed in Figure 10.10 represents randomly chosen ports for ROS nodes.

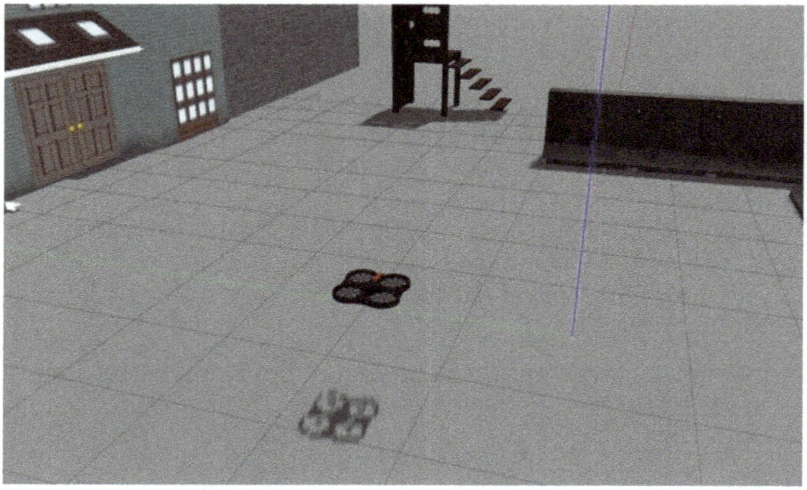

Figure 10.7 Parrot AR.Drone simulation on Gazebo

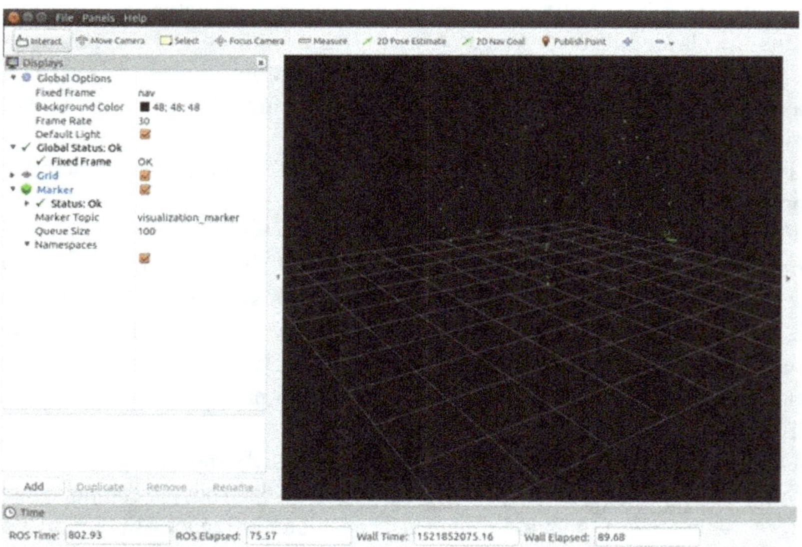

Figure 10.8 rviz flying path marker plot

This XML-RPC communication lacks privacy being performed in plain-text, which is easily eavesdropped by a malicious agent, allowing to obtain information about ROS deployment (infrastructure) and possible targets for attack (Figure 10.11).

It is easily available the information on HTTP Header about XMLRPC++ Agent and its data content on XML format. This finding can be fixed through an encryption scheme between communication nodes to avoid eavesdropping.

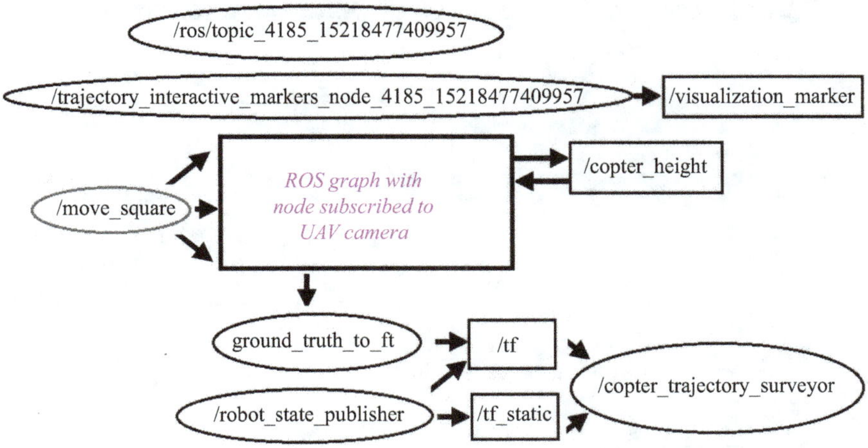

Figure 10.9 Deployed ROS infrastructure

Figure 10.10 Port scanning

As no Authentication method is implemented on ROS, neither privilege scheme (authorisation), subscribing to available nodes is easy to be implemented without any hassle for a malicious user, this can lead to a leak of data consequently loss of confidentiality, following example presents an easy subscriber to Parrot AR. Drone 2.0 front camera publisher using rviz:

On the right, the right side of the simulation on the Gazebo appears, and the left side shows the front camera topic output.

This lack of authentication and/or authorisation can be overcome with an authorisation scheme or, as previously mentioned, for ROS communication through encryption, as also presented by [17].

Another found issue relates to the lack of consistency check between nodes creation or publishing activity, as previously mentioned, the /cmd vel node is

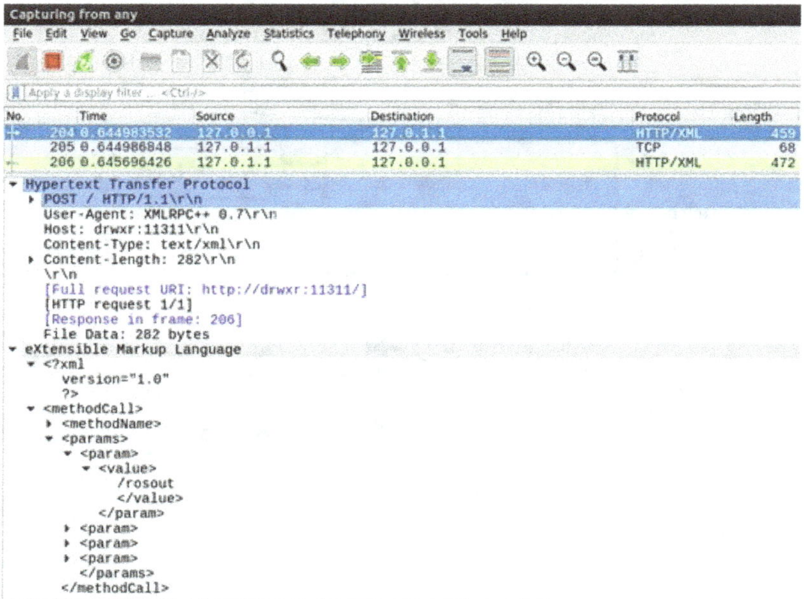

Figure 10.11 Packet sniffing and analysis

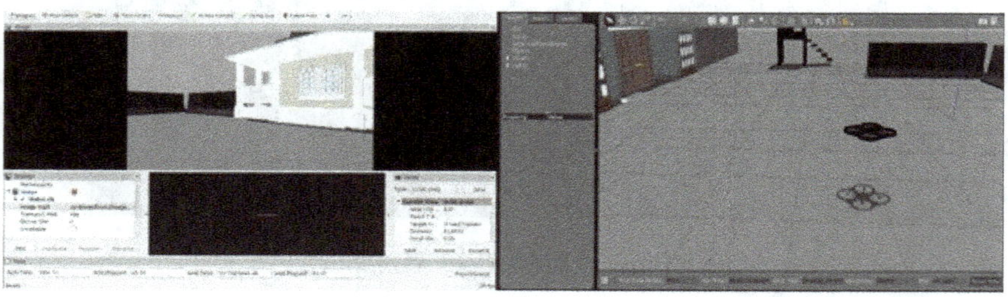

Figure 10.12 Topic eavesdropping

responsible for linear and angular movements of the UAV, due to lack of check, it is possible to create an unauthorized node that can interfere with /cmd vel proper operation. A simple malicious code is presented next (Figure 10.12).

Listing 10.1 Malicious code

```
#! /usr/bin/env python

import rospy
from geometry_msgs.msg import Twist

rospy.init_node('unauthorized')
```

```
pub = rospy.Publisher('/cmd_vel', Twist,
queue_size=1)

 rate = rospy.Rate(1)

 while True:
 pub.publish()
 rate.sleep()
```

This code is responsible for creating a node named /unauthorized which acts publishing an empty message on /cmd vel node interfering on the legit node /move square from publishing.

The evidenced behaviour on simulation relates to UAV freezing due to the fact that the last published command to /cmd vel either by /move square or /unauthorized is the response for commanding the UAV, once the /unauthorized node publishes with a higher frequency, it takes control over /move square on /cmd vel.

This malicious node appears on the architecture as another interface with /cmd vel as highlighted on the graph (Figure 10.13).

10.8 Future trends

With expected ubiquity of UAVs in a more connected environment new technological trends emerge, such the concept of Internet of Drones (IoD) [19] that take advantage of existing cellular networks and radio communication to support UAVs coordination increasing the classical ground station to vehicle communication with a vehicle to vehicle communication being an enabling technology for autonomous flight, swarm navigation and Unmanned Traffic Management (UTM) [20] (Figure 10.14).

As an emerging technology, IoD still requires research and developments [21], regarding security, solutions to overcome unauthorized access, illegal connection, eavesdropping, jamming, authentication, intrusion detection, among others.

In conjunction with Software Defined Network simulators (e.g., mininet) and Network Functions Virtualization (NFV) [22], it is possible to create simulation testbeds with ROS and Gazebo to validate secure network deployments of IoD.

10.9 Conclusion

Security is an apparent concern for UAV and a key enabling technology for wide adoption. With ROS as the supporting middleware for UAV, these security aspects shall be considered on ROS. These conclusions from UAV domain can be derived for all CPSs once ROS is glimpsed to be adopted in a wide range of domains [23–26]. This can be improved by the continuity of initiatives such as Security Robot Operating Systems (SROS) and as provided by ROS 2.

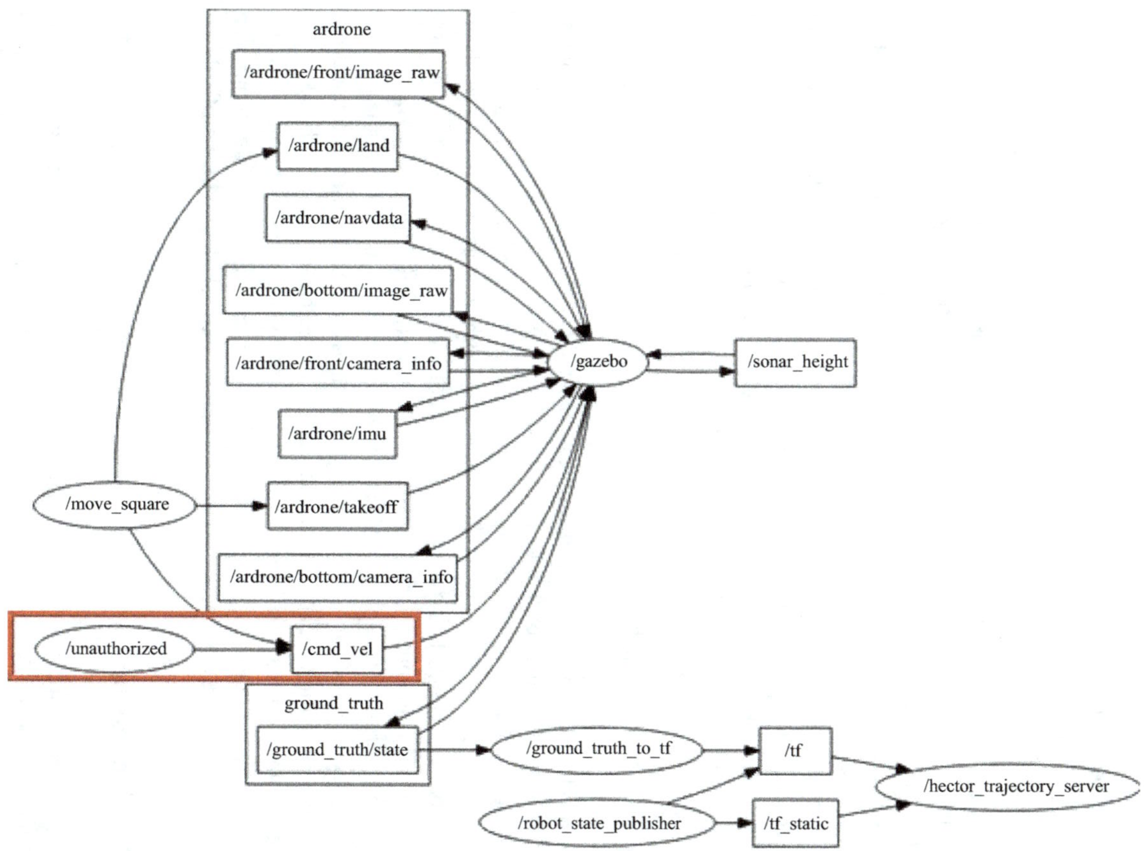

Figure 10.13 Modified architecture

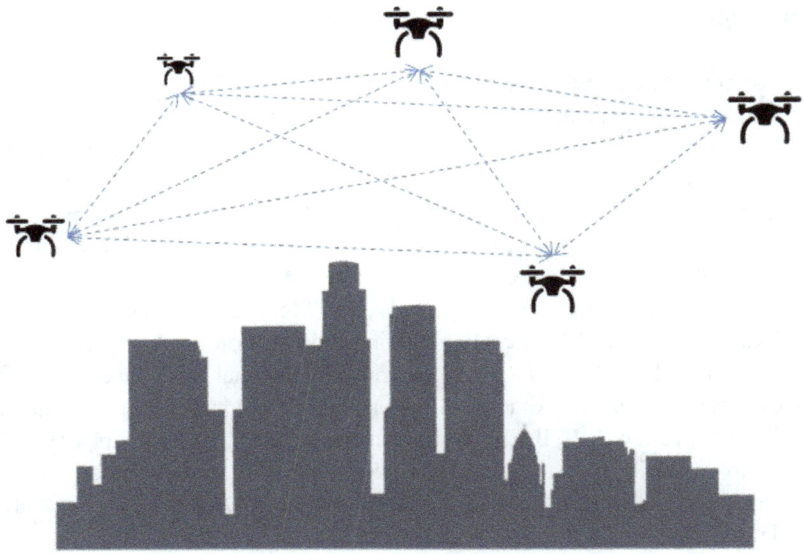

Figure 10.14 Internet of drones (IoD) concept

Despite security improvements introduced by ROS 2, it does not eliminate the need for security assessment during UAV deployments with ROS, mainly in solutions that consider the integration of multiple robots where emergence properties can arise and not expected threat conditions can be present requiring a security analysis to avoid vulnerable deployments.

A great tool to sustain design is a Security Risk Assessment that supports the decision in which risks are acceptable, due to the complexity of implementation (i.e., cost) and impact of the scenario under analysis.

Compared with the traditional approach of fixing a flaw after it occurs which in some designs can cause an earlier retirement due to high impactful re-design or fix, simulation, as presented with the Gazebo, and rviz is a great supporting tool to simulate these scenarios and provide subsidies for the security risk assessment and even design changes during early stages of development.

Current approaches consist of hardening in the application level of ROS to mitigate its findings through authentication and encryption schemes. Also, an option is to harden the operating system that runs ROS using firewall and user permission rules to access roscore, practices similar to those used on embedded systems.

Another hardening task, not mentioned in previous publications, relates to a configuration management control system, once ROS is based on the file system. This hardening consists of hash tables of components of ROS to avoid file modifications or inclusions that must be verified before every ROS application launch.

An open space to contribution relates to penetration test tools for ROS that are currently not available to the community.

References

[1] Vattapparamban E, Guvenç I, Yurekli AI, *et al.* Drones for Smart Cities: Issues in Cybersecurity, Privacy, and Public Safety. In: Proc. 2016 Int'l Wir. Comm. and Mob. Comp. Conf. (IWCMC). IEEE, 2016, Paphos, Cyprus, p. 216–221.

[2] Estrela VV, Saotome O, Loschi HJ, *et al.* Emergency Response Cyber-Physical Framework for Landslide Avoidance with Sustainable Electronics. Technologies, 6, 2018, p. 42.

[3] Khalilpour M, and Razmjooy N. A Hybrid Method for Gesture Recognition. Journal of World's Electrical Engineering and Technology, 2, 3, 2013, p. 47–53.

[4] Hooper M, Tian Y, Zhou R, *et al.* Securing Commercial WiFi-based UAVs from Common Security Attacks. In: Military Communications Conference, MILCOM 2016-2016 IEEE. IEEE; Baltimore, MD, USA, 2016, p. 1213–1218.

[5] Bisio I, Garibotto C, Lavagetto F, Sciarrone A, and Zappatore S. Blind Detection: Advanced Techniques for WiFi-Based Drone Surveillance. IEEE Transactions on Vehicular Technology, 68, 2019, p. 938–946.

[6] Lykou G, Anagnostopoulou A, and Gritzalis D. Smart airport cybersecurity: Threat mitigation and cyber resilience controls. Sensors. 2018;19(1). doi: 10.3390/s19010019.

[7] Estrela VV, and Coelho AM. State-of-the art Motion estimation in the context of 3D TV. In: Multimedia Networking and Coding. IGI Global, Hershey, PA, USA, 2013, p. 148–173. DOI:10.4018/978-1-4666-2660-7.ch006.

[8] Faria LD, Silvestre CA, Correia MA, and Roso NA. Susceptibility of GPS dependent complex systems to spoofing. Journal of Aerospace Technology and Management. 2018;10:10.

[9] Singh K, and Verma AK. Flying adhoc networks concept and challenges. In M. Khosrow-Pour, D.B.A. (Ed.), Encyclopedia of Information Science and Technology, Fourth Edition (pp. 6106–6113), 2018. Hershey, PA: IGI Global. doi:10.4018/978-1-5225-2255-3.ch530.

[10] Kaleem Z, and Rehmani MH. Amateur Drone Monitoring: State-of-the-Art Architectures, Key Enabling Technologies, and Future Research Directions. IEEE Wireless Communications, 25, 2018, p. 150–159.

[11] Singh K, and Verma AK. Threat Modeling for Multi-UAV Adhoc Networks. In: IEEE Region 10 Conf., TENCON 2017-2017. IEEE; Penang, Malaysia, 2017, p. 1544–1549.

[12] Sharma V, Kumar R, Srinivasan K, *et al.* Coagulation Attacks over Networked UAVs: Concept, Challenges, and Research Aspects. International Journal of Engineering & Technology, 7, 3.13, 2018, p. 183–187.

[13] McClean J, Stull C, Farrar C, *et al.* A preliminary cyber-physical security assessment of the robot operating system (ROS). In: Unmanned Systems Technology XV. Vol. 8741. International Society for Optics and Photonics; 2013, p. 874110.

[14] Dieber B, Kacianka S, Rass S, *et al.* Application-level security for ROS-based applications. In: 2016 IEEE/RSJ International Conference on Intelligent Robots and Systems (IROS). IEEE; Daejeon, Korea, 2016, p. 4477–4482.

[15] Lera FJR, Balsa J, Casado F, *et al.* Cybersecurity in Autonomous Systems: Evaluating the performance of hardening ROS. Malaga, Spain. 2016, p. 47.

[16] DeMarinis N, Tellex S, Kemerlis V, *et al.* Scanning the Internet for ROS: A View of Security in Robotics Research. arXiv preprint arXiv:180803322. 2018.

[17] Dieber B, Breiling B, Taurer S, *et al.* Security for the Robot Operating System. Robotics and Autonomous Systems. 98, 2017, p. 192–203.

[18] Hailong Q, Meng Z, Meng W, *et al.* Autonomous Exploration and Mapping System Using Heterogeneous UAVs and UGVs in GPS-Denied Environments. IEEE Transactions on Vehicular Technology, 68, 2019, p. 1339–1350.

[19] Gharibi M, Boutaba R, and Waslander SL. Internet of Drones. IEEE Access, 4, 2016, p. 1148–1162.

[20] Kopardekar P, Rios J, Prevot T, *et al.* Unmanned Aircraft System Traffic Management (utm) Concept of Operations. In: AIAA Aviation Forum; 2016.

[21] Choudhary G, Sharma V, Gupta T, *et al.* Internet of Drones (IoD): Threats, Vulnerability, and Security Perspectives. arXiv preprint arXiv:180800203. 2018.

[22] Rametta C, and Schembra G. Designing a Softwarized Network Deployed on a Fleet of Drones for Rural Zone Monitoring. Future Internet, 9, 1, 2017, p. 8.

[23] Estrela VV, Monteiro ACB, França RP, Iano Y, Khelassi A. and Razmjooy N. Health 4.0: Applications, Management, Technologies and Review. Medical Technologies Journal, 2, 4, 2019, p. 262–76, DOI: 10.26415/2572-004x-vol2iss4p262-276.

[24] Zheng S, Lin Z, Zeng Q, Zheng R, Liu C, and Xiong H. IAPcloud: A Cloud Control Platform for Heterogeneous Robots. IEEE Access, 6, 2018, p. 30577–30591.

[25] Motlagh NH, Taleb T, and Osama O. Low-Altitude Unmanned Aerial Vehicles-Based Internet of Things Services: Comprehensive Survey and Future Perspectives. IEEE Internet of Things Journal, 3, 2016, p. 899–922.

[26] Atat R, Liu L, Wu J, Li G, Ye C, and Yi Y. Big Data Meet Cyber-Physical Systems: A Panoramic Survey. IEEE Access, 6, 2018, p. 73603–73636.

Chapter 11
Vision in indoor and outdoor drones

Maik Basso[1] and Edison Pignaton de Freitas[1]

This chapter explores how the type of environmental impacts the CV techniques, algorithms and specific hardware to be used. Indoor environments, also known as controlled environments, generally rely on solutions based on beacons, proximity sensors and image processing for data acquisition. In this case, as the environment is controlled, the illuminance of the scene is adjusted and sensors are previously positioned, which facilitates the development and execution of these systems. In outdoor environments, generally known for uncontrolled environmental variables, frequently require solutions based on image processing techniques to provide the data acquisition. In this environment, the non-constant variation of the illuminance of the scene and the great variation of the background of images are important complicating factors for the operation of the image processing algorithms. In addition, constructions and buildings block the signal of sensors and global positioning systems making it even harder to treat the exceptions caused by these factors. Each exception being treated in a CV system has a computational cost that can be high. If this is considered in applications using embedded hardware, some projects simply become infeasible. Researchers put great effort attempting to optimise the software for high performance and better use of the hardware resources, so that less processing power is demanded as well as positively impacting the energy savings. This chapter presents a review of the main CV techniques currently used in the development of mission control software for the use in indoor and outdoor environments, providing autonomy navigation and interaction for these aerial robots.

11.1 Computer vision in unmanned aerial vehicles

Computer vision (CV) studies the development of artificial systems capable of detecting and developing the perception of the environment through image information or multidimensional data, obtained by different types of sensors [1]. It is a useful mode of robotic detection as it imitates the human visual sense allowing the extraction of measurements from the environment without contact. CV involves working with higher concepts and algorithms related to artificial intelligence,

[1]Department of Electrical Engineering, Federal University of Rio Grande do Sul, Porto Alegre, RS, Porto Alegre - RS Brazil

which involves intense programming so that it fits with the activities and requirements of the system. The CV scope encompasses several areas of knowledge, as can be seen in Figure 11.1 [2]. The CV differs from image processing because, it also handles signals obtained from sensors and other means to analyse, to understand and to provide the interaction of the system with the environment in which it is inserted.

From the engineering point of view, CV is a valuable tool to build autonomous systems that can accomplish some of the tasks that the human visual system, and, in many cases, overcome the human capabilities [1]. Currently, vision systems are widely integrated with robotic systems. Commonly, visual perception and manipulation are combined in two steps 'look' and then 'move' [3]. The accuracy of the results depends directly on the accuracy of the visual sensor. An alternative to increasing the overall system accuracy is to use a visual feedback control loop, decreasing the obtained error.

In this context, there is an increasing interest in using CV techniques in unmanned aerial vehicles (UAVs), also known as drones. Such UAV systems can be better modelled if they are treated as cyber–physical systems (CPSs) [4]. In their case, these techniques are applied to position the UAV in autonomous flight mode or to perform aerial surveillance and detection of regions of interest (ROIs) or points of interest (POIs) [5]. This trend is driven by different facts, such as the miniaturisation of electronic components, including sensors (driven by other technologies such as smartphones) [6]; the increase of computational power for the on-board CPUs; and the cost reduction of other components of this type of robotic platform [7]. With today's technology, the variety and complexity of mission tasks modern UAVs are demanding and they require higher levels of autonomy. The central part of a standalone UAV is the navigation control system and its supporting

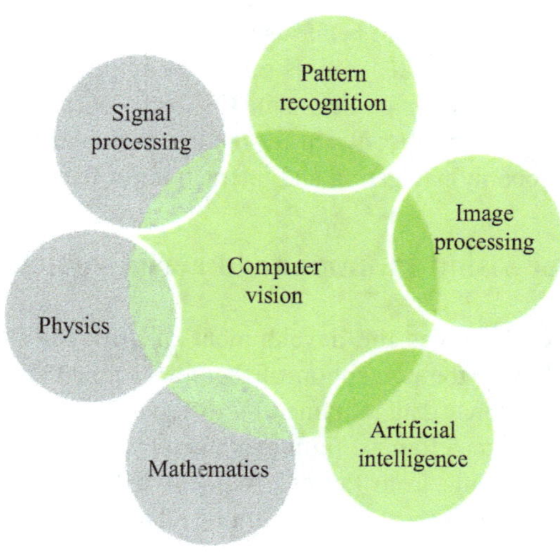

Figure 11.1 Computer vision scope [2]

subsystems, integrated to the ground control stations [8]. The ground control stations are responsible by the remote monitoring of the operations executed by the UAVs. The autonomous navigation system uses information from several subsystems to perform three main tasks: estimating the location and orientation of the UAV (location), identifying obstacles in the environment (obstacle detection), and then making decisions (decision-making). These decisions are critical to maintaining the control loop and to provide navigation in an unknown environment [9]. CV is an indispensable tool to achieve these goals.

The CV systems installed in and integrated with the architecture of the UAVs generally have similar architecture and take three steps in its operation, which are data acquisition in numerical form, data processing and data analysis. Cameras, proximity sensors, and ultrasonic sensors typically perform the data acquisition step. After data acquisition, the embedded computer performs data processing by performing algorithms with measurement techniques (variables, index, and coefficients), detection (patterns, objects or ROI/POI) or monitoring (people, vehicles or animals).

The resulting processed data are analysed and then converted into decision commands that serve as input to the autonomous robotic system [10]. Autonomy is characterised as the ability to move and act in the environment in which it is inserted, to perceive the environment through sensors, to adapt or to change the environment, to learn from past experiences, to construct representations of the environment, and to develop a process of interacting with the environment. However, the performance of a CV system has limiting and determinant factors. For example, in a visual computing system that uses image processing, some factors can affect the image acquisition, such as occlusion, blur movement, rapid pose variation, disordered background environments, and onboard mechanical vibration, among others [11].

In addition, the type of environment is a determining factor for the techniques, algorithms and specific hardware to be used. Internal environments can be controlled and most often rely on solutions based on beacons, proximity sensors and image processing for data acquisition. In this case, as the environment is controlled, the scene illumination can be adjusted, and the sensors can be pre-positioned, which facilitates the development and execution of these systems. On the other hand, these environments usually have more obstacles, and the space of navigation is much more restricted, which can lead to spending much of the processing for this verification. However, these conditions can be further hampered when the internal environments are unknown by the stand-alone system, as in the case of rescue operations [12].

In outdoor environments, often known to be variable, with uncontrolled environmental factors, often require solutions based on image processing techniques to provide data acquisition. Moreover, in outdoor operations, most navigation systems are based on the Global Positioning System (GPS) [13]. In this environment, the constant variation of the scene luminance and the large variation of the background in the acquired images are important complicating factors for the operation of the image processing algorithms. Environmental noise is also something that prevents the proper functioning of the sensors that use this form for data acquisition.

On the contrary, constructions and buildings block the signal from sensors and GPSs making it even harder to handle the exceptions caused by these factors. Each exception treated in a CV system has a computational cost that can be high, depending on the time requirements to which this system is constrained. Considering applications using embedded hardware, some projects become unfeasible. For this reason, the research in this area is focused on trying to optimise the software for high performance and better use of hardware resources, so that less processing power is required and positively impacting the energy consumption of these systems, which are usually driven by batteries.

To understand the current state of research in the area of visual computing used in UAVs, Sections 2.1.1 and 2.1.2 present a study about techniques of computational vision used in drones, along with their relation to the development of mission control software for use in internal and external environments. The text highlights the importance of this research area to provide support for autonomy in the navigation and the interaction for these airborne robots with the environment in which they are inserted.

11.1.1 Indoor environments

Differently, of the UAVs that operate in the external environment, UAVs that operate indoors usually have a reduced size, which impacts on the amount of hardware they can carry and on their autonomy. These factors also impact on a computational capacity, which is further reduced when compared to larger vehicle platforms. The GPS signal is blocked against constructions or buildings. Thus, indoor UAV cannot recognise its location in the environment [14]. This impacts on one of the most common applications developed and studied for indoor environments, which refers to how to provide localisation to these systems. In addition, indoor environments may have poor lighting, and because of this, they require that artificial lighting is installed in the hardware, which can significantly increase battery consumption.

In this context, alternative navigation techniques are required, which allow the vehicle to operate successfully in these areas. Furthermore, it is often desirable that these navigation techniques do not require external infrastructure. One of the most popular techniques for indoor navigation of air vehicles is the use of a laser with track finder, which measures the distance to the objects that are in its focus. With this equipment, it is possible to use the technique known as simultaneous localisation and mapping (SLAM) to create a three-dimensional map of the environment and to locate the vehicle in the environment. Inertial navigation with the aid of monocular vision is another paradigm applicable to internal environments. The cameras provide a wealth of information about the environment, and they are low cost and lightweight [15].

Under these circumstances, in work presented by [15], two integrated navigation systems are developed that combine the techniques of visual SLAM, using a camera and laser SLAM with an inertial navigation system. The monocular SLAM visual system had fully correlated characteristics and modelled the vehicle states. The SLAM laser system is based on a scan matching map of the Monte Carlo model and leverages the visual data to reduce ambiguities in the vehicle pose estimation. The system is validated in the simulation with 6 degrees of freedom and in real flight test. According to the authors, one of the main features of the work is that the system is validated with a controller in the navigation loop. Figure 11.2 shows the trajectory traversed by the UAV in the simulation of an exploration mission using the proposed techniques.

Although the laser scanner can provide accurate depth information, it is very expensive and cumbersome. With the use of an RGB-D camera, which can provide RGB and depth images, a simultaneous location and mapping method based on this type of camera is proposed in [16]. An RGB-D SLAM algorithm was used to locate the camera and construct the 3D map of the used test environment. The developed system can achieve the pose and the trajectory of the camera. In addition, chart optimisation and loop closure are adopted to eliminate cumulative errors. However, in this solution, when the camera moves too fast, loss of frames occurs, and few points are detected. According to the author, the performance of the algorithm and its accuracy need to be improved.

In vision-based navigation systems, the UAV path must be known a priori.

Data from the surrounding environment of the UAV's flight path is taken and analysed to identify the basic features of that path before the flight mission begins.

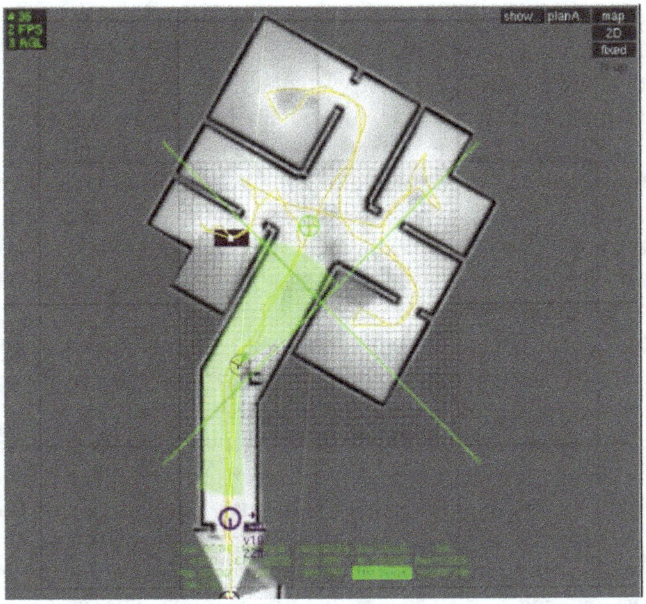

Figure 11.2 Result of the SLAM map and trajectory of the UAV while exploring a simulated indoor environment [15]

Then the real-time data taken from the on-board sensors during the flight mission is compared to the vehicle's visual memory to identify the known characteristics and estimate the UAV's motion. In this context, the work proposed by the authors in [17] describes an indoor, inexpensive and simple navigation system using three laser beams fixed to the UAV and pointed to the ground.

A proposed algorithm uses the camera to capture the laser points on the ground and to determine their coordinates. The positions of the laser points are used to obtain complete information on the position and orientation of the UAV. The proposed navigation system can be classified as a vision-based navigation system. However, it does not depend much on the quality of videos taken from the vision camera and does not require an image processing algorithm with high computational cost employed. An illustrative simulation study is conducted to demonstrate the validity of the proposed navigation system. According to the author's results, the proposed system is more efficient over a range of angles between the laser beams because it is possible to obtain the laser beam length without the need for extra sensors or estimation algorithms. This work manages to provide a location for navigation through a simple and easy method, but it is not possible to predict collisions with obstacles along the way.

Collision with obstacles is another well-known problem in indoor environments, so an indoor location assistance system is proposed by [18]. This work demonstrates the development of an indoor navigation system, specifically for industrial applications that require customised detection technologies to aid navigation. A custom sensing array, with ultrasonic transceivers, was developed to locate the drone's position in a known closed environment and to provide feedback to the navigation system. Six people were recruited to pilot the drone with and without the navigation system in an enclosed room for a predefined target at a known location. Two approaches were used, the first when the UAV was at the line of sight of the pilot, and the second with no line of sight. The duration of the flight, the number of collisions and the distance from the target were recorded and used to measure the performance of the proposed application. Using the proposed navigation system, it was possible to reduce the flight duration by an average of 19.7% during an obstructed line of sight. Moreover, UAV preventative collision detection and navigation has been improved. The navigation system provides a detection radius of 7.65 m and a position accuracy of 1 cm.

A localisation technique known as Air-SSLAM was presented in [19]. This technique uses a stereo camera setup besides two or more cameras that can be used. Figure 11.3 demonstrates the operation of the proposed system.

Stereo images are captured, and then the image features are extracted. A mapping between the features of the images is performed through a descriptor, and an initial map is generated with the depth estimate of each pair of features. This process is then repeated using the initial map as a reference. The maintenance and updating of the long-term map are continuously carried out by analysing the quality of each correspondence, as well as inserting new features in unexplored areas of the environment. According to the author, the work presents three main contributions that are the development of a new method to combine the features efficiently, the

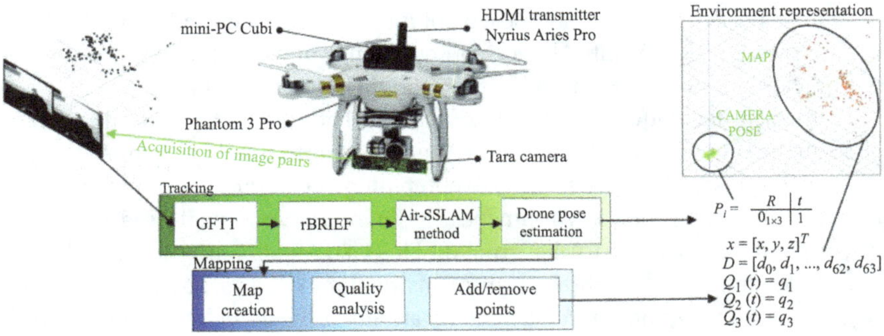

Figure 11.3 Outline of Air-SSLAM architecture [19]

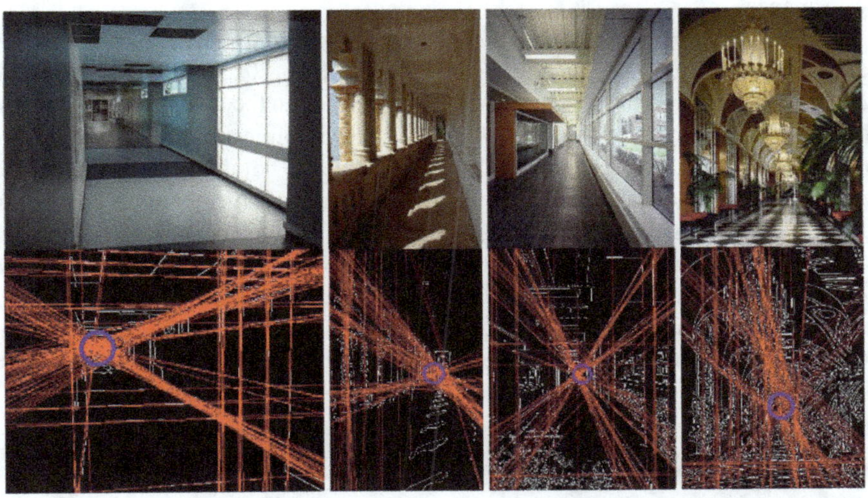

Figure 11.4 Images of the navigation system running inside a corridor [20]

development of three quality indicators to accelerate the process of mapping and the maintenance of the map with a distribution that in its technique, is performed by zones in the images. The results are promising according to the authors because the method works with a constant update of the map of the environment and with an approximate number of 200 features.

Another promising method of indoor navigation is proposed in [20]. This method does not require the construction of a 3D model of the environment. On the other hand, this technique classifies the type of internal environment in which the UAV is inserted and then uses vision algorithms based on perspective suggestions to estimate the desired direction to fly. Figure 11.4 demonstrates the system detecting the escape route in indoor images (corridors). Images are processed with a Canny edge detector, and the Hough transform is used to find the vanishing point using a proposed grid-based approach.

During the tests with the proposed solution, it was identified that these algorithms require a significantly lower computational power, allowing the UAV to react quickly and navigate through several internal environments.

Until then, techniques have been described that are generally used in uncontrolled or unknown environments. The studies reported in [21] demonstrate the development of a UAV using visual computing techniques to provide vehicle navigation in a controlled environment. The indoor environment used in the tests had coloured tracks fixed to the ground. The visual computing system detected these tracks and decoded in the form of instructions used to power the vehicle's navigation system. Important data such as attitude, speed, and acceleration of the UAV along with the real-time video, are sent as feedback to the ground station through communication links to give commands and also for monitoring purposes. The algorithms of processing and control by CV were evaluated and obtained good performance.

Using a scenario controlled with markers, the system from [22] is a completely autonomous solution used to participate in the IMAV 2013 competition (IMAV 2013), which has won the first prize of the competition to the authors. The proposed solution is a system composed of multiple UAVs without centralised coordination, whose UAV agents share their positional estimates. The ability to navigate and detect collisions is a consequence of the behaviour of each member participating in the group of UAVs. Figure 11.5 demonstrates the system execution scenario.

All processing takes place at a base station outside the vehicles. For each vehicle, there is running in the ground station an instance of the proposed architecture that communicates through Wi-Fi with the UAV. Visual markers are used to detect and map obstacles and to improve pose estimation. In the executed tests,

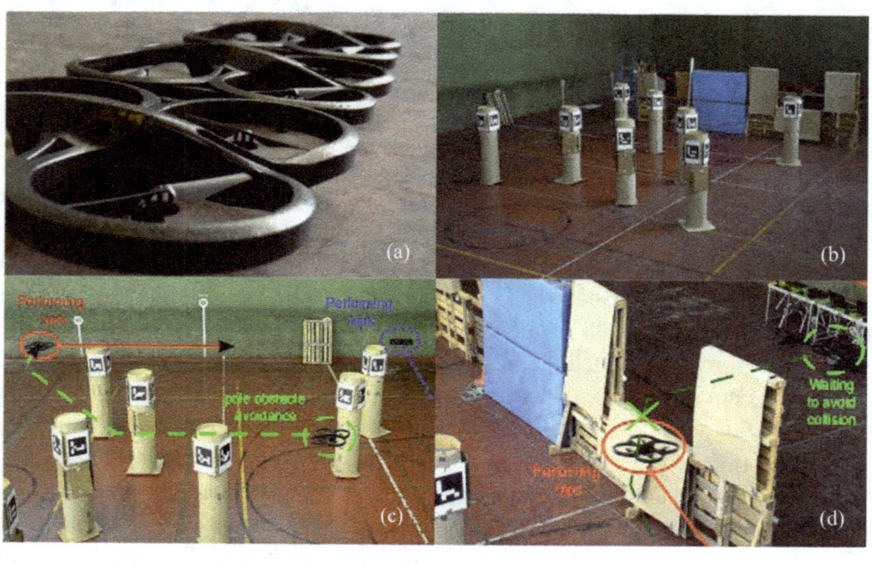

Figure 11.5 The architecture and the scenario application presented in [20]

each UAV presented the ability to navigate avoiding obstacles and even collision with other members of the system.

In the case of indoor environments, what can be observed from the use of visual computing techniques in UAVs is that most of these works are designed to develop systems with the ability to navigate and/or collision detection/avoidance in objects or obstacles. Efforts are concentrated in the development of aerial vehicles to explore the environment and interaction with it, preventing missions and optimising their flight time. Another relevant point to notice is that the research focuses on developing solutions for indoor environments that are generally uncontrolled and unknown, which can increase the level of autonomy but also the level of difficulty in implementing and merging the architecture of these systems. Another important direction of studies that must be highlighted is on platforms that consider the interaction of several vehicles with the environment (and among themselves), being collaborative or independent.

11.1.2 Outdoor environments

Unlike techniques presented in indoor environments that are restricted to the navigational problem resolution and collision detection/avoidance, the visual computing techniques employed for vehicles intended for outdoor use have a much broader application scope due to the diversity of the environment and the great number of possible applications to be developed. Because of all this diversity of outdoor environments, vision applications based on position-attitude control, pose estimation, mapping, obstacle detection, target tracking are easily found in the current literature [23].

However, these techniques tend to have a higher computational cost because they must deal with the sudden variations of the data acquired by the sensors in an uncontrolled environment. These systems also usually have more than one component in their execution loop in order to treat each of the parts of the application to which the proposed technique covers. This point requires a greater effort on the part of the researchers, who usually have to optimise their techniques extremely to be able to meet the temporal requirements of these systems.

For the localisation problem, these techniques of outdoor use most often have the availability of reference for GPSs to assist in the determination of the position of the UAV. The accuracy of these systems also depends on the number of satellites to which the vehicle has visibility. However, these GPS-based systems do not provide a reliable solution in environments such as urban areas or forests that may reduce the visibility of the vehicle to the available satellites [9].

Currently one of the main computational techniques used in UAVs is the detection of patterns. This technique is so versatile that it allows the development of a great number of applications. A good example of such applications may be the detection of fire areas. The architecture shown in Figure 11.6 characterises these applications.

As described in [25] a method for monitoring and detection of forest fire is presented. The proposed algorithm for fire detection uses colour and motion features to improve the performance of forest fire detection and increase its reliability. The motion and colour resources are extracted using an optical flow method and

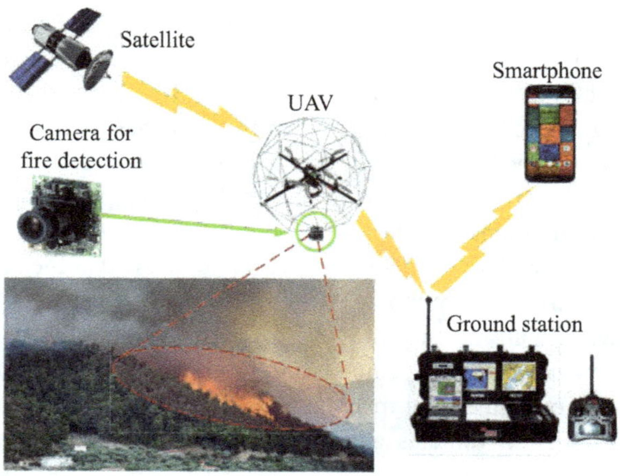

Figure 11.6 Conceptual vision-based forest fire detection system [24]

colour-based decision rules, respectively. In this work, experiments were performed using a low-cost camera installed at the bottom of the UAV to search for and detect a fire.

In the proposal reported in [24], a new method of detecting forest fire using colour and motion features for forest fire-fighting applications is presented. The first step of the software consists of the extraction of fire colour pixels as regions of fire candidates, making use of the chromatic characteristic of the fire. The second step for this proposed system is the use of an optical flow to calculate motion vectors of the candidate regions. The motion is estimated from the results of the optical flow to distinguish fire from other false positives. By applying a threshold and performing morphological operations on the motion vectors, the binary images are obtained. Then the fires are located in each binary image using the proposed method. The results obtained in the development of this work are promising, in addition, the technique presented good performance and reliability in fire detection.

Following the line of detection techniques, the work developed by [26] presents a solution for the detection and counting of cars to be used in UAVs. The proposed method does a step of screening asphalted areas in order to restrict areas where cars can be detected and thus reduce false positives. A resource extraction process based on a characteristic transformation is used to detect the key-points identified in the image. Then, it uses a classifier to determine what parts of the image are cars and what parts are not. The last step of the method is the grouping of the key-points belonging to the same car. Then the number of cars present in the scene is calculated. The results presented demonstrate the efficiency of the algorithm in the detection of cars, but it is suggested by the authors an improvement in the methodology to reduce the number of duplicate key-points to make the algorithm even faster and more efficient.

Another well-known detection technique described in the literature as Crop Row Detection (CRD) is presented in [27]. This technique performs the detection of the lines of plants on precision agriculture plantations. Combining this technique with a Line Follower algorithm, and based on software implementation in low cost embedded hardware, it represents a valuable tool for autonomous conduction of UAVs to deploy precision agriculture applications. As an auxiliary vision tool for an embedded spraying system, the author also proposes the use of a vegetation index known as Normalised Difference Vegetation Index (NDVI) for self-regulated application of agrochemicals.

Inspections also use detection techniques based on CV. As presented in [28], an autonomous UAV-based inspection system is implemented for asset evaluation and fault detection for large-scale photovoltaic systems. The detection of the defects is done through the first-order derivation of the Gaussian function and the correspondence of characteristics in the images. Two typical visible defects of photovoltaic modules are characterised, namely, snail trails and dust shading. Field experiments demonstrate that the system can perform inspection and condition monitoring tasks on large-scale photovoltaic systems in an autonomous and supervised manner with significantly improved efficiency. The performance evaluation under various conditions confirms that the proposed inspection system can adapt to different slopes within a certain flight height.

There are also several detection-based CV techniques to assist UAVs while landing. In work presented in [29], a robust system with detection and marker tracking technique is proposed. The algorithms are proposed to assist in the task of landing the UAVs. Robust tracking is done using the multiple ellipse connection with a concentric marker. The simulation results show that the algorithm is robust and very accurate.

Another system to aid the landing of UAVs in a moving vehicle with real-time image processing system based on marker detection is proposed in [30]. Marker detection is based on a colour detection algorithm with a morphological filter, and the tracking orientation is based on the relative distance between the UAV and the vehicle. The captured RGB image was transformed into the HSV colour system so that a noise filter was applied using a threshold. The target tracking algorithm is based on the relative distance between the UAV and the target. The flight test was performed with the vehicle moving in the open-air environment, and the algorithm was validated for the autonomous landing.

A standalone vision-based tracking system is presented in [31], and it is used to track a target from a UAV. To deal with the losses due to occlusion or loss of the detected object, this work uses a robust and computationally efficient visual tracking scheme in which a correlation filter and the re-detection algorithm. Target status is estimated from visual information. Extensive real-time flight experiments were performed in outdoor environments, where computing is fully implemented in the on-board computer. Experimental results illustrate that the proposed real-time vision-based tracking system achieves the tracking performance required by the application.

Another work that traces objects is proposed in [10], where an architecture that allows the user to specify an object in the image that the UAV must follow from an approximately constant distance is developed. In the event of loss of tracking of the object in the image, the system begins to hover and awaits the retrieval of tracking or a subsequent detection, which requires the use of odometry measurements for self-stabilisation. The proposed software uses the forward-facing camera images and some of the IMU data to calculate the references to the control loops. The results obtained with the work relativity show that the system was able to perform the visual detection with targets of variable size, that the system was able to follow a person with a velocity of approximately 2.5 m/s for a time of approximately 45 seconds.

A new method for detecting, locating and recognising trees with a UAV equipped with a monocular camera is presented in [32]. Trees are detected on a frame basis using the latest generation convolutional neuron networks, inspired by the recent rapid advances shown in the literature. The trees are detected, and their position is marked on the GPS. Localised trees are segmented, characterised by resource descriptors, and stored in a database that contains their GPS coordinates. The trees detected on later flights are compared with the data in the database to avoid duplication of information. The proposed method is also able to identify if the trees are absent from the expected locations with GPS marked, allowing to alert the authorities concerned about possible illegal deforestation immediately.

In addition to counting trees, other vision applications in UAVs such as counting other types of commercial plants help to generate production statistics in precision agriculture. In this context, the work presented in [33] demonstrates an economical and attractive technology for the automatic detection of palm trees to be used in UAVs. From the in-flight image, it first extracts a set of key-points using the algorithm to detect features. Afterwards, these key-points are analysed with a trained classifier in a training data set. Finally, the palm trees are identified. Then, to capture the shape of each tree, it merges the key-points with an active contour method. Finally, the texture of the obtained regions is analysed with local binary patterns to distinguish palms from other vegetation. The results of the experimental tests were acquired in two different farms and confirm the promising capabilities of the proposed structure in palms detection.

The work presented in [34] demonstrates a rigorous study on the use of embedded hardware coupled to a UAV to perform real-time image processing. A POI detection algorithm has been proposed to perform the tests on the hardware. In the article, it is shown that it is possible and feasible to use a low-cost processing board in a mini UAV for image processing. In this work, a practical application is demonstrated using image processing in UAVs; it is shown the preliminary results of an application for the detection of power lines in order to perform the autonomous guidance of the UAV on the lines for cable inspection for example. Figure 11.7 shows an image illustrating the results in the detection of three electrical power lines. Notice that the system successfully perform the detection, even considering a noise environment in the background.

Another emerging technology is the monitoring and tracking of ice for use in marine ice management applications. This technique was reported in [35], and it is aimed to be used to detect and track the movement of icebergs and other ice sheets in an arctic environment autonomously. An occupancy grid map algorithm and locations of interest generator are used, coupled with a mission control system. According to the developers of the project, one of the contributions of their work is the interface of the algorithm with a vision-based object detection module to

Figure 11.7 Electrical power line detection application [34]

generate an occupancy grid map of a real-time predefined search area using on-board processing data. A generator of places of interest has also been developed, generating locations where the UAV should investigate based on the occupation grid map. The results were obtained based on test flights performed, and it was possible to verify that the system was able to successfully create an occupation grid map based on the automatic segmentation of thermal images on-board and in real-time, in ice regions and regions without ice. However, in the tests carried out, it is possible to observe some delays in the communication and the fact that the mission controller was implemented in a ground base station.

As can be observed, the visual computing applications used in outdoor environments have a great diversity due to the number of possible applications. Advances in technology increasingly allow UAVs to be used in a variety of tasks and they are thus changing the way some processes are conducted, such as in the case of electrical transmission line inspections or even in the inspection of photovoltaic panels. However, this huge number of applications requires systems with increasingly complex and computationally costly architecture. This is one of the most significant problems in CV outdoor applications. There is a need to use several algorithms engaged in an application loop competing for processing time on a low-cost processor. To distribute tasks and processing studies from this moment on start and turn to applications available to multiple execution agents.

11.2 Other approaches handling both indoor and outdoor environments

In addition to the most common approaches, studies are being developed to perform in indoor and outdoor environments, such as the work demonstrated in [36], which presents a complete strategy of tracking a terrestrial target in complex internal and external environments with a UAV based on CV.

Navigation techniques have also gained focus on multi-environment studies, as demonstrated in [37]. The proposed architecture efficiently combines visual information from a monocular camera with measurements of inertial sensors. The presented algorithms are validated on multiple platforms in real conditions, through a 16-minute flight test, including a stand-alone landing, of a 66-kg rotorcraft UAV operating in an uncontrolled external environment without using GPS and through of a micro-UAV operating in a chaotic interior environment.

Another interesting proposal in this context is presented in work reported in [38], in which an integrated navigation sensor module, including a camera, a laser scanner, and an inertial sensor, for UAVs to fly both indoors and outdoors environments. A real-time navigation algorithm based on estimating algorithms is proposed. The algorithm merges the image features with laser track data to estimate vehicle status and position. The proposed onboard navigation system can provide real-time 3D navigation without any pre-assumptions. The experimental results demonstrate the performance of the proposed system and prove their efficiency for multiple environments.

A proposal to identify victims of natural disasters identifying them and triggering the rescue team to support these endangered people is presented in [39]. This work is in the initial phase of development, presenting the contextualisation of different applications and various scenarios in which the UAVs can act. The initial results demonstrate the level of autonomy that could be reached by using CV techniques aiming both indoor and outdoor environments, but more detailed evaluations and results are not reported yet.

To support this wide range of applications, which often require more than one UAV, advanced studies in communication and other distributed systems technologies, capable of integrating and meeting the temporal requirements of vision applications are under development.

The work proposed by [40] addresses the problem of highly mobile networks composed of UAVs as data providers in a military surveillance system. Based on a Software-Defined Networking (SDN) approach to provide the best routes for data delivery, to the goal is to achieve an enhancement in the quality of service offered by the CV applications supported by these network of UAVs. The results demonstrated by the author provide solid evidence of the usefulness of the proposed application.

The work presented by [41], addresses the problem of assigning tasks between agents represented by UAVs, in which a central entity creates the tasks, but selected for execution by each of these independent agents (UAVs). The selection partially considers data acquired by the embedded CV system that equips the UAVs. To provide the ability to decide which tasks to execute, an approach based on swarm intelligence is proposed. Experiments in a simulated environment demonstrate that the proposed method improved results compared to existing methods reported in the literature.

As can be observed, these techniques handling both indoor and outdoor environments tend to be more robust in terms of accuracy, but also have a higher computational cost because they have to deal with the disturbances and disparities in the data collected by the sensors due to the great variation of the scene and the environmental conditions. However, they can meet a more significant number of requirements for a particular mission, both indoors and outdoors.

11.3 Conclusion

This chapter examines the impact and necessities of indoor and outdoor environments on CV procedures, algorithms and hardware.

Indoor or controlled environments generally depend on solutions relying on beacons, nearness sensors and intense CV algorithms for data acquisition and decision-making. In this case, as the environment is controlled, the illuminance of the scene is adjusted and sensors are previously positioned, which simplifies the development and execution of these systems.

In outdoor environments, generally known for uncontrolled environmental variables, frequently require solutions relying on CV techniques and Geographic Information Systems (GISs) to provide the data acquisition. In this environment, the non-constant variation of the scene illuminance and the great background dissimilarity of images are important complicating factors for the operation of the CV algorithms. In addition, construction sites and buildings block the signal of sensors and GPSs making it even tougher to handle the exceptions produced by these factors. In a CV system, a high computational cost can result from treating each exception. If an application uses embedded hardware, such as an FPGA, then some projects turn out to be infeasible.

Investigators put great effort to optimise the software regarding a growth in performance and better usage of the hardware resources, to decrease the necessary computational power as well as to positively boost energy savings. These advances result from a comprehensive review of the key CV techniques now used in mission control software for indoor and outdoor cases to provide autonomy, superior navigation and interaction between these aerial robots.

Scenarios involving disaster response and people in distress would rather use a high level of indoor–outdoor interaction of geospatial information with a network data model. Nevertheless, indoor geospatial information acquisition is timewasting. Numerous research works employed the Architecture, Engineering, and Construction (AEC) paradigm to create indoor network strategies. These schemes hinge on the types of input data, and the interior characteristics of buildings and their associated objects are typically incomplete. Consequently, combining Building Information Modelling (BIM) with GIS can improve indoor–outdoor combined applications. An open standard for BIM entitled Industry Foundation Classes (IFC) is sustained by building SMART to accomplish data interoperability. Reference [42] proposed a Multi-Purpose Geometric Network Model (MGNM) depending on BIM and it uses both indoor and outdoor network connections. To attain the objectives, the IFC-to-MGNM adaptation embraces the subsequent stages: (i) obtaining edifice data from IFC, (ii) MGNM information isolation from the previously mentioned building data, and (ii) construction and accumulation the topological associations of the

MGNM into remote system Geodatabases. Moreover, an entrance-to-street tactic can connect indoor networks, entrances/paths, besides outdoor networks for meticulous route planning. The experiments point out that the MGNM can result from the BIM automatically to associate indoor and outdoor structures for multi-purpose applications [42,43]. More use-case settings will help to corroborate and advance methods to deal with applications requiring the simultaneous deployment of indoor and outdoor drones. These scenarios must take into account better strategies for path-planning that use computational intelligence to devise the lighter configuration of indoor and outdoor UAVs for a given mission [44–49] in conjunction with cloud resources [4,50,51].

References

[1] R. Szeliski, "Computer vision – Algorithms and applications," *Texts in Computer Science*. Springer, London, UK, 2011. DOI: 10.1007/978-1-84882-935-0

[2] M. Prasad, "Basic concepts of computer vision," http://maxembedded.com/2012/12/basic-concepts-of-computer-vision/, 2012, accessed 01-10-2018.

[3] F. Conticelli and B. Allotta, "Two-level visual control of dynamic look and-move systems," in *Proceedings 2000 ICRA. Millennium Conference. IEEE International Conference on Robotics and Automation*. Symposia Proceedings (Cat. No. 00CH37065), vol. 4, 2000, pp. 3784–3789.

[4] V. V. Estrela, O. Saotome, H. J. Loschi, *et al.*, "Emergency response cyber-physical framework for landslide avoidance with sustainable electronics," *Technologies*, vol. 6, p. 42, 2018. doi:10.3390/technologies6020042.

[5] P. Campoy, J. F. Correa, I. Mondragón, *et al.*, "Computer vision onboard UAVs for civilian tasks," in *Unmanned Aircraft Systems*. Springer, Berlin, 2009, pp. 105–135.

[6] M. Hentschke, E. Pignaton de Freitas, C. H. Hennig, and I. C. Girardi da Veiga, "Evaluation of altitude sensors for a crop spraying drone," *Drones*, vol. 2, no. 3, 2018. [Online]. Available: http://www.mdpi.com/2504-446X/2/3/25

[7] T. Dapper e Silva, V. Cabreira, and E. P. de Freitas, "Development and testing of a low-cost instrumentation platform for fixed-wing uav performance analysis," *Drones*, vol. 2, no. 2, 2018. [Online]. Available: http://www.mdpi.com/2504-446X/2/2/19

[8] M. Basso, I. Zacarias, C. E. Tussi Leite, H. Wang, and E. Pignaton de Freitas, "A practical deployment of a communication infrastructure to support the employment of multiple surveillance drones systems," *Drones*, vol. 2, no. 3, 2018. [Online]. Available: http://www.mdpi.com/2504-446X/2/3/26

[9] A. Al-Kaff, D. Martín, F. García, A. de la Escalera, and J. M. Armingol, "Survey of computer vision algorithms and applications for unmanned aerial vehicles," *Expert Systems with Applications*, vol. 92, pp. 447–463, 2018. [Online]. Available: http://www.sciencedirect.com/science/article/pii/S0957417417306395

[10] J. Pestana, J. L. Sanchez-Lopez, S. Saripalli, and P. Campoy, "Computer vision based general object following for GPS-denied multirotor unmanned vehicles," in *2014 American Control Conference*, June 2014, pp. 1886–1891.

[11] R. Gonzalez and P. Wintz, *Digital Image Processing*. Addison-Wesley Publishing Co., Inc., Reading, MA, 1977.

[12] T. Tomic, K. Schmid, P. Lutz, *et al.*, "Toward a fully autonomous UAV: Research platform for indoor and outdoor urban search and rescue," *IEEE Robotics & Automation Magazine*, vol. 19, no. 3, pp. 46–56, 2012.

[13] S. Ward, J. Hensler, B. Alsalam, and L. F. Gonzalez, "Autonomous UAVs wildlife detection using thermal imaging, predictive navigation and computer vision," in *2016 IEEE Aerospace Conference*, March 2016, pp. 1–8.

[14] P. Suwansrikham and P. Singkhamfu, "Indoor vision based guidance system for autonomous drone and control application," in *2017 International Conference on Digital Arts, Media and Technology (ICDAMT)*, March 2017, pp. 110–114.

[15] D. Magree and E. N. Johnson, "Combined laser and vision-aided inertial navigation for an indoor unmanned aerial vehicle," in *2014 American Control Conference*, June 2014, pp. 1900–1905.

[16] X. Liu, B. Guo, and C. Meng, "A method of simultaneous location and mapping based on RGB-D cameras," in *2016 14th International Conference on Control, Automation, Robotics and Vision (ICARCV)*, November 2016, pp. 1–5.

[17] M. K. Mohamed, S. Patra, and A. Lanzon, "Designing simple indoor navigation system for UAVs," in *2011 19th Mediterranean Conference on Control Automation (MED)*, June 2011, pp. 1223–1228.

[18] K. J. Wu, T. S. Gregory, J. Moore, B. Hooper, D. Lewis, and Z. T. H. Tse, "Development of an indoor guidance system for unmanned aerial vehicles with power industry applications," *IET Radar, Sonar Navigation*, vol. 11, no. 1, pp. 212–218, 2017.

[19] P. Araújo, R. Miranda, D. Carmo, R. Alves, and L. Oliveira, "Air-SSLAM: A visual stereo indoor slam for aerial quadrotors," *IEEE Geoscience and Remote Sensing Letters*, vol. 14, no. 9, pp. 1643–1647, 2017.

[20] C. Bills, J. Chen, and A. Saxena, "Autonomous MAV flight in indoor environments using single image perspective cues," in *2011 IEEE International Conference on Robotics and Automation*, May 2011, pp. 5776– 5783.

[21] S. K. Phang, J. J. Ong, R. T. C. Yeo, B. M. Chen, and T. H. Lee, "Autonomous mini-UAV for indoor flight with embedded on-board vision processing as navigation system," in *2010 IEEE Region 8 International Conference on Computational Technologies in Electrical and Electronics Engineering (SIBIRCON)*, July 2010, pp. 722–727.

[22] J. Pestana, J. L. Sanchez-Lopez, P. de la Puente, A. Carrio, and P. Campoy, "A vision-based quadrotor multi-robot solution for the indoor autonomy challenge of the 2013 International Micro Air Vehicle Competition," *Journal of Intelligent & Robotic Systems*, vol. 84, no. 1, pp. 601–620, 2016. [Online]. Available: https://doi.org/10.1007/s10846-015-0304-1

[23] C. Kanellakis and G. Nikolakopoulos, "Survey on computer vision for UAVs: Current developments and trends," *Journal of Intelligent & Robotic Systems*, vol. 87, no. 1, pp. 141–168, 2017. [Online]. Available: https://doi.org/10.1007/s10846-017-0483-z

[24] C. Yuan, Z. Liu, and Y. Zhang, "Vision-based forest fire detection in aerial images for firefighting using UAVs," in *2016 International Conference on Unmanned Aircraft Systems (ICUAS)*, June 2016, pp. 1200–1205.

[25] C. Yuan, K. A. Ghamry, Z. Liu, and Y. Zhang, "Unmanned aerial vehicle based forest fire monitoring and detection using image processing technique," in *2016 IEEE Chinese Guidance, Navigation and Control Conference (CGNCC)*, August 2016, pp. 1870–1875.

[26] T. Moranduzzo and F. Melgani, "Automatic car counting method for unmanned aerial vehicle images," *IEEE Transactions on Geoscience and Remote Sensing*, vol. 52, no. 3, pp. 1635–1647, 2014.

[27] M. Basso, "A framework for autonomous mission and guidance control of unmanned aerial vehicles based on computer vision techniques," Master Thesis (Master in Electrical Engineering), Federal University of Rio Grande do Sul, Porto Alegre, Brazil, 2018. [Online]. Available: http://hdl.handle.net/10183/179536

[28] X. Li, Q. Yang, Z. Chen, X. Luo, and W. Yan, "Visible defects detection based on UAV-based inspection in large-scale photovoltaic systems," *IET Renewable Power Generation*, vol. 11, no. 10, pp. 1234–1244, 2017.

[29] Y. Jung, H. Bang, and D. Lee, "Robust marker tracking algorithm for precise UAV vision-based autonomous landing," in *2015 15th International Conference on Control, Automation and Systems (ICCAS)*, October 2015, pp. 443–446.

[30] H. Lee, S. Jung, and D. H. Shim, "Vision-based UAV landing on the moving vehicle," in *2016 International Conference on Unmanned Aircraft Systems (ICUAS)*, June 2016, pp. 1–7.

[31] H. Cheng, L. Lin, Z. Zheng, Y. Guan, and Z. Liu, "An autonomous vision-based target tracking system for rotorcraft unmanned aerial vehicles," in *2017 IEEE/RSJ International Conference on Intelligent Robots and Systems (IROS)*, September 2017, pp. 1732–1738.

[32] U. Shah, R. Khawad, and K. M. Krishna, "Detecting, localizing, and recognizing trees with a monocular MAV: Towards preventing deforestation," in *2017 IEEE International Conference on Robotics and Automation (ICRA)*, May 2017, pp. 1982–1987.

[33] S. Malek, Y. Bazi, N. Alajlan, H. AlHichri, and F. Melgani, "Efficient framework for palm tree detection in UAV images," *IEEE Journal of Selected Topics in Applied Earth Observations and Remote Sensing*, vol. 7, no. 12, pp. 4692–4703, 2014.

[34] E. P. Freitas, I. A. Wieczorek, C. Pereira, and A. Vinel, "Real time embedded image processing system for points of interest detection for autonomous unmanned aerial vehicles," in *Aerospace Technology Congress*, October 2016.

[35] F. S. Leira, T. A. Johansen, and T. I. Fossen, "A UAV ice tracking framework for autonomous sea ice management," in *2017 International Conference on Unmanned Aircraft Systems (ICUAS)*, June 2017, pp. 581–590.

[36] S. Chen, S. Guo, and Y. Li, "Real-time tracking a ground moving target in complex indoor and outdoor environments with UAV," in *2016 IEEE International Conference on Information and Automation (ICIA)*, August 2016, pp. 362–367.

[37] G. Chowdhary, E. N. Johnson, D. Magree, A. Wu, and A. Shein, "GPS denied indoor and outdoor monocular vision aided navigation and control of unmanned aircraft," *Journal of Field Robotics*, vol. 30, no. 3, pp. 415–438, 2013.

[38] S. Huh, D. H. Shim, and J. Kim, "Integrated navigation system using camera and gimbaled laser scanner for indoor and outdoor autonomous flight of UAVs," in *2013 IEEE/RSJ International Conference on Intelligent Robots and Systems*, November 2013, pp. 3158–3163.

[39] L. Apvrille, T. Tanzi, and J. L. Dugelay, "Autonomous drones for assisting rescue services within the context of natural disasters," in *2014 XXXIth URSI General Assembly and Scientific Symposium (URSI GASS)*, August 2014, pp. 1–4.

[40] I. Zacarias, J. Schwarzrock, L. P. Gaspary, *et al.*, "Enhancing mobile military surveillance based on video streaming by employing software defined networks," *Wireless Communications and Mobile Computing*, vol. 2018, pp. 1–12, 2018. Available: https://doi.org/10.1155/2018/2354603

[41] J. Schwarzrock, I. Zacarias, A. L. Bazzan, R. Q. de Araujo Fernandes, L. H. Moreira, and E. P. de Freitas, "Solving task allocation problem in multi unmanned aerial vehicles systems using swarm intelligence," *Engineering Applications of Artificial Intelligence*, vol. 72, pp. 10–20, 2018. [Online]. Available: http://www.sciencedirect.com/science/article/pii/S0952197618300575

[42] T. A. Teo and K.-H. Cho, "BIM-oriented indoor network model for indoor and outdoor combined route planning," *Advanced Engineering Informatics*, vol. 30, pp. 268–282, 2016.

[43] X. Zhou, J. Wang, M. Guo, and Z. Gao, "Cross-platform online visualization system for open BIM based on WebGL," *Multimedia Tools and Applications*, Springer, Zurich, Switzerland, 2019;78: 28575–28590. https://doi.org/10.1007/s11042-018-5820-0.

[44] D. J. Hemanth and V. V. Estrela, *Deep Learning for Image Processing Applications*, Advances in Parallel Computing Series, Vol. 31, IOS Press, 2017, ISBN 978-1-61499-821-1 (print), ISBN 978-1-61499-822-8 (online).

[45] N. Razmjooy, B. S. Mousavi, M. Khalilpour, and H. Hosseini, "Automatic selection and fusion of color spaces for image thresholding," *Signal, Image and Video Processing*, vol. 8, no. 4, pp. 603–614, 2014.

[46] B. Mousavi, F. Somayeh, and F. Soleymani, "Semantic image classification by genetic algorithm using optimised fuzzy system based on Zernike moments," *Signal, Image and Video Processing*, vol. 8, no. 5, pp. 831–842, 2014.

[47] L. Yang, X. Sun, A. Zhu, and T. Chi, "A multiple ant colony optimization algorithm for indoor room optimal spatial allocation," *ISPRS International Journal of Geo-Information*, vol. 6, p. 161, 2017.

[48] M. A. de Jesus, V. V. Estrela, O. Saotome, and D. Stutz, "Super-resolution via particle swarm optimization variants," in Hemanth J. and Balas V. (eds) *Biologically Rationalized Computing Techniques for Image Processing Applications*. LNCVB, vol. 25. Springer, Cham, 2018.

[49] N. Razmjooy and V. V. Estrela, *Applications of Image Processing and Soft Computing Systems in Agriculture*, 2019.

[50] V. V. Estrela, A. C. B. Monteiro, R. P. França, Y. Iano, A. Khelassi, and N. Razmjooy, "Health 4.0: Applications, management, technologies and review," *Medical Technologies Journal*, vol. 2, no. 4, pp. 262–276, 2019.

[51] N. H. Motlagh, T. Taleb, and O. Osama, "Low-altitude unmanned aerial vehicles-based internet of things services: Comprehensive survey and future perspectives," *IEEE Internet of Things Journal*, vol. 3, pp. 899–922, 2016.

Chapter 12

Sensors and computer vision as a means to monitor and maintain a UAV structural health

Helosman Valente de Figueiredo[1], *Osamu Saotome*[1] *and Roberto Gil Annes da Silva*[1]

The aeronautical structures used in unmanned aerial vehicles (UAVs) have become more flexible and lightweight to meet the need to operate at high altitudes and for long periods in the air, such as those required in the NASA Helios project. These flexible aeroplanes, the similarity, between iron birds and natural birds. The increased structural flexibility of these aircraft reignited concerns related to aeroelastic instabilities, such as flutter. Improving the techniques and methods used in aircraft certification flights is an important concern of the aeronautical community because current standards and procedures do not provide recommendations and guidelines for aircraft with a high degree of flexibility. The techniques traditionally used in commercial aeroplanes cannot be used in UAVs, because the new UAVs have a high degree of non-linearity in their flight dynamics. Contemporary research studies indicate an increasing awareness of the importance of vision in the monitoring of UAV structural health. This chapter addresses algorithm-based solutions to real-time infrastructure damage recognition and damage accreditation for UAV systems. Damage detection and identification of effects are quantified for remediation. A vision-based UAV system may obtain visual evidence from a damaged surface of interest to detect faults, eliminate the irrelevant areas from image data, spot damage, measure the resulting consequences, record information, identify the type of fault and point to the most noticeable kind of problem present. This chapter also discusses new approaches to sense and acquire vibration data and to preprocess these data on aeroelastic certification test flights. These new approaches aim to reduce the time to identify the aeroelastic phenomenon and to reduce the size of the hardware that must be boarded in the aircraft, thus minimising the risks and costs of the vibration tests. The advance of computer vision systems enables the use of cameras as a motion tracker sensor with millimetre precision and accuracy. Non-contact sensors are suited to flutter analysis because they do not interfere with the dynamics of the aircraft. With the aid of Computer Vision algorithms, they allow the acquisition of 2D and/or 3D data, instead of one-dimensional data acquired by traditional vibration sensors, such as piezoelectric accelerometers.

[1]DCTA-ITA-IEEA, Sao Jose dos Campos, SP, Brazil

Nevertheless, it is worth noting that to capture the aeroelastic phenomena, these cameras must operate with at least five times higher frame rates than traditional ones. Therefore, this intelligent video sensor system that is capable of processing the obtained images and providing the user with the data of movements in a vector ready for use, at a reasonable cost, is an important topic under development. In addition, this work proposes modifications in the acquisition and pre-processing procedures of the signals traditionally used in flutter certification analytic, for example, modal analysis, suited to an embedded system, and near-real-time process.

12.1 Introduction

New concepts of aircraft have been investigated for application in intelligence, reconnaissance, and environmental research missions, where there is a need to operate at high altitudes and for long periods in the air, such as the NASA Helios [1]. These new designs must have efficient aerodynamic and structural design, which often results in long, flexible and lightweight structures.

These characteristics increase the risk of aeroelastic instability, such as the appearance of flutter, a phenomenon that consists of a self-powered and potentially destructive vibration, in which aerodynamic forces couple with the natural mode of structure vibration, producing periodic motion [2,3].

To avoid flutter, regulatory agencies require aeroelastic certification for new aircraft concepts or modifications of commercial aircraft. This ensures that within the flight envelope, there is no exhibited aeroelastic instability, i.e., damping (aerodynamic plus structural) should not be less than 3% for any critical mode [4,5].

For aeroelastic certification, flight tests are conducted with the expansion technique increasing the flight envelope. This consists of subjecting the aircraft to a given altitude and flight speed determined in the envelope, and, thereafter, an excitation is applied to the aircraft. The response is acquired to calculate the transfer function of the aircraft to estimate dynamic characteristics, e.g., damping and natural frequency.

The advance of computer vision systems enables the use of cameras as a motion-tracking sensor with millimetre precision and accuracy [6]. Non-contact sensors are appropriate to flutter analysis because they do not interfere with the dynamics of aircraft. With the aid of Computer Vision algorithms, it allows the acquisition of 2D and/or 3D data. Nevertheless, to capture the aeroelastic phenomena, these cameras must operate with at least five times higher frame rates than the traditional one. Thus, this intelligent video sensor system capable of processing the obtained images and providing the user with the data of movements in a vector ready for use, at a reasonable cost, is an important topic under development.

12.1.1 Case study: aeroelastic instability flutter phenomenon

A formal definition of flutter can be given as a dynamic instability of the structure associated with the interaction of aerodynamic, elastic and inertial forces. This synergy is illustrated in Figure 12.1.

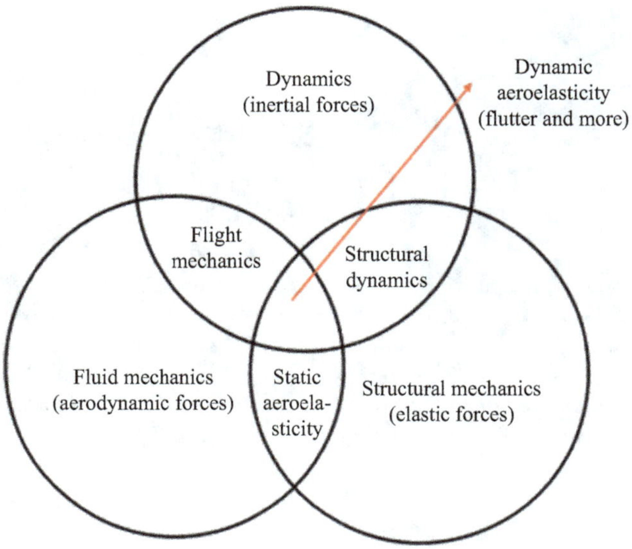

Figure 12.1 Diagram of forces generating flutter [7]

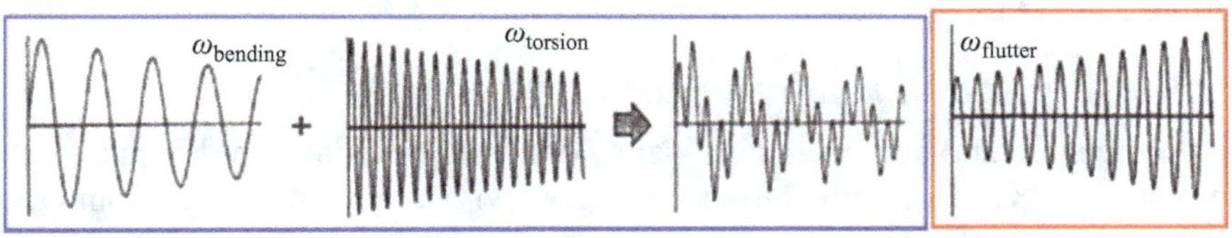

Figure 12.2 Example of a combination of modes (flexural and torsional) vibration of flutter [7]

This interaction generates a self-powered and potentially destructive vibration, in which aerodynamic forces couple with the natural vibrational mode of a structure, producing periodic motion, which can attain one or more modes of vibration. Figure 12.2 illustrates an example of these modes.

The accidents caused by flutter are catastrophic, both in the aeronautical and structural areas. Figure 12.3a illustrates the occurrence of flutter in a glider, and Figure 12.3b presents a case of a structural accident caused by aeroelastic phenomena, the Tacoma Bridge. Flutter clearance involves a combination of pre-flight analysis and flight testing.

The pre-flight analysis includes the (i) ground vibration test (GVT), (ii) structural coupling test (SCT), and (iii) aeroelastic model tests in the wind tunnel. This analysis provides guidelines for planning the flutter flight test. Flight test intention is to demonstrate that the aircraft is free from flutter and aeroservoelastic instabilities at all speeds in the flight envelope.

(a) (b)

Figure 12.3 Examples of flutter: (a) the occurrence of flutter in ultralight aircraft [8]; (b) tacoma bridge collapse [9]

12.2 Related work

This work involves two main themes of computer vision for monitoring structural health and flutter. The following sections will present these issues and how they correlate with this research.

12.2.1 Structural health monitoring

To ensure structural and operational safety and issue warnings on damage or catastrophic collapse, many structures are designed with a structural health monitoring (SHM) system. SHM requires permanently attached sensors, often combined with instrumentation, which allows frequent measurements during operation [10].

Since 1970, about 17,000 papers have been published addressing the topic of SHM. However, this great research effort has not translated into industrial applications. There are many reasons for this low applicability in the industrial sector, for example, many researchers develop their work in simple structures (bars, plates), whereas industrial structures are much more complex; as is the number of sensors required for SHM; the cost of maintaining the SHM system; and managing and processing the collected data [11].

Many papers seek to improve fault detection methods [12–15], positioning of sensors [6,16,17], use of new media, and measurement methods [18–20]. The advance in the computational systems made the use of more sophisticated techniques to detect failures possible, such as the use of neural networks, fuzzy logic, and main component analysis. Oliveira [12] work proposed to use the Savitzky–Golay filter to improve the performance of SHM systems based on neural networks.

Capellari [16] and Flynn [17] investigate methodologies to position the sensors. Capellari analyses how information changes according to the number and types of sensors used. They propose the positioning problem as an optimisation using Polynomial Chaos Expansion and the Stochastic optimisation method. On the other hand, the works from Castelini [18] and Chen [19] look for new ways to make these measures, using non-contact measurement systems. These systems have a unique ability to collect information without interfering with the structure because they do not need to couple sensors and cables to the structure.

12.2.2 Computer vision for structural health

The computer vision has been widely explored in the last years, mainly in the areas of inspection and autonomous navigation of thefts. The work on use computer vision for SHM is still in its infancy. Most of the works investigate fundamental structures such as bars and plates or large structures such as skyscrapers and bridges. The work of Jithin [21] presents a method to analyse contactless vibrations for light and thin bars, where the traditional method would be ineffective as it would alter the structural dynamics of the test body. In this work, a high-speed camera identifies and monitors the natural damping of the bar.

Structures that are difficult to the instrument can have their vibrational parameters easily estimated with computer vision; works [22] and [23] apply computer vision to estimate these characteristics. The work of Bartilson *et al.* [22] employs three types of sensors to monitor bounces in structures (accelerometers, strain gauges, images), and it proves that the computational view can be used to determine the natural frequency accurately.

12.2.3 Flutter certification

Two main approaches stand out to enhance certification flights. The first one seeks to improve the software for identification and monitoring of the aeroelastic phenomena, to improve the mathematical models of estimation and to propose new methods extract the dynamic characteristics [24–26]. Other works seek to improve this process by improving the computational methods of prediction and modelling of the phenomenon, as presented in the works of [25,27,28]. The work [29] uses a combination of signal processing and dynamic system identification techniques called Auto-Regressive Moving Average (ARMA), combined with Random Decrement (RDD) and Natural Excitation Technique (NExT). The aforementioned techniques aim to improve the accuracy of the prediction of aeroelastic instability.

The second approach deals with the processing of digital signals by searching for new means and methods for data processing. With the development of new technologies and the miniaturisation and improvement of computer systems, new architectures for processing flight data have been explored. DLR researchers have developed new procedures for vibration tests on the ground that reduce the cost of the certification process [30]. A modification to the traditional process of certification flights is also proposed, and the new approach boards all the tools and engineers on the aircraft being studied [5].

12.2.4 Computer vision and in in-flight measurements: future trends

Computer vision has been used as a measurement methodology since 1980 when an experiment was carried out by NASA titled HiMAT Aeroelastic Tailored Wing [31]. The author proposes using a set of infra-red LEDs to measure aeroelastic deformations and a set of photodiodes to measure wing displacements.

Computer vision has become increasingly attractive, as the evolution of image acquisition and processing systems has increased its applicability. Many areas are already exploring this technology such as mobile robotics, remote sensing, and health systems that help treat amputees. Aeronautical systems have many phenomena dependent on the flow, which makes it even more attractive to use a computer vision system to perform non-contact measurements. The literature shows the trend of using this type of system, and the works of [32–34] present systems of measurement deformations during flight, using computer vision. The application of this technique will also be maintained in the UAVs because as they have lighter and more flexible structures, a form of measurement that does not add mass to the structure ideal. Pang [34] already explores this methodology in a flexible UAV, using a set of stereo cameras to monitor the deformations of an X-HALE aircraft. This system can do three-dimensional tracking using a series of markings. The data is collected throughout the flight and then processed offline, thus making it possible to acquire the deformations obtained by the computational model with the actual deformations in flight.

Computer vision has become a trend, as it enables the measurements of flight displacement of a complete surface, a gap left by the measurement methods traditionally used in the aeronautical industry (strain gauges the accelerometer) [32].

12.3 Signal processing on flutter certification

The main tool in the tests for flutter certification is modal analysis. This tool describes the structure in terms of frequency, damping, and modes of vibration, which are dynamic properties [35]. Data processing for modal analysis can be divided into five steps, as shown in Figure 12.4.

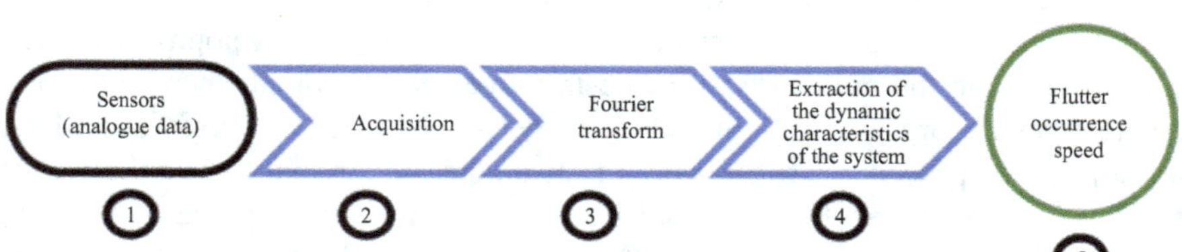

Figure 12.4 Data flow for modal analysis

12.4 Experiments and results

12.4.1 Synthetic data

To carry out this experiment, three steps were necessary. The first was to generate the data; second, pass the data to the frequency domain and third to extract dynamic characteristics. The synthetic data were generated with a mathematical model. Section 12.4.1.1 describes this model.

12.4.1.1 Model of the typical wing section

The mathematical model was developed to generate initial data for the study. A typical wing section model was developed in MATLAB®, presented in Figure 12.5. This section has two modes of vibration, associated in this case with its two degrees of freedom (DF): bending.

This model was designed to be used as a function 12.4 in MATLAB based on a set of mathematical equations, and as a theoretical reference to validate the methodology used to calculate and monitor the flutter. We used the mathematical equations of support and momentum related to bending (*L*) and torsion (*M*), respectively, with the theoretical based on [37].

$$L = pV^2\left(L_h + L_h\frac{b\dot{h}}{V} + L_\theta b\Theta L_\theta\frac{b^2\dot{\Theta}}{V}\right) \tag{12.1}$$

$$M = pV^2\left(M_h bh + M_h\frac{b^2\dot{h}}{V} + M_\theta b^2\Theta + M_\theta\frac{b^3\dot{\Theta}}{V}\right) \tag{12.2}$$

where

L and M are the synthetic data of lift and moment, respectively;
h and $\dot{\Theta}$ are the rates of bending and torsion;
L_h, L_h, L_θ and L_θ are the aerodynamic lift derivatives;
M_h, M_h, M_θ and M_θ are the aerodynamic derivatives of momentum;

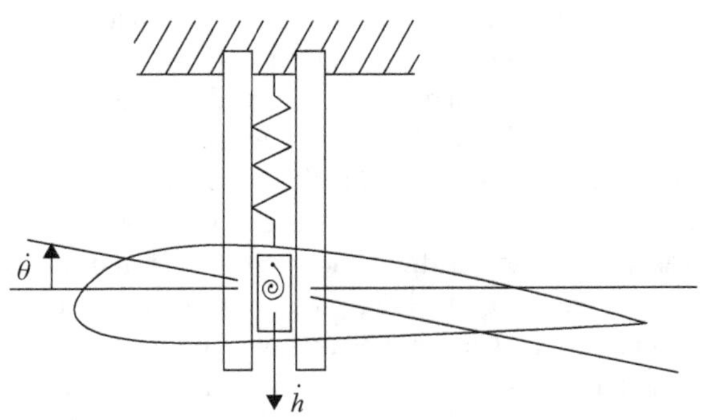

Figure 12.5 Simplified model of a typical wing section with two degrees of freedom [36]

p is air density; and
V is flow velocity.

From these equations, the matrix representation was set up to highlight the elements of the stiffness matrices and aerodynamic damping. The matrix representation is presented as:

$$\begin{Bmatrix} L \\ M \end{Bmatrix} = pV \begin{Bmatrix} bL_h & b^2L_\theta \\ b^2M_h & b^3M_\theta \end{Bmatrix} \begin{Bmatrix} \dot{h} \\ \dot{\Theta} \end{Bmatrix} + pV^2 \begin{Bmatrix} L_h & bL_\theta \\ bM_h & b^2M_\theta \end{Bmatrix} \begin{Bmatrix} h \\ \Theta \end{Bmatrix} \tag{12.3}$$

With the matrix representation, one a term is proportional to the bending velocities ($\dot{h}$) and torsion ($\dot{\Theta}$), while the other term is proportional to the bending (h) and torsion (Θ) displacements. Thus, aerodynamic forces (L and M) acting on an aerodynamic profile subjected to oscillatory movement can be considered as damping and stiffness behaviour in a structure. Thus, B and C are called the damping and rigidity aerodynamic matrices, respectively. These matrices depend on the flight condition, for example, the flow rate.

To apply the aeroelastic study, the aerodynamic forces (L and M) are considered together with the structural equations, and this leads to the classical equations of motion:

$$\mathbf{A} \begin{Bmatrix} \ddot{h} \\ \ddot{\Theta} \end{Bmatrix} (pV\boldsymbol{B} + \boldsymbol{D}) \begin{Bmatrix} \dot{h} \\ \dot{\Theta} \end{Bmatrix} + (pV^2\boldsymbol{C} + \boldsymbol{E}) \begin{Bmatrix} h \\ \Theta \end{Bmatrix} = 0 \tag{12.4}$$

where

A is the matrix of structural inertia;
B is the aerodynamic damping matrix;
C is the aerodynamic stiffness matrix;
D is the structural damping matrix; and
E is the matrix of structural rigidity.

The output of the function is the accelerations of bending ($\ddot{h}$) and torsion ($\ddot{\theta}$) movements in the time domain, and the input is the type of air gust (a type of excitation) and airflow velocity, which is submitted in this section. These signals in the time domain compose the generation of synthetic data to simulate steps 1 and 2 for the theoretical validation of the proposed solution to calculate and monitor the flutter of Figure 12.4. Some experiments with this dynamic model were planned and carried out.

Figure 12.6 shows the results obtained with the initial response test of this dynamic model, which was taken as the standard for the development of flutter analysis and monitoring algorithms. In this initial simulation, the first positive cycle of the cosine function was used as input (a), to constitute the intensity of force of this gust of air. This cycle simulates the passage through a high-pressure zone, which can excite the flutter under normal flight conditions. The bending and twisting outputs are presented in 1.6b considering a constant velocity at 55 % of the flutter speed and 95 % in 1.6c. The objective of this experiment is to demonstrate that when approaching the speed and situation of occurrence the flutter damping decreases and the excitation energy dissipate, thus proving the correspondence with the model.

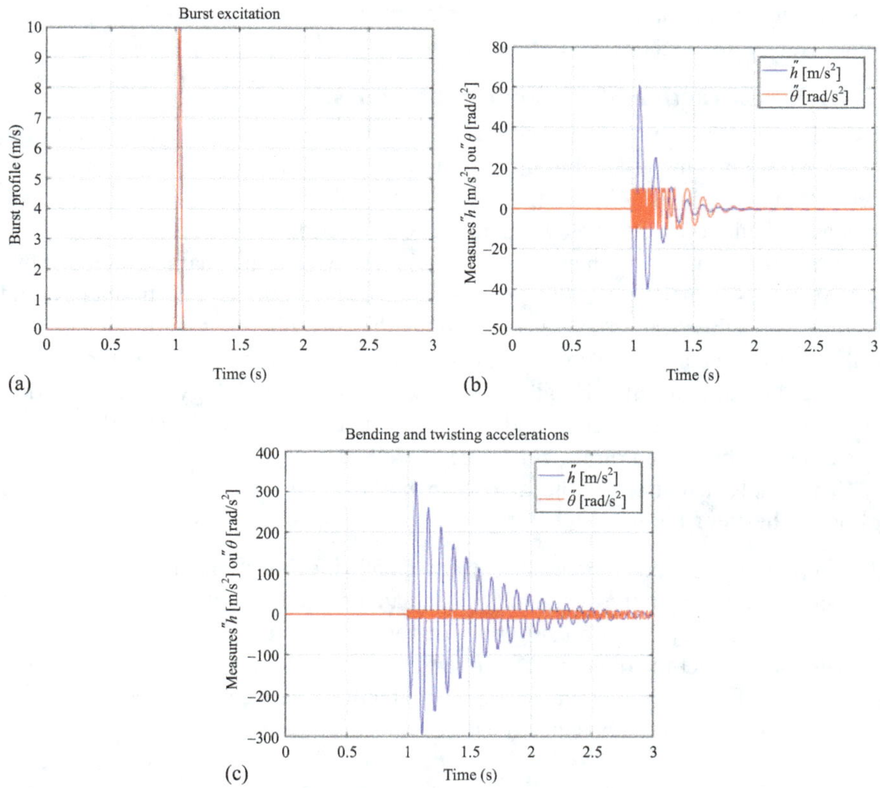

Figure 12.6 Model excitation and responses for 55% and 95% of flutter: (a) excitation signal; (b) response to 55% of flutter speed; (c) response to 95% of flutter speed

12.4.1.2 Pre-processing

The test data analysis helps to obtain the aircraft flutter characteristics [38–40]. Since the measured signal, in general, tends to have low quality, the pre-processing is crucial ameliorate the accuracy of subsequent flutter data analysis. Several types of filtering can be introduced to improve data processing of the flutter structure. For instance, the effectiveness of the morphological filter depends on the different operation types performed and the structural elements used. After this pre-processing, the usual Flutter Boundary Prediction (FBP) technique helps to validate and confirm the feasibility of the pre-processing stage through the numerical simulations, and the test of the flight flutter data. The results can be examined employing the Signal-to-Noise Ratio (SNR) metric of structural response from the physical assessment regarding the improvement of the prediction accuracy in actual applications [41,42]. Among pre-processing algorithms, there are basic ones, like least-squares variants, maximum likelihood variations, spectral analysis, Kalman filtering, and more contemporary methodologies such as subspace methods [43], and computational intelligence algorithms [44–47].

12.4.1.3 Extraction of dynamic characteristics

The literature contains several methods to extract dynamic characteristics in the measured signal, including highlight peak-amplitude method, inverse or line-fit method, and residuals [48]. In this work, the circle-fit method was investigated, because calculation methodology facilitates the embedded process.

From the frequency domain transformation, two dynamic characteristics of this system are estimated using the peak extraction method: natural frequency and damping of each vibration mode. This method has two phases: first, the location of peaks and estimation of natural frequency for each vibration mode; the second estimate the damping of each vibration mode. This methodology is performed in step 4 of the proposed solution and was also implemented in MATLAB for theoretical validation, based on the references [37,49–51].

The implementation of the algorithm based on the peak extraction method followed the steps below:

Steps 1–2: Synthetic torsion and flexural data in the time domain of experiment 1;

Step 3: Transform the synthetic torsion data and bending of the time domain to the frequency domain using the Fourier transform;

Step 4: Extract the dynamic characteristics of a system. Calculate the response in the frequency domain of *H*(*w*). In practice, the frequency response function (FRF) can be calculated as:

$$H(\omega) = \frac{X(\omega)}{F(\omega)}, \tag{12.5}$$

where $X(\omega)$ is an input Fourier transform for output, and $F(\omega)$ is a Fourier transform for input. From the considerations of digital signal processing, we can write that:

$$S_{XX}(\omega) = |H(\omega)|^2 S_{ff}(\omega), \tag{12.6}$$

where $S_{XX}(\omega)$ is the autocorrelation of input, and $S_{ff}(\omega)$ is the output autocorrelation. Next, plot the Nyquist diagram and locate the points of half-arc farthest from the origin. Determine the adjacent frequencies ω_a, ω_b and ω_R (farthest from the origin). Calculate the angle that represents the angle between these adjacent frequencies (Figure 12.7).

For each mode of vibration, estimate the corresponding damping as follows:

$$\zeta = \frac{\omega_b^2 - \omega_a^2}{2\omega_R\left[\omega_a ran\left(\frac{\alpha}{2}\right) + \omega_b tan\left(\frac{\alpha}{2}\right)\right]} \tag{12.7}$$

Step 5: Monitor the damping (ζ) for each mode of vibration, estimate the speed and monitor the flutter.

Figure 12.8 illustrates the flowchart for implementing these steps. The results of implementing this algorithm are presented in Section 12.4.2.1.

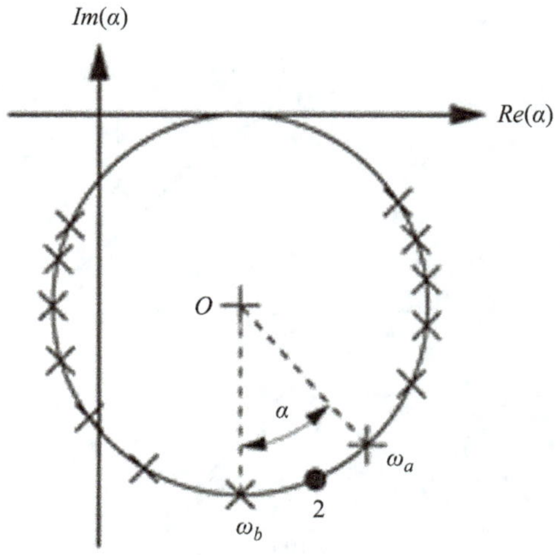

Figure 12.7 Frequencies of interest

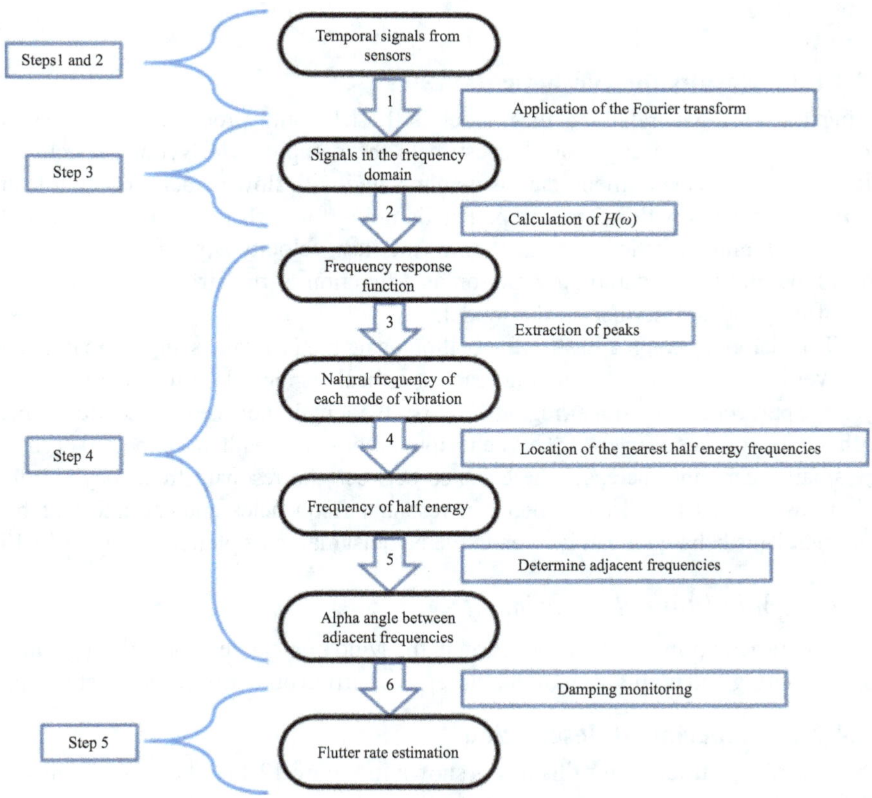

Figure 12.8 Flowchart of implementation of the steps for the calculation and monitoring of flutter via the Nyquist diagram method

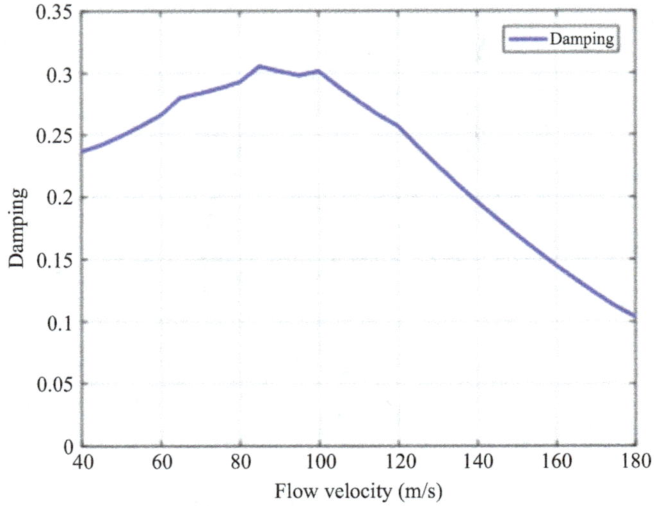

Figure 12.9 Damping from bending as a function of airflow velocities

12.4.1.4 Results for synthetic data

Using the data generated by the model in 12.4.1.1 and procedure described in Section 12.4.1.3, it was possible to estimate the damping via Nyquist (circle fit) diagram. For this experiment, data were generated with flow velocity ranging from 40 m/s to 180 m/s with a 5 m/s step. The damping for each flow rate is estimated, and the damping variation as a function of the flow velocity is plotted. Figure 12.9 shows the plot of the damping variation as a function of the airflow velocities, for one of the modes described in the model.

This damping graph allows extrapolation using the last values and estimating the flow velocity at which the damping reaches critical region (close to zero) when the system approaches the critical region ($\zeta = 0$), the length increases of the arc formed with the frequencies of interest. This behaviour is consistent with that expected because the smaller damping increases the distance between the resonant frequency and the origin, as well as the distance between resonant frequencies and the adjacent frequencies. This behaviour can be seen in the Nyquist diagrams plotted in Figure 12.10.

12.4.2 Wind tunnel experiment

The experimentation and data collection in the wind tunnel to evaluate the algorithm for calculating and monitoring of the flutter was carried out with an aeroelastic wing.

12.4.2.1 Experiment description

The assembled structure for this test is shown in Figure 12.11. The experiment was performed in the Laboratorio de Estruturas Aeroespaciais (LAB-ESP) of the Instituto Tecnologico de Aeronautica (ITA) and used the ADS1800-SV signal acquisition system from Lynx.

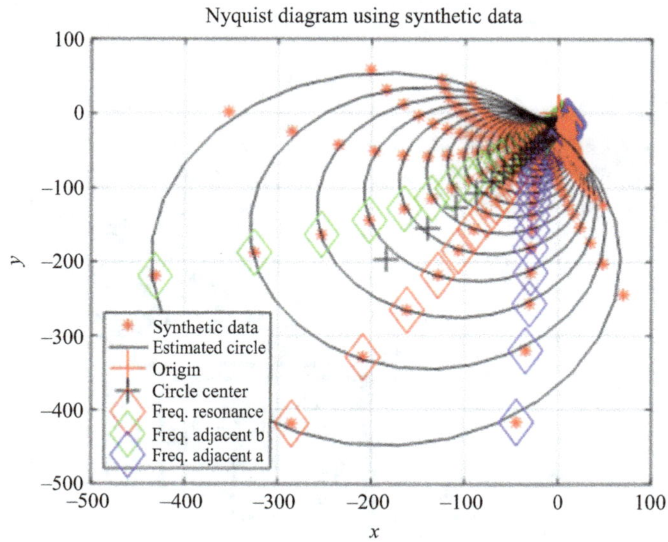

Figure 12.10 Nyquist diagram as a function of airflow velocities

The aeroelastic wing was mounted in a low-speed wind tunnel operating at low subsonic speed, with a maximum velocity of 33 m/s, whose measurements were performed with two degrees of freedom, bending, and twisting. For this, we used:

- Three ICP accelerometers M352C68 at the root of the aeroelastic wing, according to the figure;
- Two differential pressure meters connected to a Pitot tube to measure the airflow velocity. The meters used were a testo512 model manometer and the MPXV7002DP differential pressure sensor;
- A PCB 086C03 impact hammer for excitation (force application) on the aeroelastic wing;
- Lynx ADS1800-SV acquisition system.

For this test, the flow rate was manually increased until the flutter speed. For each speed, an excitation was performed on the wing, and during this process flow rate, and the measured accelerations was monitored. Another parameter that influences the speed of flutter in this wing is the position of the ballast about the elastic axis. This test evaluated two conditions; in the first configuration, the centre of the ballast was positioned 5 mm in front of the elastic axis of the wing, then the distance from the centre of the ballast was modified to 10 mm in front of the elastic axis of the wing. Figure 12.12 illustrates the ballast positions for this test. As expected, the change in velocity of the flutter was observed when changing the position of ballast when the centre of the ballast is positioned 10 mm in front of the elastic axis of the wing, the speed required to reach the flutter increases. Figure 12.12 also shows the positions for wing excitation.

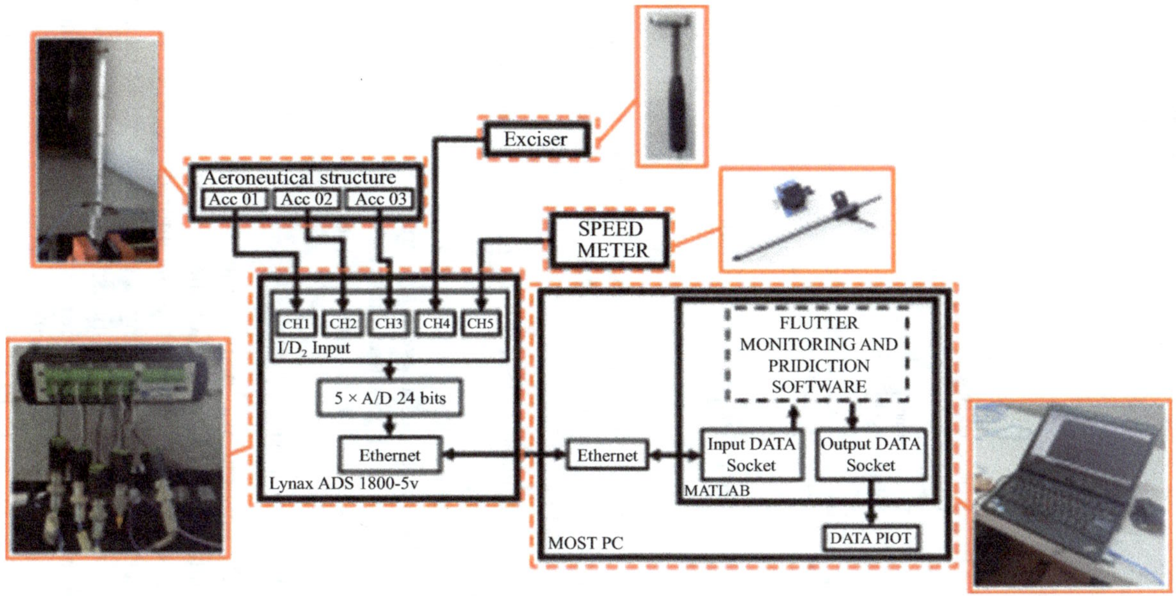

Figure 12.11 Structure of the aeroelastic wing test in a wind tunnel

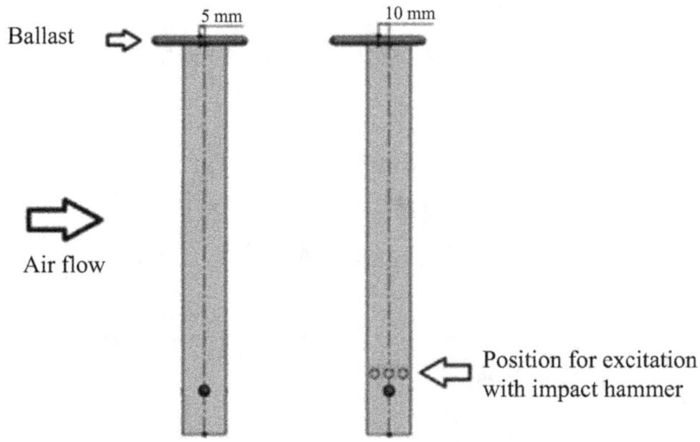

Figure 12.12 Ballast positioning and excitation positions in the aeroelastic wing

Signal acquisition was performed using a sampling rate of 48,000 samples/s, at a time interval of the 60 s. The sampling rate of Lynx's ADS1800-SV signal acquisition system was higher than the accelerometer response rate, which is 10 kHz.

This sampling rate allows monitoring of the entire accelerometer operating range, guaranteeing four times the maximum working frequency of the accelerometers and attending to the Nyquist sampling theorem.

For this experiment, the flow velocity was manually increased from 0 to 10.18 m/s by a potentiometer connected to the frequency inverter of the wind tunnel. For each velocity, applications of excitation forces were performed in the positions indicated in Figure 12.12, following the order: trailing edge, elastic axis, and leading-edge. The change in damping can be seen in Figures 12.13 and 12.14. In Figure 12.13, the system was excited and did not enter into a flutter. In Figure 12.14, the system fluttered at the moment the excitation was performed on the elastic axis, around 15s.

12.4.2.2 Results for experimental data

For the estimation of flutter velocity with these experimental data, the proposed solution and the methods described in Section 12.4.1.3 were applied. Figure 12.15 shows the estimated damping as a function of airflow velocity and a third-order polynomial approximation. The experimental data collected in the wind tunnel tests with the aeroelastic wing described in Section 12.4.2.1 were used.

As can be seen in Figure 4.31, with increasing airflow velocity, damping (ζ) tends to decrease. The number of points for the experimental data was lower because the flow velocities at which the differential pressure sensor (*MPXV*7002*DP*) presented reasonable accuracy when velocities were greater than 2 m/s. The sensor does not have sufficient sensitivity to measure speeds less than 2 m/s. The damping coefficient values were between 0.014 and 0.005. These values are proportional to the mass and structure of the aeroelastic wing used. The Figure shows that the flutter condition occurred with the proximity of airflow velocity at 10.2 m/s.

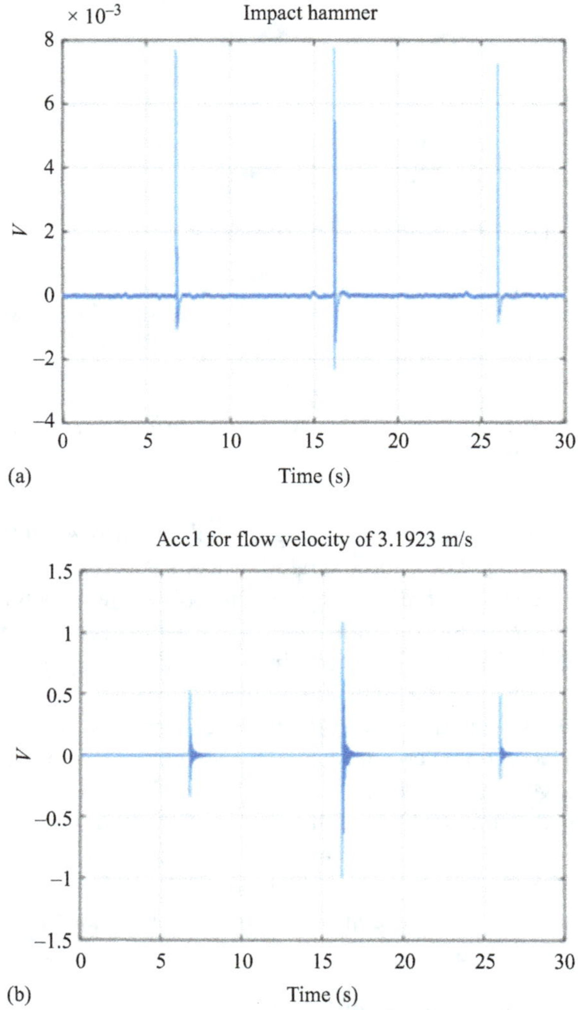

Figure 12.13 Acceleration and excitation measurements for airflow velocity of 6.52 m/s. (a) Impact Hammer. (b) Data collected by Acc1 with a flow velocity of 6.52 m/s

12.5 Discussion

The experiments carried out proved the technical feasibility of a real-time monitoring system for structural health. The developed software was able to estimate the damping for each flight condition. However, the implementation of this concept in real systems still requires some development of technologies, since the cost of instrumentation and the processing of these measures in real-time are still impediments.

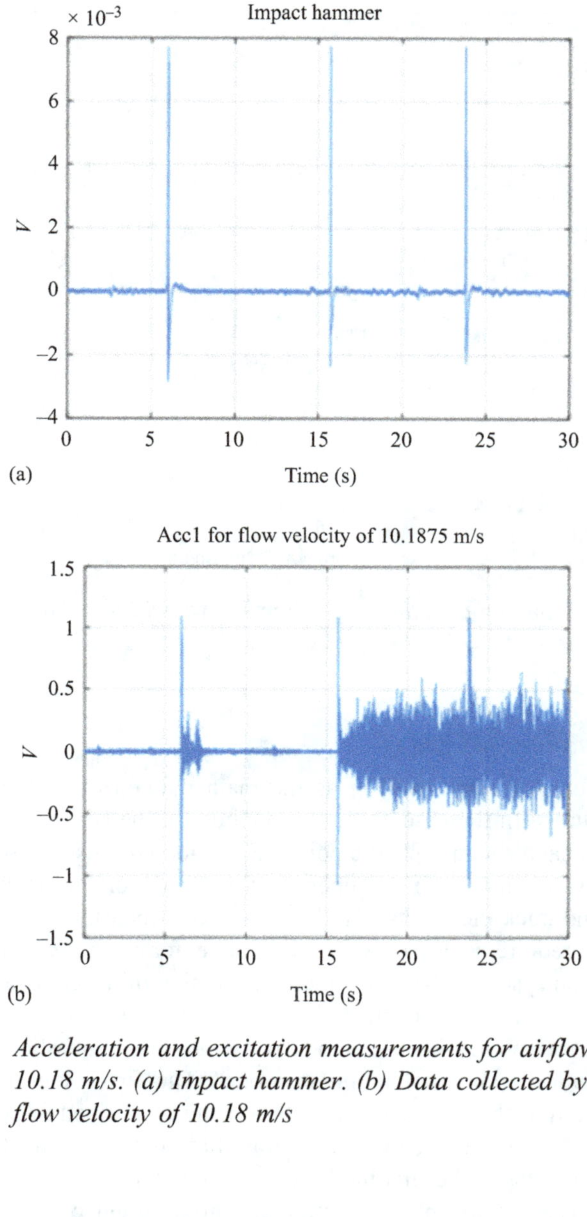

Figure 12.14 Acceleration and excitation measurements for airflow velocity of 10.18 m/s. (a) Impact hammer. (b) Data collected by Acc1 with a flow velocity of 10.18 m/s

In the development of new aircraft, the prototypes must pass a sequence of tests, and in this scenario, the time available for ground and flight vibration tests is restricted. Agile instrumentation optimises vibration testing, which can be achieved through the use of contactless measurements based on computer vision. In the aeronautical area, no commercial solutions have been found for this instrumentation problem. Currently, only the test aircraft used in certification flights are instrumented.

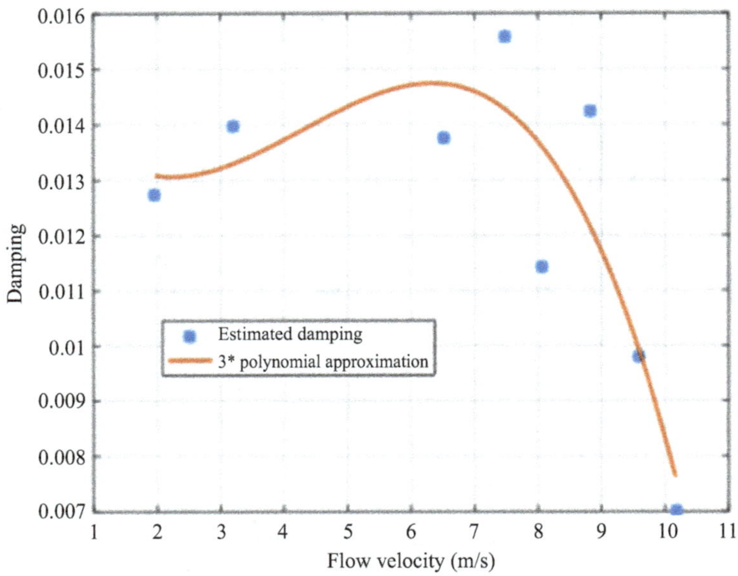

Figure 12.15 Damping as a function of airflow velocities, according to the data collected with the Lynx system

12.5.1 Computer vision

Computer vision has been growing as the main method of data acquisition for complex structures. In the case of flutter monitoring, the computer vision system must be able to handle some specific points. For a good discussion, see the video in [52]. The software must be able to operate in different conditions, with a variable background, and track the points in at least two coordinates.

With these requirements, some questions arise, for example, which is the ideal image acquisition system, what is the ideal positioning of the cameras, how to treat the relative movement of the camera, and what is the performance of the tracking algorithm for the mobile background. To answer these questions, a series of experiments will be presented in future works. The first experiment will be a comparative study with the sensor acquisition system described in Section 12.4.2.1 with the computer vision. The purpose of this study is to confirm if the computer vision system can detect the first modes of bending and torsion. The setup described in Section 12.4.2.1 will be used with the addition of video camera operating at 60 FPS. Figure 12.16 shows this initial computer vision setup.

This experiment will prove and identify the main challenges for the implementation of this SHM system in flight. Some studies have dealt with this problem, although preliminary, for well-controlled conditions, such as the works [21,22,53], so the trend of using computer vision for vibration measurements is verified.

With the verification of the proposed measurement concept, we propose the development of a measurement system based on computer vision that will track the

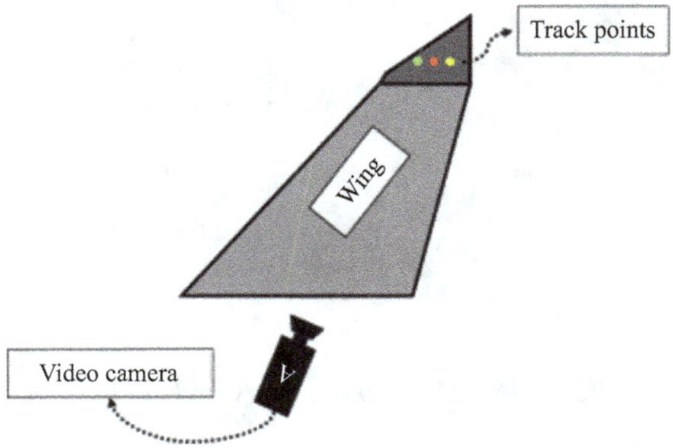

Figure 12.16 Setup for video acquisition system

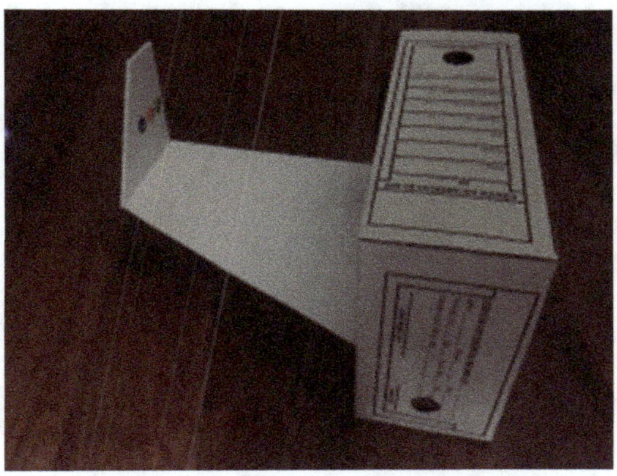

Figure 12.17 Setup used for concept validation

wing movements of a model aeroplane. Aiming at the precepts of INCOSE (International Council of Systems Engineering Handbook), other works can be developed in this system to characterise the degree of reliability, thus determining the feasibility for application, calibrations, and test parts in wind tunnels. The system will track and measure the torsional and bending movements of aeronautical structures of a model aeroplane with the aid of a camera and MATLAB software. The setup for validation of this concept is presented in Figure 12.17, using a personal computer, a cell phone camera (Galaxy S8), and a wing-like structure with track points.

For this setup, the frequency of acquisition was 240 fps, with 1280 × 720. Through this relationship, the camera will be able to capture the components of harmonic motion safely: the starting point of equilibrium, maximum positive amplitude, negative, and return to an equilibrium point.

The division into three targets has the purpose of discriminating the movements through analysis according to Figure 12.18. Displacements capable of forming an

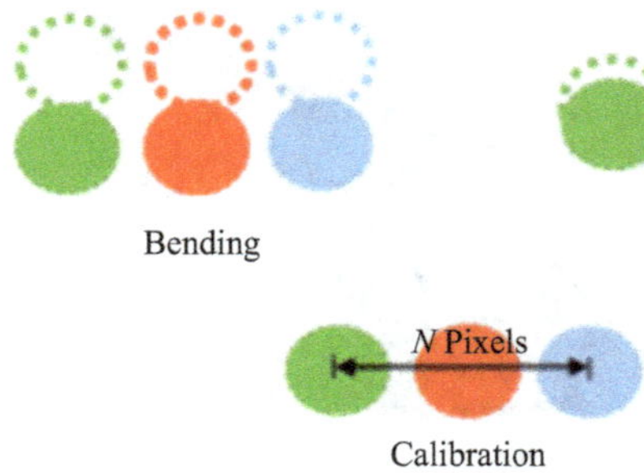

Figure 12.18 Marks movements and calibration

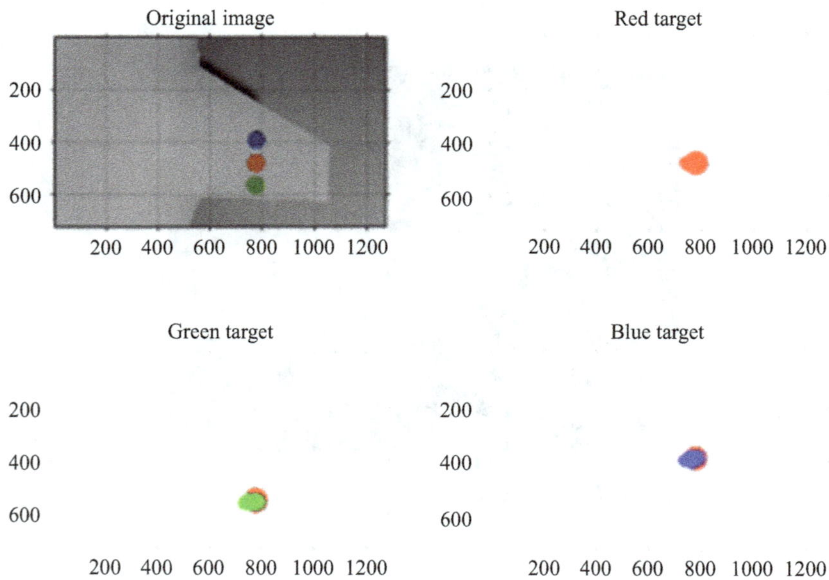

Figure 12.19 Marks subtract

angle with abscissa axis or normal inclination of the wing characterise a rotation. Knowing the distances between targets and the camera, the relation between pixels and real distance is set.

The targets were made in the colours red, green, and blue (RGB) because of an image recognised by the software is formed by the sum of contributions of these primary colours (RGB image), thus avoiding secondary colours. Once each target is identified, the contributions of blue and red colour matrices can be subtracted to obtain only the green colour, and thus respectively, to identify each target as shown in Figure 12.19. This software will track the markings throughout the video as shown in Figure 12.20.

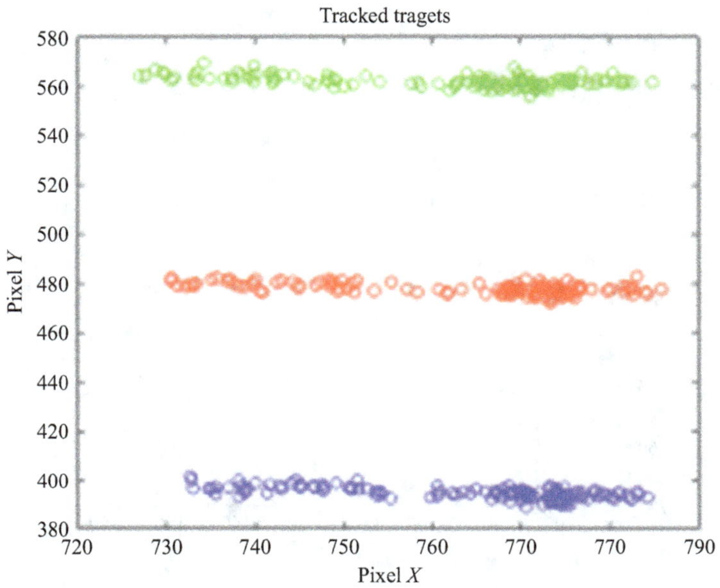

Figure 12.20 Marks subtracted

Then convert those measurements into bending (Figure 12.21) and torsional displacements (Figure 12.22). After the conversion of the pixel measurements to displacements, they will be processed by the algorithms developed in the previous phases of this work.

To verify if the proposed system was able to capture the frequency of vibration of the bending movement, the FFT (Figure 12.23) of the signal was taken with the DC signal average removed to verify if the peak was in the position expected.

The oscillation period was verified in the time signal used, thus verifying if the vibration mode was at the correct frequency.

12.6 Final remarks

This chapter discusses new approaches to sense and acquire vibration data and to pre-process these data on aeroelastic certification test flights. These new approaches aim to reduce the time to identify the aeroelastic phenomenon and to reduce the size of hardware that must be boarded in the aircraft, thus minimising the risks and costs of the vibration tests.

The presented experiments construct a way to develop a non-contact measurement system for flight vibration tests in the aircraft certification process. These experiments have shown that the techniques used today for in-flight trials will be obsolete in the near future, as the aeronautical structures are becoming lighter every day, thus not admitting any additional mass for instrumentation in-flight trials.

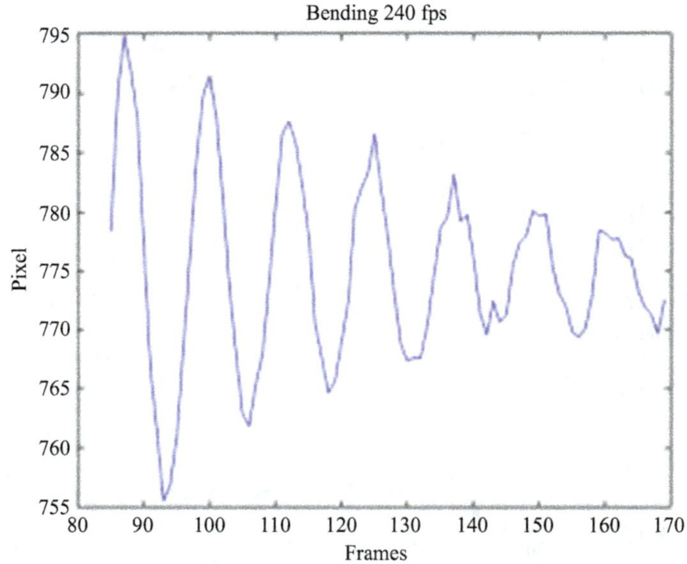

Figure 12.21 Bending estimation for 240 fps

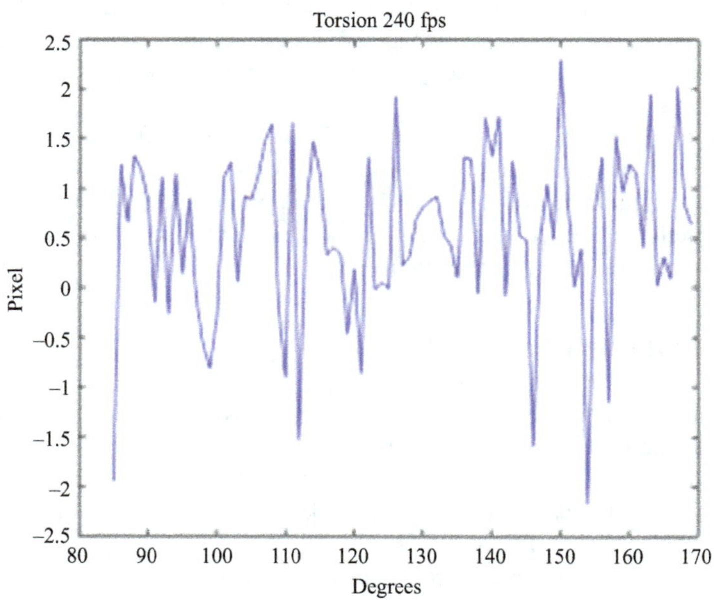

Figure 12.22 Torsion estimation for 240 fps

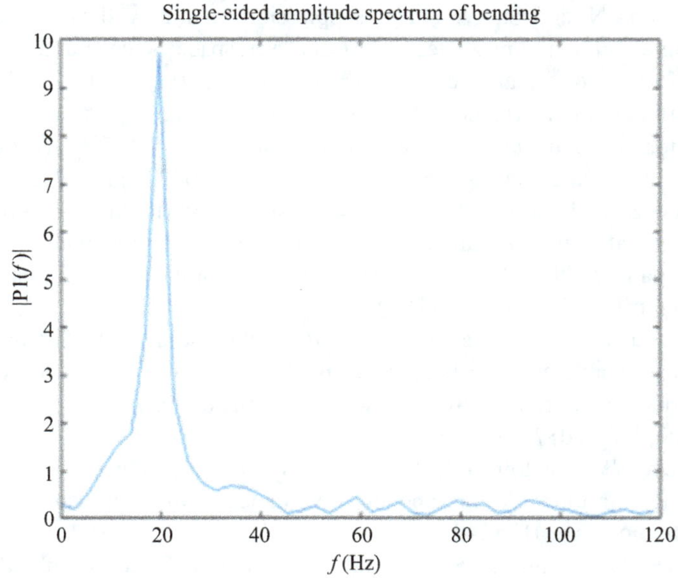

Figure 12.23 FFT of the bending signal

The proposed computer vision experiment is based on works that have already proven this type of measurement system can accurately determine natural frequencies and structural damping, thus opening the way for investigation of natural frequencies and damping resulting from aeroelastic phenomena.

The main advantages of the computer vision for this case are that it does not interfere with the structure, low complexity to be replicated, low instrumentation cost, and easy to use on difficult access structures (wings). Works have already been obtaining excellent results with commercial cameras and suggest that the results can be further improved by using cameras with higher resolutions and acquisition frequencies [22].

In addition, this work proposes modifications in the procedures of acquisition and pre-processing of the signals traditionally used in flutter certification analytic, for example, modal analysis, suited to an embedded system and near-real-time process.

References

[1] Noll TE, Ishmael SD, Henwood B, *et al.* Technical findings, lessons learned, and recommendations resulting from the helios prototype vehicle mishap. Security. Technical Report RTO-MP-AVT-145. NASA Langley Research Center Hampton, VA 23681 USA 2007:1–17. Available from: http://stinet.dtic.mil/oai/oai?fn&gverb=getRecordfn&gmetadataPrefix=htmlfn&gidentifier=ADA478771.

[2] Tsushima N, and Su W. Flutter suppression for highly flexible wings using passive and active piezoelectric effects. Aerospace Science and Technology. 2017;65:78–89. Available from: http://dx.doi.org/10.1016/j.ast.2017.02.013.

[3] Kayran A. Flight flutter testing and aeroelastic stability of aircraft. Aircraft Engineering and Aerospace Technology. 2007;79(5):494–506. Available from: http://www.emeraldinsight.com/doi/abs/10.1108/00022660710732707.

[4] Saeed S, and Salman S. Flutter analysis of hybrid metal-composite low aspect ratio trapezoidal wings in supersonic flow. Chinese Journal of Aeronautics. 2017;30(1):196–203. Available from: http://linkinghub.elsevier.com/retrieve/pii/S1000936116302357.

[5] Sinske J, Jelicic G, Buchbach R, *et al.* Flight testing using fast online aeroelastic identification techniques with DLR research aircraft. In Proc. 17th International Forum on Aeroelasticity and Structural Dynamics (IFASD 2017), Como, Italy 2017;1:1–19.

[6] Estrela VV, Saotome O, Loschi HJ, *et al.* Emergency response cyber-physical framework for landslide avoidance with sustainable electronics. Technologies, 2018;6:42.

[7] Names B. 5 things you should know about flutter—LinkedIn; 2016. Available from: https://www.linkedin.com/pulse/5-things-you-should-know-flutter-ben-names/.

[8] AIRBOYD. Ressonancia Aeroelastica - Efeito Flutter—YouTube. Available from: https://www.youtube.com/watch?v=3CMlXyV2XnE.

[9] Sez S. Tacoma narrows bridge collapse 1940—Symon Sez; 2010. Available from: https://symonsez.wordpress.com/tag/tacoma-narrows-bridge-collapse-1940/.

[10] Ko JM, and Ni YQ. Technology developments in structural health monitoring of large-scale bridges. Engineering Structures. 2005;27(12 SPEC. ISS.):1715–1725.

[11] Cawley P. Structural health monitoring: Closing the gap between research and industrial deployment. Structural Health Monitoring: An International Journal. 2018;17:1225–1244, 147592171775004. Available from: http://journals.sagepub.com/doi/10.1177/1475921717750047.

[12] de Oliveira M, Araujo N, da Silva R, *et al.* Use of Savitzky Golay filter for performances improvement of SHM systems based on neural networks and distributed PZT sensors. Sensors. 2018;18(1):152. Available from: http://www.mdpi.com/1424-8220/18/1/152.

[13] Pozo F, and Vidal Y. Damage and fault detection of structures using principal component analysis and hypothesis testing; In Advances in Principal Component Analysis, Springer, Berlin, Germany, 2017, 137–191.

[14] Flynn EB. A Bayesian experimental design approach to structural health monitoring with application to ultrasonic guided waves. University of California; 2010. Available from: https://escholarship.org/uc/item/9m15r6ph.

[15] Coelho AM, de Assis JT, and Estrela VV. Error concealment by means of clustered blockwise PCA, in *Proceedings of 2009 IEEE Picture Coding Symposium*, Chicago, IL, USA, 2009. DOI: 10.1109/PCS.2009.5167442

[16] Capellari G, Chatzi E, and Mariani S. Optimal sensor placement through bayesian experimental design: Effect of measurement noise and number of sensors. Proceedings. 2016;1(3):41. Available from: http://www.mdpi.com/2504-3900/1/2/41.

[17] Flynn EB, and Todd MD. Optimal placement of piezoelectric actuators and sensors for detecting damage in plate structures. Journal of Intelligent Material Systems and Structures. 2010;21(3):265–274.

[18] Castellini P, Martarelli M, and Tomasini EP. Laser Doppler Vibrometry: Development of advanced solutions answering to technology's needs. Mechanical Systems and Signal Processing. 2006;20(6):1265–1285.

[19] Chen JG, Wadhwa N, Cha YJ, *et al.* Modal identification of simple structures with high-speed video using motion magnification. Journal of Sound and Vibration. 2015;345:58–71. Available from: http://dx.doi.org/10.1016/j.jsv.2015.01.024.

[20] Estrela VV, and Coelho AM. State-of-the-art motion estimation in the context of 3D TV, in *Multimedia Networking and Coding*. IGI Global, 2013. 148–173. DOI:10.4018/978-1-4666-2660-7.ch006.

[21] Jithin TV, Sudheesh Kumar N, and Gopi Krishna N. Vibration analysis using machine vision system. In Proc. Seventh International Conference on Theoretical, Applied, Computational and Experimental Mechanics, ICTACEM 2017, Kharagpur, India, 2017;1:1–12.

[22] Bartilson DT, Wieghaus KT, and Hurlebaus S. Target-less computer vision for traffic signal structure vibration studies. Mechanical Systems and Signal Processing. 2015;60:571–582. Available from: http://dx.doi.org/10.1016/j.ymssp.2015.01.005.

[23] Silva S, Bateira J, and Caetano E. sistema de visao artificial para monitorizacao de vibracoes em tirantes de pontes. Revista da Associacao Portuguesa de Analise Experimental de Tensoes, Lisbon, Portugal, ISSN;1646:7078.

[24] Brandt A. Advantages of using long DFT computation for signal processing in operational modal analysis, in *International Conference on Structural Engineering Dynamics*; (EURODYN 2017) Rome, Italy, 2017.

[25] Conti E, Saltari F, Eugeni M, *et al.* Modal parameter estimate of time varying system using operational modal analysis based on Hilbert transform. 2017;1–14.

[26] Zeng J, and Kukreja SL. Flutter prediction for flight/wind-tunnel flutter test under atmospheric turbulence excitation. Journal of Aircraft. 2013;50(6):1696–1709. Available from: http://arc.aiaa.org/doi/abs/10.2514/1.C031710.

[27] Raveh DE. Assessment of advanced flutter flight test techniques and flutter boundary prediction. J.A. Aircraft 2017;55(5):1–26.

[28] Ahmed F, and Kandagal SB. Modal identification of aircraft wing coupled heave-pitch modes using wavelet packet decomposition and logarithmic decrement. Procedia Engineering. 2016;144:435–443. Available from: http://dx.doi.org/10.1016/j.proeng.2016.05.153.

[29] Tamayama M, Saitoh K, Yoshimoto N, *et al.* Effect of vibration data preprocessing for flutter margin prediction, in *JAXA Special Publication:*

Proceedings of the First International Symposium on Flutter and its Application, Tokyo, Japan *ł*; 2017. p. 97.
[30] Schwochow J, Jelicic G, and Govers Y. Recent developments in operational modal analysis. In Proceedings EURODYN 2017, Paris, France 2005;1:1–19.
[31] DeAngelis VM. In-flight deflection measurement of the HiMAT aeroelastically tailored wing. J.A. Aircraft, 1982;19(12):1088–1094.
[32] Bakunowicz J, and Meyer R. In-flight wing deformation measurements on a glider. Aeronautical Journal. 2016;120(1234):1917–1931.
[33] Kurita M, Koike S, Nakakita K, *et al.* In-flight wing deformation measurement. *51st AIAA Aerospace Sciences Meeting Including the New Horizons Forum and Aerospace Exposition*. Grapevine, Texas, USA, 2013;(January):1–7.
[34] Pang ZY, Cesnik CES, and Atkins EM. In-flight wing deformation measurement system for small unmanned aerial vehicles. *55th AIAA/ASMe/ASCE/AHS/SC Structures, Structural Dynamics, and Materials Conference*. National Harbor, Maryland, USA, 2014;(January):1–13.
[35] Avitabile P. Experimental modal analysis. Sound and Vibration. 2001; 35(1):1–15.
[36] Wentz JPTG, Medeiros EB, and Duarte MLM. Determinacao da Velocidade de Flutter em Asa de Aeronaves Ultraleves de Construcao Mista.
[37] Wright JR, and Cooper JE. Introduction to aircraft aeroelasticity and loads; 2007. Available from: http://books.google.com/books?hl=enfn&glr=fn&gid=BUfn gtRQaz9gICfn&gpgis=1.
[38] Razmjooy N, Ramezani M, Estrela VV, Loschi HJ, and do Nascimento DA. Stability analysis of the interval systems based on linear matrix inequalities, in Y. Iano *et al.* (eds), *Proceedings of 4th Brazilian Technology Symposium (BTSym'18)*, Smart Innovation, Systems and Technologies, vol. 140. Springer, Cham, Campinas, SP, Brazil, 2019. DOI: 10.1007/978-3-030-16053-1_36
[39] Razmjooy N, Khalilpour M, Estrela VV, and Loschi HJ. World cup optimization algorithm: An application for optimal control of pitch angle in hybrid renewable PV/wind energy system. In M. Quiroz *et al.* (eds), *NEO 2018: Numerical and Evolutionary Optimization*, Cancun, Mexico, 2019.
[40] Razmjooy N, Ramezani M, and Estrela VV. A solution for Dubins path problem with uncertainties using world cup optimization and Chebyshev polynomials, in Y. Iano *et al.* (eds), *Proceedings of 4th Brazilian Techn. Symp. (BTSym'18). BTSym 2018*. Smart Innovation, Systems and Technologies, vol 140. Springer, Cham, Campinas, SP, Brazil, 2019. DOI: 10.1007/978-3-030-16053-1_5
[41] Hua ZX, Junhao L, and Shiqiang D. A preprocessing method for flutter signals based on morphological filtering. *2018 9th International Conference on Mechanical and Aerospace Engineering (ICMAE)*, 2018, p. 430–434.
[42] Bo Z, and Jian-Jun L. Denoising method based on Hankel matrix and SVD and its application in flight flutter testing data preprocessing. Journal of Vibration and Shock; 2009;2:162–166.

[43] Bucharles A, Cumer C, Hardier G, *et al.* An overview of relevant issues for aircraft model identification. Aerospacelab Journal, 2012;4:13–33.

[44] de Jesus MA, and Estrela VV. Optical flow estimation using total least squares variants. Oriental Journal of Computer Science and Technology. 2017:10:563–579.

[45] Razmjooy N, and Estrela VV. *Applications of Image Processing and Soft Computing Systems in Agriculture*, IGI Global, Hershey, PA, USA, 2019. DOI:10.4018/978-1-5225-8027-0

[46] Alcalay G, Seren C, Hardier G, Delporte M, and Goupil P. Development of virtual sensors to estimate critical aircraft flight parameters. IFAC-PapersOnLine. 2017:50:1:14174–14179 DOI: 10.1016/j.ifacol.2017.08.2083

[47] Antoniadou I, Dervilis N, Papatheou E, Maguire AE, and Worden K. Aspects of structural health and condition monitoring of offshore wind turbines. Philosophical Transactions. Series A, Mathematical, Physical, and Engineering Sciences. 2015;373:1–14.

[48] Ewins DJ. *Modal Testing: Theory, Practice and Application*. Wiley, New Jersey, United States, 2000; p. 562.

[49] Ewins DJ. *Modal Testing: Theory and Practice*; Research Studies Press Ltd, Taunton, England, 1984.

[50] Inman DJ. *Engineering Vibration*. Fourth Ed. Pearson, Elsevier; 2014.

[51] Rao SS. *Mechanical Vibrations*. vol. 67; 2010.

[52] Noakes A. (54) Boeing 747-400 wing flexing - YouTube; 2010. Available from: https://www.youtube.com/watch?v=1URyA7-3PSQ.

[53] Kohut P, and Giergiel M. Optical measurement of amplitude of vibration of machine. Mechanics and Mechanical Engineering, 2008;12(2):147–156.

Chapter 13

Small UAV: persistent surveillance made possible

Ohood Al Nuaimi[1], *Omar Almelhi*[1], *Abdulrahman Almarzooqi*[1], *Abdulla Al Saadi Al Mansoori*[1], *Slim Sayadi*[1] *and Issacniwas Swamidoss*[1]

In typical intelligence surveillance and reconnaissance (ISR) missions, persistent surveillance is commonly defined as the exercise of automatic intelligence discovery by monitoring a wide area coverage for hours of operation at a high altitude leveraging aerial platforms (manned or unmanned). The platform can be large enough to carry a matrix of high-resolution sensors and a rack of high-performance computing equipment to process in real-time all sensors' feeds. With the current ISR growing in capability, engineering and optics-based aerial surveillance to find a suitable design solution became a design challenge. More onboard processing is desired for an increasing fidelity/resolution sensors' feed, while matching a constraining SWaP (size, weight, and power) budget requirements in a bandwidth-constrained operating theatre. The advent in small unmanned aerial vehicle (sUAV) technology, able to carry sophisticated optics payloads and to take aerial images from strategic viewpoints has become unavoidable in nowadays battlespace contributing in moving forward the ISR capabilities. The constrained on-board processing power in addition to the strict limit in the flying time of sUAV are amongst the serious challenges to overcome to enable cost-effective persistent surveillance based on sUAV platforms. All previous examples show that tailoring the sensors to match the platforms' environment is a challenging endeavour and therefore architects have shifted their design methodology to be based on hardware and software open architectures as a centrepiece of their approach in building cost-effective surveillance solution design. This chapter is a brief introduction to hardware and software building blocks for developing persistent surveillance systems. In our context, the focus is in particular on Electro-Optic (EO, visual spectrum) and Infrared (IR) integrated solutions leveraging computer vision techniques for surveillance missions.

[1]Tawazun Technology and Innovation (TTI), Abu Dhabi, United Arab Emirates

13.1 Introduction

Currently, in our connected digital world, there are many ways to track, watch and collect data on individuals. Such a thing is termed 'surveillance', and one method of surveying targets is by using cameras, which is a type of visual surveillance [1]. While there are many other types of surveillance, visual surveillance is heavily used in different areas in the world. From roads to workplaces, crowds to individuals, cameras that image in the large spectrum of light with varying resolutions and fields of view are used daily to collect data about multiple targets. Surveillance systems can be as simple as a single fixed camera or complex like a network of connected cameras over a huge area [1].

In our current times, using aerial systems fitted with cameras to acquire imagery is becoming more and more prevalent, and one specific application of these imaging systems is the wide-area surveillance system (WASS) [2], which is concerned about covering a wide area, either by imagery or by signals, and detecting multiple targets in this broad area. The added benefit of wide-area coverage is that it eliminates the 'soda straw' view that most narrow-field surveillance systems have when embedded on aerial platforms. WASS provides the user with the opportunity to track multiple targets while not losing any data in the area of interest. However, tracking moving objects in a video sequence is a challenging task in WASS applications, and it poses severe problems for the unmanned aerial vehicle (UAVs) due to their cyber-physical nature [2–4] especially for small UAVs (sUAVs). In addition, WASS requires typically big and heavy imaging and processing components that make their integration into the smaller aerial vehicle a challenging task.

In general, sUAVs are small, agile and flexible, facilitating the design of easy, fast, and cost-efficient solutions for a variety of day-to-day problems. Recently, sUAV with visual sensors attained proper attention for a broad range of applications such as visual surveillance, border control, search and rescue missions, wildlife monitoring, crowd detection and monitoring [5]. Leveraging sUAV for persistent surveillance is by itself an evolutionary solution for smarter aerial surveillance. Developments in sUAV technologies caused a rapid growth of their application areas such as disaster management, tactical military area surveillance and major events surveillance. Although there are still lots of possible sUAV applications, most of the existing applications suffer from limited flight time endurance (mostly less than 3 h) [6,7]. Hence, while there are numerous ongoing studies for improving sUAV flying time [8,9], they appear to be unsuitable platforms for persistent surveillance which requires more extended operating hours.

In [10] an approach to tether sUAV to a ground power source was promoted as a feasible solution. This is due to the recent market availability of commercial tethered aerial platforms that have gotten very satisfactory feedbacks in terms of flight time and operational consistency [11–14]. This approach proposes that the multi-rotor type sUAV continuously receive power during a flight from a ground power source at the cost of limiting their freedom of movement. For some applications such as persistent surveillance, the duration of the operation is more important than the freedom of the movement. In this chapter, we will baseline our hardware and software recommendations based on an exemplary sUAV persistent surveillance that leverage computer vision techniques to achieve its surveillance missions.

13.2 System view

13.2.1 System description

Typical sUAV surveillance systems consist of three major components; aerial platform, gimballed image acquisition system and a high-performance computing system for motion [15,16] imagery called processing and exploitation unit (PEU). The aerial platform is responsible for achieving and maintaining the flight altitude needed in the desired area in order for the imaging system to operate. The gimballed system can host a visual spectrum's optics camera, an Infrared (IR) camera, or any type and number of cameras needed for the application. It stabilises the stream of imageries that are fed to the PEU in both day and night conditions. The PEU processes in real-time stream of imageries generating a wealth of operational intelligence based on the track, and geo-locating moving objects within the surveyed area. Through a provision for a recording and playing back capabilities, sUAV surveillance system provides an intelligence surveillance and reconnaissance (ISR) infrastructure for post-mission investigation including patterns of life analysis and anomalies detection.

13.2.2 Hardware components

In this section, the detailed components of the sUAV surveillance system are explained (in Figure 13.1). The high-level specification of each component is presented, including a summary of their advantages and disadvantages that can aid in the selection of parts for any specific application.

Aerial platform (Drone): It is envisioned that the ultimate sUAV surveillance system will conduct a mission lasting for a couple of hours to ensure the persistence of surveillance. Hence, the drone's endurance should satisfy this projected mission time which by itself is a challenging requirement for two factors:

- The drone has limited power to ensure long flying period (e.g., up to 2 h).
- The drone datalink has limited bandwidth and throughputs while the PEU system would expect the streaming of high-resolution imageries.

Both previously cited factors could be addressed with the tether technology that would transport the power to the drone, as well as eventually act as a data link carrier linking the drone to its ground segment. Nevertheless, the performance of the aerial platform will be impacted by the extra weight of the tether, and it will suffer from the forces exercised by the tether. In addition, the flying altitude is drastically reduced, and the tether now constrains the manoeuvrability of the drone. Also, to our knowledge, tether technology is limited to carry only power

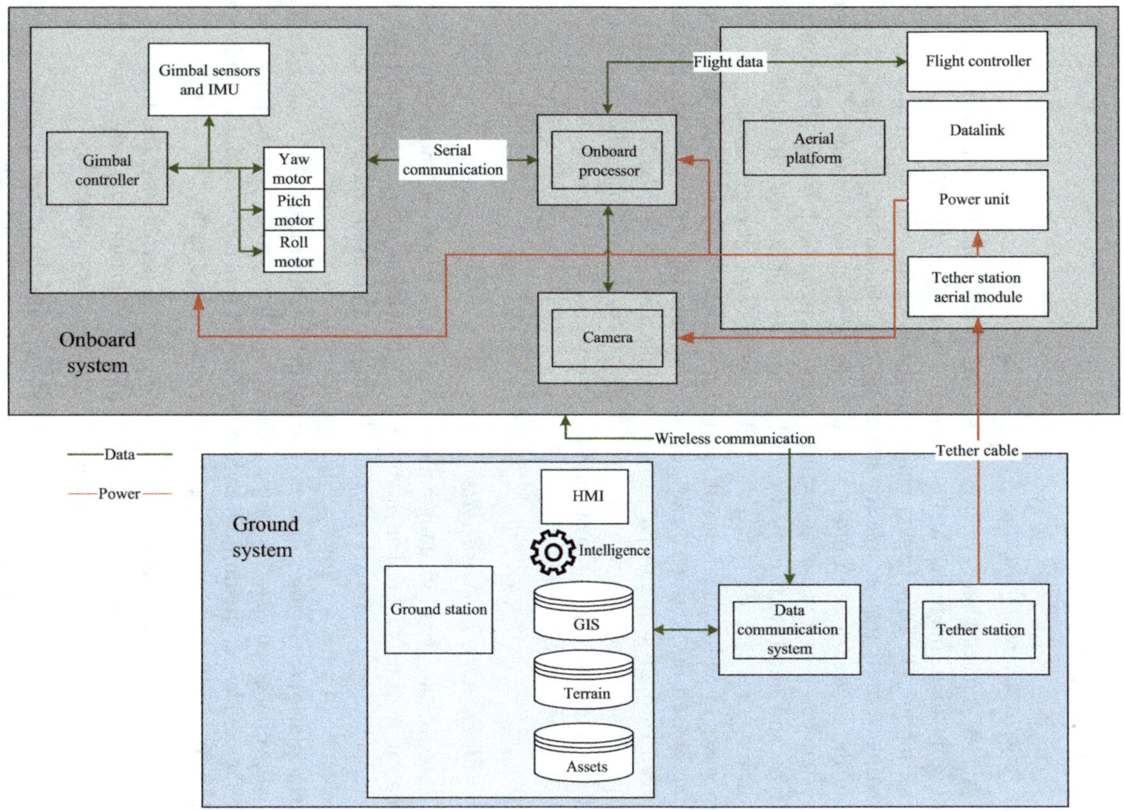

Figure 13.1 System view

(no datalink support) to the drone and therefore, the data link component selection becomes a crucial factor in the feasibility of our design.

In summary, the drone technology choice should ensure to carry the imaging system and to fly with a tether to achieve the height and duration needed for capturing 'quality' images, while being able to downlink them in real-time. The drone presents the advantage of flying in closed spaces and between city buildings without obstructions. The tether helps in supplying the drone with the power to keep it operational for a longer period. Nevertheless, since it is tethered, the length of the tether limits the height performance of the drone, and the payload weight budget has to account for the pull exercised by the tether.

Imaging system: Generically, we refer to the imaging system as the component that includes at the same time the gimbal device with all its attached optics and sensors. The gimbal hosts the cameras and is responsible for stabilising the acquired images. The gimbal should be small enough to be attached to the selected drone, but big enough to accommodate the imaging equipment. The gimbal has to withstand the effects of altitude, humidity, vibration and temperatures' variation. To ensure the tracking capability of moving targets, the gimbal design should offer at least 2-axes of movement. The main advantages of the gimbal reside in providing the observation and targeting capabilities in addition to offering the stabilisation and control functions. However, since all drone is attached parts are competing for the same payload weight all gimbal's sub-components need to be as small as possible and might have their functionalities downgraded to fit the size and weight budget.

One of the most critical sub-components of the gimbal is the cameras. Various cameras sensitive to different wavelength can be considered. As such, electro-optic (EO, visual spectrum), IR and hyperspectral (HS) are all options that are suitable for sUAV surveillance. Moreover, the multi-cameras configuration can be considered in order to achieve wider fields of view or to gain multiple spectrums of imagery. However, the size and weight of these cameras have to be selected with great attention, aligned with the aerial platform's overall weight and performance.

Data link: The data link allows in parallel the control of the aerial platform navigation, and the downlink of the images of interest to the ground operator. Depending on where the PEU is deployed (on the ground or on-board), the data link system is exercised differently. With a PEU on the ground, the data link has to handle real-time streaming of the images to avoid any side effects to the image processing pipeline running in the PEU. With a PEU on board, the data link is freed up from the image streaming. Having a datalink with bigger bandwidth allows the use of higher quality imagery from a bigger coverage area and thus resulting in better system performance. However, there is always a limited bandwidth no matter which datalink is chosen. Therefore, both the frame rate and resolution of the images transmitted have to be selected carefully.

PEU: The PEU is the component responsible for receiving and processing the images obtained by the imaging system, and sent by the data link. The size, placement, and performance of the PEU are solely dependent on the requirements of the user. As such, multiple placement and design options can be adopted for the PEU.

13.2.3 Components recommendation

Before building any sUAV system, one should consider the operational limits of the system. The altitude, flight time, coverage area, resolution, ambient conditions, and speed of image transmission should all be known in order to build the system according to the needs of the user. In this section, some recommendations for the components are mentioned. It is vital to have the components checked in sequence, as to eliminate any compatibility problems among them.

Aerial platform (drone): Constrained by the tether, the aerial platform for the sUAV surveillance system should be a rotary-wing satisfying the following criteria:

- Fly higher than the desired altitude;
- Have a payload weight that allows for the carrying of the imaging system and all its components;
- Be compatible with the tether system;
- Provide the housing and interface for the power module of the tether; and
- Operate in the ambient temperature of its deployment location.

Tether station: The tethering station provides power to the aerial platform. It consists of a tethering cable connected to an on-board power module. Some recommendations when designing the tether are as follows:

- Be compatible with the aerial platform;
- Have a tether cable measuring slightly greater than the maximum desired altitude;
- Have an optimised weight to fit within the allocated payload weight budget; and
- Operate in the ambient temperature of its deployment location.

Imaging system: It is defined as the generic component that represents both the mechanical gimbal device with all its attached optics and sensors. We attribute to the gimbal the main critical function of stabilising the cameras while the flying drone experiences its natural navigation movement. Therefore, the imaging system recommendations are as follows:

- It should have an optimised weight to fit within the allocated payload weight budget.
- It should support standard power requirements compatible and achievable by the aerial platform.
- It should provide a stabilisation mechanism.
- It should support tracking by aiming at a certain geographic location.

Within the motorised gimbal is attached at least a single camera capturing motion imageries in the visual spectrum of light (referred in this chapter as EO camera) or in the IR spectrum. EO cameras are responsible for taking images mostly in the daytime. Selecting the adequate EO camera for sUAV surveillance is a challenging endeavour, due to conflicting camera settings that affect the quality of the photos, and hence might impact the images processing pipeline taking place in the PEU. As such, we recommend that the imaging system should:

- have an optimised weight to fit within the allocated gimbal weight budget;
- be compatible with the data link and PEU components in order to transfer and consume the images successfully;
- have a configurable optimised frame rate (FPS) compatible with the PEU's expectations (e.g., track people and vehicles);
- have a resolution that achieves a Ground Sampling Distance (GSD) compatible with the PEU's expectation at the maximum flying altitude; and
- have a field of view (FOV) big enough to cover the surveyed area.

Of particular interest to the tracking capability of the sUAV surveillance is the GSD and FOV criteria. A higher GSD will result in pixilation of the object and might result in losing its track. The FOV dictates the limit of the surveyed geographic area. Governed by the following equations and based on typical camera specifications, the sUAV surveillance system designer can easily deduce the expected performance of the selected imaging system. The equations are describing the imaging area and quality of a single-camera gimbal system.

Imaging system variables:

- $xsensor$: width of the sensor, in mm
- $ysensor$: height of the sensor, in mm
- $focallen$: focal length of lens, in mm
- $altitude$: height, in m
- $xgimbal$: x-axis gimbal tilt angle, in degree
- $ygimbal$: y-axis gimbal tilt angle, in degree
- $xres$: x-axis sensor resolution, in pixels
- $yres$: y-axis sensor resolution, in pixels

Equations:

$$\text{Field of view wide} = 2\tan^{-1}\left(\frac{xsensor}{2 \times focallen}\right), \text{ and} \tag{13.1}$$

$$\text{Field of view tall} = 2\tan^{-1}\left(\frac{ysensor}{2 \times focallen}\right). \tag{13.2}$$

Equations (13.1) and (13.2) provide the camera's horizontal and vertical Fields of View (FOV), which helps in determining the coverage area generated on the ground.

$$\text{From drone to bottom of picture} = altitude \times \tan\left(ygimbal - \frac{1}{2} \times \text{Field of view wide}\right) \tag{13.3}$$

$$\text{From drone to top of picture} = altitude \times \tan\left(ygimbal + \frac{1}{2} \times \text{Field of view wide}\right) \tag{13.4}$$

$$\text{From drone to the left of picture} = altitude \times \tan\left(xgimbal - \frac{1}{2} \times \text{Field of view tall}\right) \tag{13.5}$$

$$\text{From drone to the right of picture} = altitude \times \tan\left(xgimbal + \frac{1}{2} \times \text{Field of view tall}\right) \tag{13.6}$$

$$\text{Height of photo footprint} = \text{From drone to the right of picture} - \text{From drone to the left of picture} \tag{13.7}$$

$$\text{Width of photo footprint} = \text{From drone to top of picture} - \text{From drone to bottom of picture} \tag{13.8}$$

Equations (13.3) through (13.6) show the position of the drone relative to the edges of the imaging area, while equations (13.7) and (13.8) give the dimensions of the coverage area on the ground-based on altitude, tilt of the gimbal, and FOV.

Under normal conditions (with no tilt), the GSD for the camera is calculated by the following equation:

$$GSD = \frac{altitude \times xsensor}{focallen \times xres}. \tag{13.9}$$

where GSD denotes how much distance on the ground is 1 pixel in the image.

Under no-tilt conditions, the x-axis GSD and the y-axis GSD are the same. However, when the camera is tilted, a different GSD is obtained for each tilt angle, and it is as follows:

$$GSD_{tiltx} = \frac{\text{Width of photo footprint}}{xres}. \tag{13.10}$$

$$GSD_{tilty} = \frac{\text{Height of photo footprint}}{yres}. \tag{13.11}$$

The sUAV surveillance system designer should be mindful about the GSD values while designing the system, as a GSD higher than the wanted amount will result in pixilation of the target which could affect the tracking process.

Datalink: Datalinks are used to transfer imagery, commands and various data back and forth between the ground segment and the platform. The recommendations below are specifically for data links that are used to transfer imagery only:

- Have an optimised weight to fit within the allocated payload weight budget;
- Be compatible with the aerial system in terms of power and operational system;

- Satisfy the transmission frequency regulations of the country where the solution is deployed;
- Have the transmission range greater than the maximum flying distance;
- Embed standard encryption mechanism; and
- Have a bandwidth and throughputs aligned with the PEU's expectation.

PEU: The PEU is a central piece of the sUAV surveillance system. The PEU provides the computing power required to ingest the images from the sensors and build in real-time any useful intelligence to the user. Therefore, the main mission of the PEU is to run the image processing and computer vision algorithms [17–20] efficiently.

Image processing is a branch of computer knowledge which is concerned with processing a digital signal representing images taken with a digital camera or scanned by a scanner [15,16,21,22].

The specification of the PEU is dictated by various requirements such as the resolutions of the images, the frame rate, the numbers of sensors, the flying platform, etc. Therefore, the PEU solution should pay particular attention to the following:

- Scalability: achieved by leveraging distributed architecture, multithreading and GPU programming.
- Performance: achieved by leveraging, for example, special-purpose parallel computing libraries such as the ones mentioned below.
  - Open Multi-Processing (OpenMP) can be used for distributing the workload over multiple CPUs and CPU cores;
  - Message Passing Interface (MPI) can be used for inter-workstation and inter-process communication;
  - Compute Unified Device Architecture (CUDA) can be considered for General Purpose computing on Graphics Processing Units (GPGPUs) computations.

13.3 Software components

Having the appropriate software architecture is a key enabler for an agile and scalable solution. The objective is to design a system:

1. Scalable enough to support a wide range of sensing components of different specifications (resolution, focal length, etc.).
2. Agile enough to allow the substitutability of hardware and software components with minimal development and integration effort.

The PEU is of central importance since it translates raw information captured by the sensors to actionable intelligence by discovering, in real-time, all objects of interest from the streamed images. By leveraging a plugins architecture in which logical software components are grouped under common interfaces, multiple implementations of different algorithms can be swapped easily. This is crucial to swap 'seamlessly' various hardware baselines and/or to support a progressive integration of state of the art surveillance algorithms. In our context, the major software interfaces of the image-processing pipeline comprise the following elements (Figure 13.2).

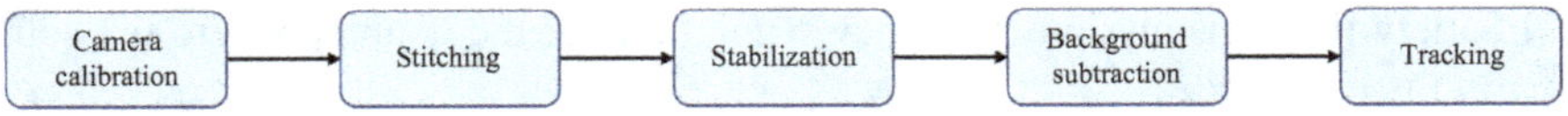

Figure 13.2 PEU image processing pipeline

13.3.1 Camera calibration

The process of geometric camera calibration is a fundamental step to minimise the differences that occur between an ideal lens model and camera lens model in the camera system. Calibration is needed for removing the radial and/or tangential distortion as well as to measure the amount of distortion introduced by the lenses [23]. It outputs the camera's intrinsic and extrinsic parameters, which constitute to rotation and translation vectors needed to translate a point in the 3D world coordinates to a 2D image coordinate point. A review of various camera calibration techniques is studied for the vision systems in [24]. The camera calibration techniques can be classified into the following three techniques: traditional calibration, self-calibration techniques, and motion-based calibration techniques. Traditional calibration methods such as the two-step calibration method [25] and Zhang's method [26,27] use the relationship between the 2D image coordinates and the 3D world coordinates of the feature points. Due to the high accuracy of Zhang's calibration method, it is mainly used in many camera calibration applications. Self-calibration techniques have the advantage of calibrating the system without the need for using a known scene. Its disadvantage is the less guaranteed accuracy, and its terrible lack of robustness. In motion-based calibration, the cameras are calibrated through some procedures such as pure translation, pre-rotation, and a combination of both. Since this method can provide high accuracy in the calibration target, it can be used for automated calibration. In general, there are many camera calibration techniques in the literature, but in our sUAV-based surveillance systems, only limited studies have involved images acquired on sUAV [28]. We cite in particular the techniques that have been explored for short object distances in [29–33] and long object distance in [34] and [35]. In most cases, the camera calibration procedure is done in a laboratory condition to estimate the camera parameters as the prerequisite through close-range calibration.

13.3.2 Image stitching

Stitching process creates a panoramic image by combining separate partially overlapping images provided simultaneously from multiple cameras or by combining a sequence of partially overlapping images provided by a single camera. For successful surveillance, mission stitching is a crucial process since the stitched images provide a large field of view of the surveyed area. For the construction of the stitched image, the transformations occurred between the images have to be estimated based on overlapping areas among these images. Several techniques have been explored using area-based or feature-based for finding the overlapping areas and for modelling the transformation between the images. For example, a fast Fourier transform (FFT) stitching method has been studied in [36]. In area-based image stitching, a window of an image is used for matching subsequent images by calculating similarity metrics such as the normalised cross-correlation [37], and entropy and mutual information [38]. Feature-based matching techniques have also been employed, such as the Scale-Invariant Feature Transform (SIFT) [39–41], Harris points [42] and Speeded Up Robust Features (SURF) [43–45]. The designer selecting the appropriate stitching technique for sUAV surveillance system should consider surveying the area of interest, accounting for the frequency of image acquisition, and the required percentage of the overlapping area [46].

13.3.3 Stabilization

With a moving camera, it will not be possible to detect and track moving objects based on processing motion imagery. Stabilisation is the process of transforming the mosaicked (stitched) motion images into a new one where the camera motion has been cancelled out. Moreover, sUAV with streaming video cameras suffer from the undesired shaking motion due to atmospheric turbulence, jittery flight control emphasising the need for more and more efficient stabilisation implementation techniques. These implementations should take into consideration the following points:

- Feature detection and description is an expensive (w.r.t. time) operation. Therefore image scaling needs to be supported in this phase;
- Consecutive images do not contain the view of the identical region on the ground plane:
  - The angle of view/perspective changes due to the aircraft movement, and
  - Even when the aircraft movement and gimbal movement is optimised to view a fixed point on the ground, the visible areas change among different frames.

There is a limited number of works related to the stabilisation of videos taken from sUAV [47–51]. Generally, UAV video stabilisation algorithms include the following three main steps: (1) motion estimation, (2) motion compensation and (3) image composition. Many studies attempted to find the 2D motion model (such as homography) for estimating the global motion trajectory and then applied a low-pass filter to the trajectory for removing the high-frequency jitters. Finally, the low-frequency parameters are used onto frames via warping techniques, which is more effective for scenes with very little dynamic movement taken by UAVs. In [47], a video stabilisation algorithm using a circular block to search and match key places is proposed. In [48], a smoothing method utilises the Lucas–Kanade tracker [49] to detect interest points. In [50], a three-step video stabilisation method is proposed for UAVs. Initially, a Features from Accelerated Segment Test (FAST) corner detector is used to locate the feature points in the frames, and then the matched key-points are used to estimate the affine transform for reducing the false matches. Finally, the affine model based on motion estimation and the compensation is performed.

13.3.4 Background subtraction

Background Subtraction is a technique that is utilised to extract the foreground of an image. Then, this is used for further processing, such as object tracking or recognition. The majority of background subtraction techniques (as shown in Table 13.1), assume that the images supplied to the algorithm have a static background, in which moving targets are detected.

Table 13.1 Various background subtraction techniques

Methods	Description	Advantages	Disadvantages
Frame Differencing [51]	The difference between the current frame and 1-3 other frames is calculated to detect moving objects. A threshold is set to distinguish foreground pixels from background pixels.	• Computation time is fast • Easy to apply • Low memory requirements	• Does not perform well with noisy background or background with moving objects.
Eigen Background [52]	The background is modelled by an eigenspace to segment moving objects. The dimensionality of the eigenspace is then reduced by using Principal Component Analysis.	• Good performance for unstable backgrounds	• Does not perform well with dynamic scenes.
Gaussian Mixture Model [53]	All the background pixels are modelled as a Gaussian mixture. Pixels are then matched with the existing models set. A new Gaussian is initialised by the least model pixel value, which it then replaced if a match is not found.	• Low memory requirements • Good with slow lighting variations. • Performs well with noisy backgrounds	• Demanding computations are involved
Approximated Median Filter [54]	The background is calculated as the median of N previous frames. Background pixels are incremented by 1 if the corresponding pixel from the current frame is bigger, it is decremented by 1 if it is smaller. Eventually converging to a new estimated median.	• Very robust • Low complexity level • Low memory requirements	• Results are not smooth, as the background model could still be affected by previous frames.

Considerations for the background subtraction process are as follows:

- Robustness in case of stabilisation errors which are inevitable due to the scale of the input imagery,
- Being able to cope with insufficient information (small amount of pixels available for foreground objects, low-intensity differences with foreground objects and neighbouring background pixels),
- Being able to cope with object shadows which are common in wide-area imagery,
- Combining the outcomes of multiple background subtractions to obtain results that are better than a single subtractor.

13.3.5 Object tracking

Object tracking is a method where a moving object is located (spatial) throughout a set of images captured over time (temporal). Tracking allows estimating the trajectory of the moving objects within the monitored area. Information such as speed and size of the object is utilised to track the objects and distinguish them from other objects in the same frame.

The tracking process needs to be:

- Robust against false alarms caused by stabilisation errors due to large extensive area imagery,
- Robust against false alarms caused by changes in the angle of view (i.e., high objects like buildings and towers move since they are viewed from different angles at different times),
- Robust in case of occlusion,
- Robust in case of reflection and to problems due to light conditions,
- Able to cope with insufficient information available for foreground objects (i.e., different objects that appear to be similar, objects that are hardly visible over the roads).

The following mechanisms are developed in order to address the considerations above:

- Motion models that reflect real-world objects are designed in order to be able to classify incorrect detections (i.e., false alarms) in the background subtraction phase,
- Coasting methodology is designed so that in case of a detection failure, a track from the previous frame is not found in the current frame; a track is marked as suspect before it is allowed to die (i.e., occlusion),
- Appearance-based techniques are combined with motion-based techniques to provide a more robust association (i.e., reflection, insufficient information),
- As a safeguard when all of the above mechanisms fail, merging is designed where each track that was allowed to die in previous frames are correlated with tracks that are birthed after the death of those tracks.

Some of the tracking methods useful for UAV-based tracking as given in Table 13.2.

Table 13.2 Tracking methods for UAVs applications

Tracking method	Description	Methods	Advantages	Disadvantages
Kernel Tracking [55,56]	Moving objects are represented as an undeveloped object section, where their shapes, features, representation and appearance, are what this tracking method is based upon.	Layering-based Tracking	• Good accuracy • Great occlusion handling • Multiple objects can be tracked	• Computation time is high
		Support Vector Machine	• Good accuracy • Good occlusion handling	• Training is needed beforehand
Point Tracking [57,58]	Moving objects that are being tracked are represented as points that can hold information such as trajectory and position. However, in order to detect those points in the first place, another external mechanism is required.	Particle Filter	• High accuracy • Good occlusion handling • Good with complex backgrounds	• Computation time is high due to a large number of calculations. • Not good for real time applications
		Kalman Filter	• Performs well with noisy images	• Low accuracy
Silhouette Tracking [59]	In cases where objects have a complicated shape, this method can be used as it uses an accurate outline of the moving object from past frames.	Shape matching	• High accuracy • Performs well to object silhouette changes • Good occlusion handling	• Computation time is high • Training is needed beforehand
		Contour matching	• Flexible to a wide range of shapes • Good accuracy	• Time is needed to find the best configuration approximation.

13.3.6 Geo-location pointing

Geo-location pointing is a method used in gimbal systems to allow the user to point in a desired geographic location based on the drone and gimbal locations and altitude. This pointing method allows for keeping an eye on a certain location for a long time, which is essential in persistent surveillance. The pointing is created by sending commands from the PEU to the gimbal motors to specify the angle needed. The pipeline for such a process is explained in Figure 13.3 below.

The PEU reads the needed data from the gimbal and drone and incorporates the user's desired location into the algorithms. From the algorithms, the pitch and yaw of the gimbal are obtained and sent to the gimbal as angle commands to point it in the wanted location. The algorithms and variables used in the PEU are provided below.

Variables:

$\varphi_{A,B}$	Latitude of points *A* and *B*, where *A* is the drone point, and *B* is the user-defined point. Assume north is positive.
$L_{A,B}$	Longitude of points *A* and *B*, where *A* is the drone point, and *B* is the user-defined point. Assume east is positive.
β	Bearing angle
D	Distance between two points on a flat plane
R	Radius of Earth in a spherical model
σ	Drone heading, clockwise from north
x	The horizontal distance from the centre of the drone to the centre of the camera
z	The vertical distance from the centre of the drone to the centre of the camera
H	Drone height from the ground
γ	Camera yaw angle. Clockwise is positive
α	Camera pitch angle from the horizontal position, downward is positive

If there are two points *A* and *B* on the surface of Earth (assumed spherical with radius *R*), and the two points are close together (within a few kilometres of each other), then the following formulas can be used to calculate the distance between the two points [60].

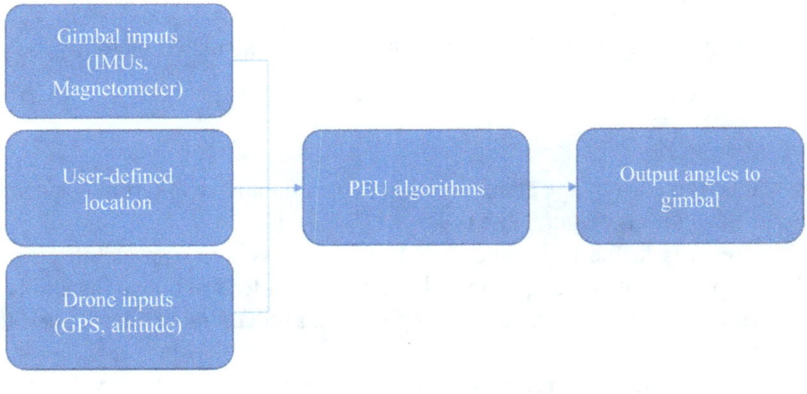

Figure 13.3 Geo-location pointing pipeline

An angle θ can be calculated using the haversine as follows:

$$\text{hav}(\theta) = \text{hav}(\Delta\varphi) + \cos(\varphi_A)\cos(\varphi_B)\text{hav}(\Delta L), \tag{13.12}$$

where $\Delta\varphi$ is the difference in latitudes, ΔL is the difference in longitudes, and haversine function is:

$$\text{hav}(\theta) = \sin^2\frac{\theta}{2}.$$

Therefore, we have:

$$\theta = 2\,\sin^{-1}\left(\sqrt{\text{hav}(\theta)}\right), \text{ and } D = R\theta. \tag{13.13}$$

For the bearing, it will be measured eastward from the north. The bearing of point B, as seen from point A, can be determined by first calculating two quantities S and C as follows:

$$S = \cos(\varphi_B)\sin(\Delta L), \tag{13.14}$$

$$C = \cos(\varphi_A)\sin(\varphi_B) - \sin(\varphi_A)\cos(\varphi_B)\cos(\Delta L). \tag{13.15}$$

Then, the bearing angle β can be calculated by:

$$\beta = \tan^{-1}\left(\frac{S}{C}\right).$$

However, this cannot be solved if C is zero, thus the atan2 function can be used in the code or any equivalent function.

The distance between the drone and the point of interest and the bearing angle can be calculated from the above equations. Referring to Figure 13.4, the yaw angle of the camera can be calculated as follows:

$$\gamma' = \tan^{-1}\left(\frac{D\sin\beta - x\sin\sigma}{D\cos\beta - x\cos\sigma}\right), \text{ and}$$
$$\gamma = \gamma' - \sigma.$$

Then, referring to Figure 13.2, the pitch angle can be calculated as follows:

$$\alpha = \tan^{-1}\left(\frac{H - z}{D - x}\right),$$

which can be replaced by the atan2 function to prevent any errors from occurring in the angle calculations (Figure 13.5).

With regards to the roll angle, it is used in the system as a stabiliser angle, which shall prevent the gimbal from rolling, which can lead to image instability. Of course, the value can be set manually if needed, but remaining at a horizontal level is better for stabilization.

If the used gimbal has a limited range of movement, it is better to specify it in the program to prevent the gimbal from getting damaged by huge ranges of motion.

As an example, the current ranges for yaw, pitch and roll angles are specified.

$$-90^\circ \leq \gamma \leq 90^\circ,$$
$$0^\circ \leq \alpha \leq 180^\circ, \text{ and}$$
$$-90^\circ \leq \beta \leq 90^\circ.$$

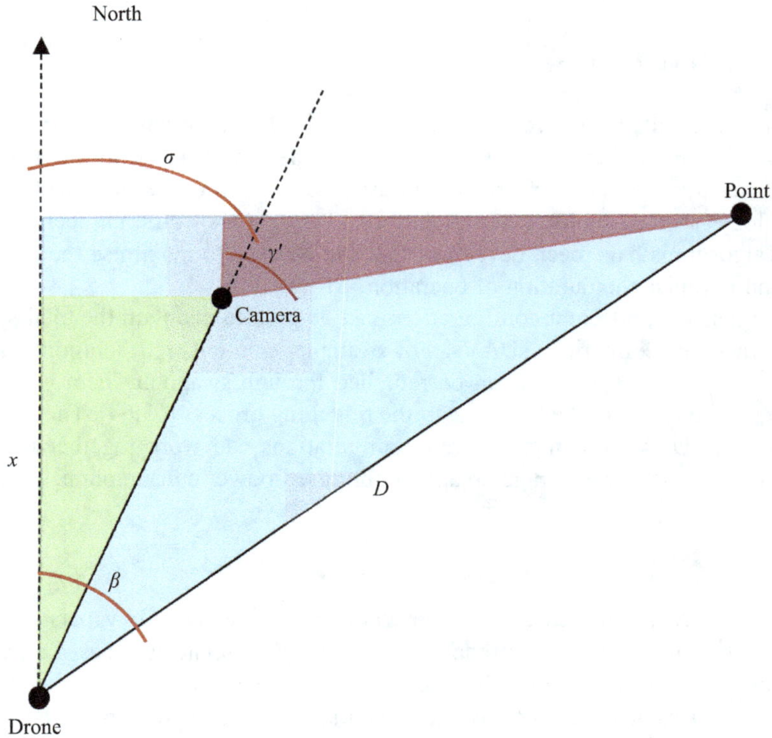

Figure 13.4 Top view of the drone and the point of interest

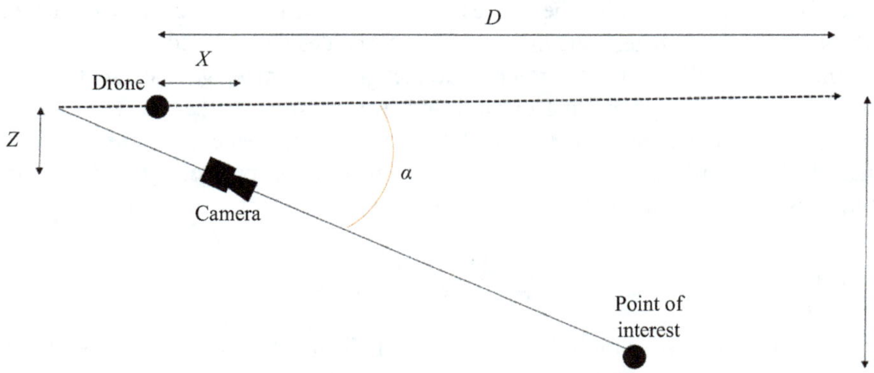

Figure 13.5 Side view of the pitch angle calculations

Of course, such values can be changed or modified based on the gimbal system used and its capabilities. The last thing to do would be to check the bit rate of the gimbal controller. As most gimbals' motors work in encoder counts, the designer should find the number of counts corresponding to the angle value. This is to allow the algorithm to output the angle in encoder counts in order to be sent to the gimbal controller. Of course, if the gimbal can read normal angles immediately, then there is no need to use such a transformation.

13.4 Future trends

Persistent surveillance is evolving into a major field when it comes to drone technologies. Many different applications require persistent surveillance to either extend security or monitor critical infrastructure. Further development of this technology is to use multiple aerial platforms to cover a specific number of areas. Many algorithms have been developed that can be used to maximise the coverage area and optimise the duration of operation [61–68].

Furthermore, different configurations can be used to maintain the total system power in case of multiple sUAVs. For example, non-tethered solutions can be utilised, and aerial platforms can be refuelled through swapping them with other platforms and return to the base to start the refuelling process [63,64]. The concepts of multiple sUAVs and non-tethered configurations can work together to cover multiple areas, larger coverage area, and optimised power consumption.

13.5 Conclusion

In this chapter, we presented the system view of a referential sUAV persistent surveillance system. Such a system view enables an open architecture based on COTS hardware components. Within this architecture, there are two types of supported payloads; EO payload and IR payload which allow the system to operate day and night. The communication between on-board and ground stations is done through datalink systems. The gimbal controller controls the gimbal IMU sensors and motors in order to stabilise the gimbal. A key component in this system view is the PEU, which is in charge of running the images processing pipeline required for the tracking capability. The PEU ingests the imageries and schedules the required processing such as stitching, stabilisation, background subtraction, and tracking. Seeking for scalability, ease of deployment and extensibility, we recommend that the PEU should adhere to a distributed/multithreaded architecture running, whenever deemed possible, on embedded hardware with GPU parallelisation in mind promoting software substitutability consequently with minimum development and integration effort.

References

[1] Ren J., Xu M., Smith J., Zhao H., and Zhang R. 'Multi-view visual surveillance and phantom removal for effective pedestrian detection'. *Multimedia Tools and Applications*. 2017;77(14):18801–18826.

[2] Sommer L.W., Teutsch M., Schuchert T., and Beyerer J. 'A survey on moving object detection for wide area motion imagery'. *Proceedings of the IEEE Winter Conference on Applications of Computer Vision*; Lake Placid, NY, USA, 2016. pp. 1–9.
[3] Palaniappan K., Rao R., and Seetharaman G. 'Wide-area persistent airborne video: Architecture and challenges'. *Distributed Video Sensor Networks: Research Challenges and Future Directions*, Springer, 2010; pp. 349–371.
[4] Blasch E., Seetharaman G., Suddarth S., *et al.* 'Summary of methods in wide-area motion imagery (WAMI)'. *Proceedings of SPIE 9089*; Baltimore, Maryland, USA, 2014. p. 90890C.
[5] Cai G., Dias J., and Seneviratne L. 'A survey of small-scale unmanned aerial vehicles: Recent advances and future development trends'. *Unmanned Systems.* 2014;2(02):175–199.
[6] Batista da Silva L.C., Bernardo R.M., de Oliveira H.A., and Rosa P.F.F. 'Unmanned aircraft system coordination for persistent surveillance with different priorities'. *Proceedings of the IEEE 26th International Symposium on Industrial Electronics*; Edinburgh, UK, 2017. pp. 1153–1158.
[7] Khandani H., Moradi H., and Yazdan P.J. 'A real-time coverage and tracking algorithm for UAVs based on potential field'. *Proceedings of the 2nd RSI/ISM International Conference on Robotics and Mechatronics*; Tehran, Iran, 2014. pp. 700–705.
[8] Logan M.J., Chu J., Motter M.A., Carter D.L., Ol M., and Zeune C. 'Small UAV research and evolution in long endurance electric powered vehicles'. *Proceedings of the AIAA Infotech Aerospace Conference and Exhibition*; Rohnert Park, California, USA, 2007. pp. 1–7.
[9] Zeng Y., Zhang R., and Lim T.J. 'Wireless communications with unmanned aerial vehicles: Opportunities and challenges'. *IEEE Communications Magazine*, 2016;54(5):36–42.
[10] Ohood A.A., Omar A., Abdulrahman A., Abdulla A.A., Slim S., and Issacniwas S, 'Persistent surveillance with small unmanned aerial vehicles (SUAVs): A feasibility study'. *Proceedings of SPIE 10796*; Electro-Optical Remote Sensing XII; Berlin, Germany, 2018. p. 107960K.
[11] https://www.ntpdrone.com/ [Accessed 28 July 2018].
[12] http://tethered.solutions/ [Accessed 01 Feb 2018].
[13] http://www.unmannedsystemstechnology.com/2017/05/mmc-announces-new-tethered-drone-power-supply-system/ [Accessed 28 Jan 2018].
[14] https://hoverflytech.com/defense/ [Accessed 05 Feb 2018].
[15] Estrela V.V., and Galatsanos N.P. 'Spatially-adaptive regularized pel-recursive motion estimation based on cross-validation'. In *Proceedings 1998 International Conference on Image Processing. ICIP98 (Cat. No. 98CB36269)*; Chicago, IL, USA, 1998 October; Vol. 2. pp. 200–203. IEEE.
[16] Coelho A.M., and Estrela V.V. (2016). EM-based mixture models applied to video event detection. arXiv preprint arXiv:1610.02923.

[17] Razmjooy N., Mousavi B.S., Soleymani F., and Khotbesara M.H. 'A computer-aided diagnosis system for malignant melanomas'. *Neural Computing and Applications*. 2013;23(7–8):2059–2071.

[18] Estrela V.V., and Herrmann A.E. 'Content-based image retrieval (CBIR) in remote clinical diagnosis and healthcare'. In *Encyclopedia of E-Health and Telemedicine*, IGI Global, 2016, pp. 495–520.

[19] Razmjooy N., Ramezani M., and Ghadimi N. 'Imperialist competitive algorithm-based optimization of neuro-fuzzy system parameters for automatic red-eye removal'. *International Journal of Fuzzy Systems*. 2017;19(4): 1144–1156.

[20] de Jesus M.A., Estrela V.V., Saotome O., and Stutz D. 'Super-resolution via particle swarm optimization variants.' Biologically *Rationalized Computing Techniques For Image Processing Applications*. Lecture Notes in Computational Vision and Biomechanics. Springer, Zurich, Switzerland, 2018;25: 317–337.

[21] Somayeh Mousavi B., and Soleymani F. 'Semantic image classification by genetic algorithm using optimised fuzzy system based on Zernike moments'. *Signal, Image and Video Processing*. 2014;8(5):831–842.

[22] Razmjooy N., Mousavi B.S., Khalilpour M., and Hosseini H. 'Automatic selection and fusion of color spaces for image thresholding'. *Signal, Image and Video Processing*. 2014;8(4):603–614.

[23] Emilia G.D., and Gasbarro D.D. 'Review of techniques for 2D camera calibration suitable for industrial vision systems'. *Journal of Physics: Conference Series*. 2017;841:012030.

[24] Yusuf A.R., Ariff M.F.M., Khairulnizam M.I., Zulkepli M., and Albert K.C. 'Camera calibration accuracy at different UAV flying heights'. *The International Archives of the Photogrammetry, Remote Sensing and Spatial Information Sciences*, vol. XLII-2/W3, *3D Virtual Reconstruction and Visualization of Complex Architectures*. 2017; pp. 1–3.

[25] Tsai R.Y. 'A versatile camera calibration technique for high-accuracy 3D machine vision metrology using off the-shelf TV cameras and lenses'. *IEEE Transactions on Robotics and Automation*. 1987;3(4):323–344.

[26] Zhang Z.Y. 'A flexible new technique for camera calibration'. *IEEE Transactions on Pattern Analysis*. 2000;22(11):1330–1334.

[27] Zhang Z. 'Flexible camera calibration by viewing a plane from unknown orientations'. *Proceedings of the 7th IEEE International Conference on Computer Vision*; Kerkyra, Greece, 1999. pp. 666–673.

[28] Pérez M., Agüera F., and Carvajal F. 'Digital camera calibration using images taken from an unmanned aerial vehicle'. *The International Archives of the Photogrammetry, Remote Sensing and Spatial Information Sciences*, Zurich, Switzerland, Vol. XXXVIII, Part 1/C22, 2011; pp. 167–171.

[29] Chiang K.-W., Tsai M.-L., and Chu C.-H. 'The development of an UAV borne direct georeferenced photogrammetric platform for ground control point free applications'. *Sensors*, 2012;12:9161–9180.

[30] Sauerbier M., and Eisenbeiss H. 'Investigation of UAV systems and flight modes for photogrammetric applications'. *The Photogrammetric Record*, 2011;26(136):400–421.

[31] Deng D.W., and Li B.A. 'Large Unmanned Aerial Vehicle Ground Testing System'. *In Applied Mechanics and Materials*, 2015;719:1244–1247.

[32] Jimenez P.L., and Agudelo D. 'Validation and calibration of a high-resolution sensor in unmanned aerial vehicles for producing images in the IR range utilizable in precision agriculture'. *American Institute of Aeronautics and Astronautics (AIAA) SciTech*, 2015.

[33] Tahar K.N. 'Aerial terrain mapping using unmanned aerial vehicle approach'. *International Archives of the Photogrammetry, Remote Sensing and Spatial Information Sciences*, Vol. XXXIX-B7, XXII ISPRS Congress, Melbourne, Australia, 2012. pp. 493–498.

[34] Mohamed M.R.M., and Schwarz K.P. 'An autonomous system for aerial image acquisition and georeferencing'. *American Society of Photogrammetry and Remote Sensing Annual Meeting*, 1999;pp. 17–21.

[35] Liu P., Xi C., and Yang L. 'An approach of system calibration for UAV photogrammetry'. *Proceedings of SPIE*, 2011;p. 8200.

[36] Xie H., Hicks N., Keller G.R., Huang H., and Kreinovich V. 'An IDL/ENVI implementation of the FFT-based algorithm for automatic image registration'. *Computers & Geosciences*. 2003;29:1045–1055.

[37] Zhao F., Huang Q., and Gao W. 'Image matching by normalized cross-correlation'. *Proceedings of the IEEE International Conference on Acoustics, Speech and Signal Processing*, 2006.pp. II.

[38] Brown M., and Lowe D.G. 'Automatic panoramic image stitching using invariant features'. *International Journal of Computer Vision*. 2007;74:59–73.

[39] De Césare C., Rendas M.-J., Allais A.-G., and Perrier M. 'Low overlap image registration based on both entropy and mutual information measures'. *OCEANS 2008. IEEE*, 2008. pp. 1–9.

[40] Jia Y., Su Z., Zhang Q., Zhang Y., Gu Y., and Chen Z. 'Research on UAV remote sensing image mosaic method based on SIFT'. *International Journal of Signal Processing, Image Processing and Pattern Recognition*, 2015;8 (11):365–374.

[41] Cruz B.F., de Assis J.T., Estrela V.V., and Khelassi A. 'A compact SIFT-based strategy for visual information retrieval in large image databases.' Medical Technologies Journal, 2019;3(2):402–412, doi:10.26415/2572-004X-vol3iss2p402-412.

[42] Zagrouba E., Barhoumi W., and Amri S. 'An efficient image-mosaicing method based on multifeature matching'. *Machine Vision and Applications*. 2009;20:139–162.

[43] Wang J., and Watada J. Panoramic image mosaic based on SURF algorithm using OpenCV, *9th International Symposium on Intelligent Signal Processing*, IEEE, Siena, Italy, 2015; pp. 1–6.

[44] Rong W., Chen H., Liu J., Xu Y., and Haeusler R. Mosaicing of microscope images based on SURF, *24th International Conference on Image and Vision Computing New Zealand*, IEEE, Wellington, New Zealand, 2009; pp. 271–275.

[45] Geng N., He D., and Song Y. 'Camera image mosaicing based on an optimized SURF algorithm'. *TELKOMNIKA Indonesian Journal of Electrical Engineering*. 2012;10:2183–2193.

[46] Zhang Y., Xiong J., and Hao L. 'Photogrammetric processing of low-altitude images acquired by unpiloted aerial vehicles'. *The Photogrammetric Record*, 2011;26:190–211.

[47] Shen H., Pan Q., Cheng Y., and Yu Y. Fast video stabilization algorithm for UAV. In *IEEE International Conference on Intelligent Computing and Intelligent Systems, 2009. ICIS 2009*, Shanghai, China, 2009, vol. 4, pp. 542–546.

[48] Vazquez M., and Chang C. 'Real-time video smoothing for small RC helicopters'. *Proc. IEEE International Conference on Systems, Man and Cybernetics*, San Antonio, Texas, USA, 2009;pp. 4019–4024.

[49] Lucas B.D., and Kanade T., 'An iterative image registration technique with an application to stereo vision'. *Proceeding of the 7th International Joint Conference on Artificial Intelligence*. San Francisco, USA, 1981; 81:674–679.

[50] Wang Y., Hou Z., Leman K., and Chang R. 'Real-time video stabilization for unmanned aerial vehicles'. Proc. IAPR Conference on Machine Vision Applications, 2011, Nara, JAPAN, 2011:336–339.

[51] Issacniwas S., Slim S., and Maya A.H. 'Parallax rectification and stabilization technique for multiple objects tracking in wide area surveillance system'. *Proc. SPIE 10649, Pattern Recognition and Tracking XXIX*, Orlando, Florida, USA, 2018;p. 1064914.

[52] Srinivasan K., Porkumaran K., and Sainarayanan G. 'Improved background subtraction techniques for security in video applications'. *Proc. IEEE 3rd International Conference on Anti-counterfeiting, Security and Identification in Communication*, Hong Kong, China, 2009.

[53] Oliver N.M., Rosario B., and Pentland A.P. 'A Bayesian computer vision system for modeling human interactions'. *IEEE Transactions on Pattern Analysis and Machine Intelligence*. 2000;22(8):831–843.

[54] Stauffer C., and Grimson W. 'Adaptive background mixture models for real-time tracking'. *Proc. IEEE Computer Society Conference on Computer Vision and Pattern Recognition*. Fort Collins, CO, USA, 1999.

[55] McFarlane N.J.B., and Schofield C.P., 'Segmentation and tracking of piglets in images'. *Machine Vision and Applications*, 1995;8:187–193.

[56] Wei L., Jianhu W., and Qin L. 'Study on moving object tracking algorithm in video images'. *Proc. 8th IEEE International Conference on Electronic Measurement and Instruments*, Xian, China, 2007;pp. 810–816.

[57] Shai A. 'Support vector tracking'. *IEEE Transactions on Pattern Analysis and Machine Intelligence*. 2004;26(8):1064–1072.

[58] Hu W., Tan T., Wang L., and Steve M. 'A survey on visual surveillance of object motion and behaviors'. *IEEE Transactions on Systems, Man, and Cybernetics Applications and Reviews*. 2004;34(3):334–352.

[59] Javed O., and Shah M. 'Tracking and object classification for automated surveillance'. *Proceedings of the 7th European Conference on Computer Vision-Part IV*, Copenhagen, Denmark, 2002;pp. 343–357.

[60] Karasulu B. 'Review and evaluation of well-known methods for moving object detection and tracking in videos'. *Journal of Aeronautics and Space Technologies*. 2010;4(4):11–22.

[61] Jeong B., Ha J., and Choi H. 'MDP-based mission planning for multi-UAV persistent surveillance'. *14th International Conference on Control, Automation and Systems*. Seoul, South Korea, 2014;pp. 831–834.

[62] Nigam N., and Kroo I. 'Persistent surveillance using multiple unmanned air vehicles'. *2008 IEEE Aerospace Conference*, Big Sky, MT, USA, 2008; pp. 1–14.

[63] Hartuv E., Agmon N., and Kraus S. Scheduling spare drones for persistent task performance under energy constraints. 2008 International Conference on Autonomous Agents and Multiagent Systems AAMAS, Stockholm, Sweden, 2018.

[64] Mitchell D., Corah M., Chakraborty N., Sycara K.P., and Michael N. 'Multi-robot long-term persistent coverage with fuel constrained robots'. *Proc. 2015 IEEE International Conference on Robotics and Automation (ICRA)*, Seattle, WA, USA, 2015;pp. 1093–1099.

[65] Peters, J.R., Wang, S.J., Surana, A., and Bullo, F. 'Cloud-supported coverage control for persistent surveillance missions'. *Journal of Dynamic Systems Measurement and Control*. 2017;139(8):1–12.

[66] Razmjooy N., Estrela V.V., and Loschi H.J. 'A survey of potatoes image segmentation based on machine vision.' *Applications of Image Processing and Soft Computing Systems in Agriculture*. IGI Global, Hershey, PA, USA, 2019;1–38. doi:10.4018/978-1-5225-8027-0.ch001

[67] Nigam N. 'The multiple unmanned air vehicle persistent surveillance problem: A review'. Machines. 2014;2(1):13–72.

[68] Mahmood S., Afsharchi M., and Weindling A.M. 'A Markovian decision process analysis of experienced agents joining ad-hoc teams'. *2018 21st Euromicro Conference on Digital System Design (DSD)*, Prague, Czech Republic, 2018; 691–698.

Chapter 14
Conclusions

Vania V. Estrela[1], Jude Hemanth[2], Osamu Saotome[3], George Nikolakopoulos[4] and Roberto Sabatini[5]

The current awareness in UAVs has prompted not only military applications but also civilian uses. Aerial vehicles' requirements aspire to guarantee a higher level of safety comparable to see-and-avoid conditions for piloted aeroplanes. The process of probing obstacles in the path of a vehicle, and to determine if they pose a threat, alongside measures to avoid problems, is known as see-and-avoid or sense and-avoid involves a great deal of decision-making. Other types of decision-making tasks can be accomplished using computer vision and sensor integration since they have great potential to improve the performance of UAVs. Macroscopically, Unmanned Aerial Systems (UASs) are cyber-physical systems (CPSs) that can benefit from all types of sensing frameworks, despite severe design constraints such as precision, reliable communication, distributed processing capabilities, and data management.

The first chapter overviews some main concepts and it calls attention to several issues still undergoing discussions in UAV-CPSs. Several trends and needs are discussed to foster the criticism from readers to the upcoming chapters and provide food for thought.

Some of the significant UAV-CPSs advantages are their capability to withstand the duties and objectives of human beings while accomplishing chores for them. Simultaneously, they also take some decisions and execute some actions independently. Therefore, people and machines must collaborate. Even though these capabilities offer noteworthy paybacks, there is still a tremendous amount of effort necessary to fully master suitable ways of assisting human–machine interaction.

Using low-cost, open-source components together with multiple sensors and actuators is quite a challenge in terms of effort and cost. Hence, it is desirable to employ Open Source Software (OSS) and Hardware (OSH) in UAV-CPSs.

[1]Telecommunications Department, Universidade Federal Fluminense, RJ, Brazil
[2]ECE Department, Karunya University, Coimbatore, India
[3]DCTA-ITA-IEEA, Sao Jose dos Campos, SP, Brazil
[4]Department of Computer Science, Luleå University of Technology, Luleå, Sweden
[5]Department of Aerospace Engineering and Aviation, RMIT University, Bundoora, Australia

Whenever possible, available OSH and OSS should be used in designs independently of the type of hardware framework and s operating system.

The performance of an Orthogonal Frequency Division Multiplexing (OFDM) UAV-CPS can be enhanced by adding channel coding (i.e., error correction code) to identify and correct the errors that happen throughout data transmission. The massive bandwidth of UWB channels can originate new effects as compared to conventional wireless channel modelling in several MAV applications. Ultra-Wide Band (UWB) technologies based on IEEE 802.15.4a have various uses when there is a need for highly accurate localisation with various sensors for stabilisation and navigation. Still, absolute indoor positioning challenges UAVs.

Flying a UAV in unstructured settings with changeable conditions is challenging. To support the development of better algorithms, a multipurpose dataset for low-altitude UAV flights in a given Brazilian environment is proposed as a benchmark for positioning and other avionics tasks for assessing computer vision procedures in terms of robustness and generalisation with a baseline for depth estimation with and without landmarks. This stage of development can help the advancement of future integration with remote-sensing (RS) modules that will bring more spectral information to the analyses.

UAV-CPSs involve a great deal of knowledge on networking – more specifically – on Flying Ad-hoc Networks (FANETs). The fact that the traffic of high-dimensional multimedia data streams through UAV-CPSs tends to grow exponentially raises several issues and point towards future research directions.

The texture is an important feature to recognise objects or regions of interest (ROIs) in any images, and it has been commonly used for image classification from satellite images to assess biomass for instance. UAV imagery takes advantage of ultra-high spatial resolution, which shows that texture is also a paramount source of knowledge. Nevertheless, texture in the UAV imagery was seldom used in surveillance. Moreover, merging ground hyperspectral data could compensate for the limited bands of UAV sensors and increase the estimation precision of analyses. Consequently, the goals of this chapter are (i) to explore UAV-based multispectral imagery and (ii) to improve several types of estimation accuracy through hyperspectral information.

The application of camera-equipped UAVs for visually observing ecological balance, construction and operation of hydric bodies, buildings, bridges, forest reservations and other types of infrastructure arrangements has grown exponentially. These UAV-CPSs can frequently examine all sorts of sites, monitor work-in-progress, generate documents/reports for safety, and scrutinise existing structures, chiefly for hard-to-reach regions.

Regardless of the sophistication of the on-board sensors, the Cloud, RS, computational intelligence, and communication advances, super-resolution (SR) will be in demand for quite a long time. This will continue to happen in contexts where acquiring imageries is expensive and troublesome such as in healthcare, astronomy, and disaster relief.

Quality of Service (QoS) and Quality of Experience (QoE) (besides other qualitative performance metrics) will play a pivotal role to impulse further improvements in all stages of a UAV-CPS.

This book intends to be a reference for current and forthcoming applications of UAV-CPSs. It will display fundamental aspects, ongoing research efforts, accomplishments, and challenges faced when it comes to the deployment of imaging capabilities and sensor integration in UAVs.

Succinctly, this book addresses the challenges, role, technical issues, and applications of computer vision/image processing frameworks, plus sensors in UAV design. The book chapters help the readers to focus on and apprehend the most critical factors that relate a subject to computer vision and sensing.

Index

www.ingramcontent.com/pod-product-compliance
Lightning Source LLC
LaVergne TN
LVHW011842200926
848484LV00002B/386
* 9 7 8 1 7 8 5 6 1 6 4 2 6 *